学术引领系列

国家科学思想库

中国学科发展战略

新能源材料

中国科学院

科学出版社
北京

内 容 简 介

本书涵盖能量转换材料，储能材料，新型节能材料，新能源材料发展的新概念、新应用、新机遇等四个专题。全书针对各专题特点，介绍其研究对象与内容，分析评述了国内外研究现状和发展趋势，并针对我国在这些领域的未来发展提出了相关战略建议与措施。

本书适合高层次的战略和管理专家、相关领域的高等院校师生、科研院所的研究人员阅读，是科技工作者洞悉学科发展规律、把握前沿领域和重点方向的重要指南，也是科技管理部门重要的决策参考，同时是社会公众了解新能源材料学科发展现状与趋势的重要读本。

图书在版编目（CIP）数据

新能源材料 / 中国科学院编. —北京：科学出版社，2023.3
（中国学科发展战略）
ISBN 978-7-03-072434-2

Ⅰ. ①新…　Ⅱ. ①中…　Ⅲ. ①新能源－材料技术－研究
Ⅳ. ①TK01

中国版本图书馆 CIP 数据核字（2022）第094358号

丛书策划：侯俊琳　牛　玲
责任编辑：张　莉　姚培培 / 责任校对：韩　杨
责任印制：师艳茹 / 封面设计：黄华斌　陈　敬　有道文化

科学出版社 出版
北京东黄城根北街 16 号
邮政编码：100717
http://www.sciencep.com

北京中科印刷有限公司 印刷
科学出版社发行　各地新华书店经销
*
2023年3月第　一　版　开本：720 × 1000　1/16
2024年1月第二次印刷　印张：22 1/2
字数：400 000

定价：158.00元

（如有印装质量问题，我社负责调换）

中国学科发展战略

指 导 组

组　　长：侯建国
副 组 长：高鸿钧　包信和
成　　员：张　涛　朱日祥　裴　钢
　　　　　郭　雷　杨　卫

工 作 组

组　　长：王笃金
副 组 长：苏荣辉
成　　员：钱莹洁　赵剑峰　薛　淮
　　　　　王　勇　冯　霞　陈　光
　　　　　李鹏飞　马新勇

中国学科发展战略·新能源材料

项目专家组

负责人： 邹志刚

成　员（以姓氏笔画为序）：

芮筱亭　李述汤　汪卫华　沈保根　张清杰
陈维江　周孝信　段文晖　宣益民　都有为
郭万林　郭烈锦　唐叔贤

项目工作组

负责人： 邹志刚

成　员（以姓氏笔画为序）：

于振涛　王镜喆　田　浩　冯建勇　任斐隆
闫世成　麦立强　李朝升　吴聪萍　张　策
张清杰　陈　泓　罗文俊　周　勇　姚　伟
姚颖方　都有为　郭万林　唐叔贤　黄　林
熊　威

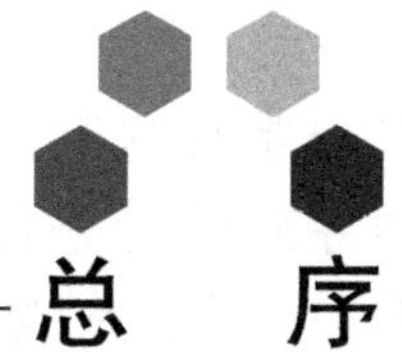

总　序

九层之台，起于累土[①]

白春礼

近代科学诞生以来，科学的光辉引领和促进了人类文明的进步，在人类不断深化对自然和社会认识的过程中，形成了以学科为重要标志的、丰富的科学知识体系。学科不但是科学知识的基本的单元，同时也是科学活动的基本单元：每一学科都有其特定的问题域、研究方法、学术传统乃至学术共同体，都有其独特的历史发展轨迹；学科内和学科间的思想互动，为科学创新提供了原动力。因此，发展科技，必须研究并把握学科内部运作及其与社会相互作用的机制及规律。

中国科学院学部作为我国自然科学的最高学术机构和国家在科学技术方面的最高咨询机构，历来十分重视研究学科发展战略。2009 年 4 月与国家自然科学基金委员会联合启动了“2011～2020 年我国学科发展战略研究”19 个专题咨询研究，并组建了总体报告研究组。在此工作基础上，为持续深入开展有关研究，学部于 2010 年底，在一些特定的领域和方向上重点部署了学科发展战略研究项目，研究成果现以“中国学科发展战略”丛书形式系列出版，供大家交流讨论，希望起到引导之效。

根据学科发展战略研究总体研究工作成果，我们特别注意到学

① 题注：李耳《老子》第 64 章：“合抱之木，生于毫末；九层之台，起于累土；千里之行，始于足下。”

科发展的以下几方面的特征和趋势。

一是学科发展已越出单一学科的范围，呈现出集群化发展的态势，呈现出多学科互动共同导致学科分化整合的机制。学科间交叉和融合、重点突破和“整体统一”，成为许多相关学科得以实现集群式发展的重要方式，一些学科的边界更加模糊。

二是学科发展体现了一定的周期性，一般要经历源头创新期、创新密集区、完善与扩散期，并在科学革命性突破的基础上螺旋上升式发展，进入新一轮发展周期。根据不同阶段的学科发展特点，实现学科均衡与协调发展成为了学科整体发展的必然要求。

三是学科发展的驱动因素、研究方式和表征方式发生了相应的变化。学科的发展以好奇心牵引下的问题驱动为主，逐渐向社会需求牵引下的问题驱动转变；计算成为了理论、实验之外的第三种研究方式；基于动态模拟和图像显示等信息技术，为各学科纯粹的抽象数学语言提供了更加生动、直观的辅助表征手段。

四是科学方法和工具的突破与学科发展互相促进作用更加显著。技术科学的进步为激发新现象并揭示物质多尺度、极端条件下的本质和规律提供了积极有效手段。同时，学科的进步也为技术科学的发展和催生战略新兴产业奠定了重要基础。

五是文化、制度成为了促进学科发展的重要前提。崇尚科学精神的文化环境、避免过多行政干预和利益博弈的制度建设、追求可持续发展的目标和思想，将不仅极大促进传统学科和当代新兴学科的快速发展，而且也为人才成长并进而促进学科创新提供了必要条件。

我国学科体系系由西方移植而来，学科制度的跨文化移植及其在中国文化中的本土化进程，延续已达百年之久，至今仍未结束。

鸦片战争之后，代数学、微积分、三角学、概率论、解析几何、力学、声学、光学、电学、化学、生物学和工程科学等的近代科学知识被介绍到中国，其中有些知识成为一些学堂和书院的教学内容。1904 年清政府颁布“癸卯学制”，该学制将科学技术分为格致科（自然科学）、农业科、工艺科和医术科，各科又分为诸多学科。1905 年清朝废除科举，此后中国传统学科体系逐步被来自西方的新学科体系取代。

民国时期现代教育发展较快，科学社团与科研机构纷纷创建，现代学科体系的框架基础成型，一些重要学科实现了制度化。大学引进欧美的通才教育模式，培育各学科的人才。1912 年詹天佑发起成立中华工程师会，该会后来与类似团体合为中国工程师学会。1914 年留学美国的学者创办中国科学社。1922 年中国地质学会成立，此后，生理、地理、气象、天文、植物、动物、物理、化学、机械、水利、统计、航空、药学、医学、农学、数学等学科的学会相继创建。这些学会及其创办的《科学》《工程》等期刊加速了现代学科体系在中国的构建和本土化。1928 年国民政府创建中央研究院，这标志着现代科学技术研究在中国的制度化。中央研究院主要开展数学、天文学与气象学、物理学、化学、地质与地理学、生物科学、人类学与考古学、社会科学、工程科学、农林学、医学等学科的研究，将现代学科在中国的建设提升到了研究层次。

中华人民共和国成立之后，学科建设进入了一个新阶段，逐步形成了比较完整的体系。1949 年 11 月中华人民共和国组建了中国科学院，建设以学科为基础的各类研究所。1952 年，教育部对全国高等学校进行院系调整，推行苏联式的专业教育模式，学科体系不断细化。1956 年，国家制定出《十二年科学技术发展远景规划纲要》，该规划包括 57 项任务和 12 个重点项目。规划制定过程中形成的“以任务带学科”的理念主导了以后全国科技发展的模式。1978 年召开全国科学大会之后，科学技术事业从国防动力向经济动力的转变，推进了科学技术转化为生产力的进程。

科技规划和“任务带学科”模式都加速了我国科研的尖端研究，有力带动了核技术、航天技术、电子学、半导体、计算技术、自动化等前沿学科建设与新方向的开辟，填补了学科和领域的空白，不断奠定工业化建设与国防建设的科学技术基础。不过，这种模式在某些时期或多或少地弱化了学科的基础建设、前瞻发展与创新活力。比如，发展尖端技术的任务直接带动了计算机技术的兴起与计算机的研制，但科研力量长期跟着任务走，而对学科建设着力不够，已成为制约我国计算机科学技术发展的“短板”。面对建设创新型国家的历史使命，我国亟待夯实学科基础，为科学技术的持续发展与创新能力的提升而开辟知识源泉。

反思现代科学学科制度在我国移植与本土化的进程，应该看到，20 世纪上半叶，由于西方列强和日本入侵，再加上频繁的内战，科学与救亡结下了不解之缘，中华人民共和国成立以来，更是长期面临着经济建设和国家安全的紧迫任务。中国科学家、政治家、思想家乃至一般民众均不得不以实用的心态考虑科学及学科发展问题，我国科学体制缺乏应有的学科独立发展空间和学术自主意识。改革开放以来，中国取得了卓越的经济建设成就，今天我们可以也应该静下心来思考“任务”与学科的相互关系，重审学科发展战略。

现代科学不仅表现为其最终成果的科学知识，还包括这些知识背后的科学方法、科学思想和科学精神，以及让科学得以运行的科学体制、科学家的行为规范和科学价值观。相对于我国的传统文化，现代科学是一个“陌生的”“移植的”东西。尽管西方科学传入我国已有一百多年的历史，但我们更多地还是关注器物层面，强调科学之实用价值，而较少触及科学的文化层面，未能有效而普遍地触及到整个科学文化的移植和本土化问题。中国传统文化以及当今的社会文化仍在深刻地影响着中国科学的灵魂。可以说，迄20 世纪结束，我国移植了现代科学及其学科体制，却在很大程度上拒斥与之相关的科学文化及相应制度安排。

科学是一项探索真理的事业，学科发展也有其内在的目标，探求真理的目标。在科技政策制定过程中，以外在的目标替代学科发展的内在目标，或是只看到外在目标而未能看到内在目标，均是不适当的。现代科学制度化进程的含义就在于：探索真理对于人类发展来说是必要的和有至上价值的，因而现代社会和国家须为探索真理的事业和人们提供制度性的支持和保护，须为之提供稳定的经费支持，更须为之提供基本的学术自由。

20 世纪以来，科学与国家的目的不可分割地联系在一起，科学事业的发展不可避免地要接受来自政府的直接或间接的支持、监督或干预，但这并不意味着，从此便不再谈科学自主和自由。事实上，在现当代条件下，在制定国家科技政策时充分考虑“任务”和学科的平衡，不但是最大限度实现学术自由、提升科学创造活力的有效路径，同时也是让科学服务于国家和社会需要的最有效的做法。这里存在着这样一种辩证法：科学技术系统只有在具有高度创造活力的情形下，才能在创新型国家建设过程中发挥最大作用。

在全社会范围内创造一种允许失败、自由探讨的科研氛围；尊重学科发展的内在规律，让科研人员充分发挥自己的创造潜能；充分尊重科学家的个人自由，不以“任务”作为学科发展的目标，让科学共同体自主地来决定学科的发展方向。这样做的结果往往比事先规划要更加激动人心。比如，19 世纪末德国化学学科的发展史就充分说明了这一点。从内部条件上讲，首先是由于洪堡兄弟所创办的新型大学模式，主张教与学的自由、教学与研究相结合，使得自由创新成为德国的主流学术生态。从外部环境来看，德国是一个后发国家，不像英、法等国拥有大量的海外殖民地，只有依赖技术创新弥补资源的稀缺。在强大爱国热情的感召下，德国化学家的创新激情迸发，与市场开发相结合，在染料工业、化学制药工业方面进步神速，十余年间便领先于世界。

中国科学院作为国家科技事业“火车头”，有责任提升我国原始创新能力，有责任解决关系国家全局和长远发展的基础性、前瞻性、战略性重大科技问题，有责任引领中国科学走自主创新之路。中国科学院学部汇聚了我国优秀科学家的代表，更要责无旁贷地承担起引领中国科技进步和创新的重任，系统、深入地对自然科学各学科进行前瞻性战略研究。这一研究工作，旨在系统梳理世界自然科学各学科的发展历程，总结各学科的发展规律和内在逻辑，前瞻各学科中长期发展趋势，从而提炼出学科前沿的重大科学问题，提出学科发展的新概念和新思路。开展学科发展战略研究，也要面向我国现代化建设的长远战略需求，系统分析科技创新对人类社会发展和我国现代化进程的影响，注重新技术、新方法和新手段研究，提炼出符合中国发展需求的新问题和重大战略方向。开展学科发展战略研究，还要从支撑学科发展的软、硬件环境和建设国家创新体系的整体要求出发，重点关注学科政策、重点领域、人才培养、经费投入、基础平台、管理体制等核心要素，为学科的均衡、持续、健康发展出谋划策。

2010 年，在中国科学院各学部常委会的领导下，各学部依托国内高水平科研教育等单位，积极酝酿和组建了以院士为主体、众多专家参与的学科发展战略研究组。经过各研究组的深入调查和广泛研讨，形成了“中国学科发展战略”丛书，纳入“国家科学思想库—学术引领系列”陆续出版。学部诚挚感谢为学科发展战略研究付出心血的院士、专家们！

按照学部“十二五”工作规划部署，学科发展战略研究将持续开展，希望学科发展战略系列研究报告持续关注前沿，不断推陈出新，引导广大科学家与中国科学院学部一起，把握世界科学发展动态，夯实中国科学发展的基础，共同推动中国科学早日实现创新跨越！

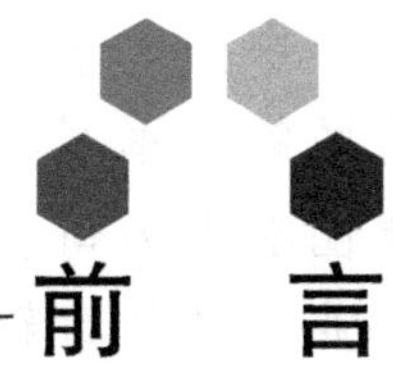

前　言

能源是人类文明进步的基础和动力，能源问题是当今社会面临的重要问题之一。面对气候变化、环境风险挑战、能源资源约束等日益严峻的全球问题，寻找一种或几种新型能源已成为国际研究的热点。2021年全国两会，碳达峰和碳中和被首次写入《政府工作报告》。

习近平总书记指出，实现碳达峰、碳中和是一场广泛而深刻的经济社会系统性变革①。在这场重大战略变革中，能源结构转型是重中之重，尤其通过光伏、储能、燃料电池发电等碳替代路径的新能源技术发展，是实现国家“双碳”目标和社会可持续发展的关键环节。

新能源材料科学与技术伴随着新能源的变革和进步不断发展壮大，是新能源利用的重要基础和支撑，其发展也是提升我国能源科技自主创新能力的根本推动力。近年来，新能源材料研究取得了快速发展，研究领域不断拓展，研究内容不断深入，已成为自然科学的一个前沿研究方向。但是，在新能源开发利用与日俱增的形势下，国内新能源材料开发能力不足的问题日益凸现，面临诸多亟待解决的“卡脖子”技术难题。大量如太阳能发电、氢燃料电池等关键核心材料及核心部件与国外先进水平差距明显，在一定程度上制约了我国能源产业的健康发展。因此，解决新能源发展中的核心材料与部件的关键性科学技术难题，是国家未来能源健康发展的重要保障。

在此背景下，中国科学院技术科学部决定设立“新能源材料学科发展战略研究”专题咨询项目。项目组由来自南京大学、南京航空航天大学、南京理工大学、南方科技大学、清华大学、武汉理工大学、西安交通大学、香港城市大学、香港中文大学（深圳）等高

① 中华人民共和国中央人民政府. 习近平主持召开中央财经委员会第九次会议. http:www.gov.cn/xinwen/2021-03/15/content_5593154.htm.

校，中国科学院物理研究所、中国科学院金属研究所、中国科学院工程热物理研究所、中国人民解放军军事科学院防化研究院等科研院所，以及中国钢研科技集团有限公司、国家电网有限公司、中国航天科技集团有限公司等企业长期从事新能源材料各分支学科研究的科学家组成。项目组瞄准国际上新能源材料领域研究的热点和难点，根据新能源材料学科的分支、历史沿革及其发展态势，系统梳理了新能源材料学科的发展历程，提出了我国在这一研究领域所面临的重要挑战，同时也凝练了关键科学问题，明确了未来的发展方向和目标，形成了本书。

虽然新能源材料在本质上属于材料学的分支学科范畴，但是多种学科，如物理学、化学、生物科学、环境科学等，皆能在这一学科领域进行广泛的交叉融合。鉴于这一特点，本书将以科学问题为依据划分新能源材料学科各分支方向。全书共四章，依次为能量转换材料，储能材料，新型节能材料，新能源材料发展的新概念、新应用、新机遇。各章撰写分工为：第一章由田虎、唐叔贤、姚颖方、任斐隆、罗文俊、冯建勇、李朝升、闫世成、周勇、王镜喆、黄林、吴聪萍编写；第二章由麦立强、张清杰编写；第三章由都有为编写；第四章由张策、姚伟、陈泓、于振涛、熊威、邹志刚、郭万林编写。全书由姚颖方整理。

需要指出的是，本书涉及能源、环境、材料、生物、物理、化学等多个方向和领域，内容难免存在疏漏，不可能完全涵盖目前新能源材料学科研究的各个方面。另外，由于新能源材料学科发展日新月异，本书中提出的建议也存在一定的局限性。他山之石，可以攻玉。东壁余光，或许裨益子明。期冀本书能给大家提供一个参考，以促进我国新能源材料研究的发展。

本书在编写过程中得到了中国科学院学部工作局的大力支持，本书成稿后得到了项目组邀请专家的宝贵修改意见和建议。在此，谨向所有为本书的完成付出辛勤劳动的同行表示衷心的感谢！

邹志刚

2021年3月30日

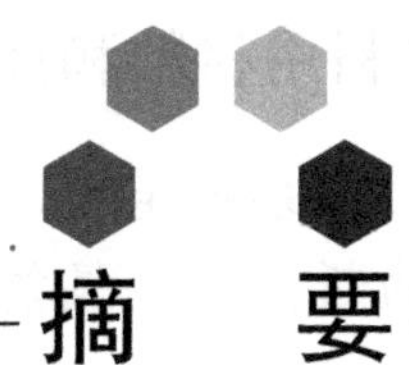

摘　要

21 世纪是能源革命的时代。面对世界范围内的资源和环境制约，很多国家都把能源结构的低碳化提到首要战略地位，并致力于确立未来大比例的新型可再生能源的战略目标，努力构建以太阳能为基础，包括基于太阳能发电、二氧化碳转换、氢燃料电池的能量转换系统，以及基于氢能和锂电的储能系统等在内的以其他新能源为补充的新型综合可持续能源体系。新能源的发展是事关我国社会主义现代化建设全局、中华民族伟大复兴和长远发展重大而紧迫的战略任务，是我国建设创新型国家、建成世界科技强国的重要支柱。新能源材料科学研究是新能源事业发展的基石，它的发展是提升我国能源科技自主创新能力的根本推动力。

新能源材料是材料学的重要组成部分，包括能量转换材料、储能材料、新型节能材料等。新能源材料的发展决定了世界能源驱动的走向和能源科技的进步，是完成我国 2030 年碳排放达峰目标、推进清洁可持续发展和智慧能源技术的重要保障，也是信息、交通运输、航空航天、生物医学等高技术领域和国防建设不可或缺的关键支撑。支撑未来产业变革的新型能源供给方式必须满足几个条件：①能源转换系统可以从太阳能中获得能量，确保能源供应源源不断；②能量载体必须闭环循环，不需要大规模消耗其他资源，实现环境和能源的可持续发展；③能源载体必须是地球高丰度元素，且热值显著高于现有的化石燃料，足以支撑急剧上升的能源需求；④必须发展快速、高效、高密度、安全的储能系统。

一、氢能转换与高效利用材料和技术

以氢能为载体的能源循环系统是可以满足以上条件的能源利用体系之一，作为零碳绿色的新能源，氢能具有能量密度大、转化效率高、储量丰富和适用范围广等特点，可实现从开发到利用全过程的零排放、零污染，是最具发展潜力的高效替代能源。氢能是二次能源，燃料电池是氢能清洁利用的最佳工具之一，氢能的利用横跨电力、供热和燃料动力三个领域，因此氢能与燃烧电池的技术发展在未来全球能源结构变革中居于重要的地位，具有重要的作用。

我国在转变能源结构和开展节能减排过程中，也把氢能利用提升到国家战略高度。2016 年 4 月，国家发展和改革委员会和国家能源局印发了《能源技术革命创新行动计划（2016—2030 年）》，并同时发布了《能源技术革命重点创新行动路线图》，将“氢能与燃料电池技术创新”列为 15 项重点任务之一。2018 年 8 月 8 日，《“十三五”国家科技创新规划》明确要求发展氢能、燃料电池这类引领产业变革的颠覆性技术。与此同时，地方政府也开始积极布局氢能与燃料电池产业。2017 年 9 月，上海市率先从地方政府层面发布了《上海市燃料电池汽车发展规划》，计划到 2020 年，氢燃料电池产业链在上海能创造 150 亿元的产值，建成 5～10 个加氢站，建成两个乘用车示范区，运行规模达到 3000 辆。随后，西安市、广州市、武汉市、佛山市等地陆续发布了氢燃料电池产业发展规划。神华集团有限责任公司、中国石油化工集团有限公司等能源化工集团相继宣布在全国布局加氢站等基础设施，它们的加入使得氢能产业发展开始提速。2018 年初，在科学技术部、工业和信息化部、财政部、交通运输部、国务院国有资产监督管理委员会、国家能源局、国家开发银行、中国科学院、中国工程院、中国科学技术协会、国家标准化管理委员会的指导下，跨学科、跨行业、跨部门的中国氢能及燃料电池产业创新战略联盟正式成立，它的成立也预示着中国氢能及燃料电池产业开始进入规范与加速发展新时期。

在氢能高效利用技术相对完备的态势下，2019 年“推动充电、加氢等设施建设”首次被写入《政府工作报告》。在现有化石能源制氢的基础上，从源头加速氢燃料的清洁制备技术的发展，直接关系到国家能源战略计划的部署，是未来国家实现经济可持续发展的关键之一。以光催化材料直接分解海水制氢为重点，可有效实现太阳能的高效转换、存储一体化，同时利用地球上最丰富的海水资源作为能源的物质来源，大幅度降低了氢能制备过程中对清洁淡水资源的需求。通过材料微结构调控，提高光催化材料的太阳能转化率和反应速率，构建效率15%以上的太阳能光催化分解海水制氢装置已成为清洁能源利用、能源绿色化产业建设和可持续发展的重点方向。

二、太阳能高效转化和利用

太阳能是重要的清洁能源之一。我国的太阳能资源十分丰富，每年可供开发利用的太阳能约为 1.6×10^{15} W，是实现能源可持续发展的基础。从长远来看，太阳能的有效开发与利用对优化我国能源结构具有重大意义。太阳能在使用过程中不会对环境产生污染，而且其取之不尽、用之不竭的显著特点备受人们关注。目前太阳能利用的主要方式是太阳能电池，又称光电池。太阳能电池通过光—电或光—热—电的途径将太阳光能转化成电能。但是，目前大多数太阳能电池存在制造成本高、废弃污染、能量转换效率低和使用稳定性差等缺点，导致其在实际应用中受到极大的限制。因此，开发经济、环境友好型材料用以制备高效且稳定的太阳能电池成为研究的重点。

太阳能等新型可再生能源存在能量密度低、分布不均匀、昼夜与季节变化大、不易储存等问题，导致清洁能源发电面临的并网消纳问题日益严重。其中，弃水、弃风、弃光等“三弃”问题是清洁能源电力发展中的热点和难点，严重制约了可再生能源发电的发展。可再生能源发电往往与电力厂网不协调，需要耗费大量的人力、物力资源来架设电力系统和输送通道。因此，发展新型可靠的电化学能量转换与存储体系，将可再生能源转化成便于存储、运输和再次利用的二次能源，是解决太阳能高效转化利用瓶颈问题的当务之急。

通过太阳能光催化技术将太阳能转换为二次能源，如氢能等，可以有效解决可再生能源发电中的“三弃”问题。目前氢能的利用技术逐渐趋于成熟，以氢气为燃料的燃料电池已开始实用化，氢能汽车和氢能汽轮机等一些绿色产品已开始投入市场。高效氢能利用技术的成熟提高了对制氢技术快速发展的要求。采用太阳能高效转化的方式实现低成本、大规模的清洁制氢技术的开发必将成为满足“氢经济”时代迫切氢能需求的突破性技术。

三、新型储能材料与系统

储能材料与技术是实现分布式发电、节能减排目标的重要举措，也是集中式发电的有效补充。储能技术在电力系统发、输、配、用各个环节均有广泛用途，备受国内外业界人士的高度关注。“十三五”开局之年，国家发展和改革委员会、国家能源局也陆续颁布了《关于促进电储能参与“三北”地区电力辅助服务补偿（市场）机制试点工作的通知》《能源技术革命创新行动计划（2016—2030年）》《关于在能源领域积极推广政府和社会资本合作模式的通知》等系列政策，极大地激励了我国的储能产业化发展，建设百兆瓦级储能电站已经上升到国家战略层面。

在强有力的政策引导下，主要类型的化学储能技术获得了迅速发展。兆瓦级化学储能技术已进行了多种示范运行，锂离子电池储能方式在我国已开始商业化运行。继张北国家风光储输示范工程之后，陆续涌现出了一批有代表性的示范性储能电站和光储一体化电站，如辽宁卧牛石 5 MW 储能电站、甘肃酒泉兆瓦级储能电站、青海 15 MW 储能电站等；在用户侧，深圳宝清兆瓦级储能电站、欣旺达居民园区兆瓦级调频移动式储能系统也得到了示范应用。当前储能技术正逐步从试验示范阶段迈向商业化推广阶段，大规模储能系统的广泛应用需重点突破高安全、低成本、长寿命、高能量转换效率等储能装备的关键技术。

四、新能源材料发展的新概念、新应用与新机遇

基于太阳能的新能源材料发展，日益呈现出多元化、低碳化、智能化和分布式等特征。诸多新奇现象的发现、新概念的提出以及传统材料在新领域的突破，为新能源材料的发展带来了巨大挑战与生机。未来，新能源材料在航空航天领域的应用、模拟自然光合成的研究，以及水伏发电材料体系的构建中，均有可能取得重大突破，引发新能源科技革命，推动人类的生产生活方式发生深刻变革。

（一）地外人工光合成材料与技术的提出

随着人类探索疆域的拓展，重返月球、载人火星探索等极具挑战性的航天任务已具备可实施性，“拓展人类在太阳系的存在，更好地理解我们在宇宙中的位置”已成为人类的共同目标。2018 年 1 月，由十四国航天局组成的国际太空探索协调组（International Space Exploration Coordination Group，ISECG）提出分步实施月球轨道“深空之门”（Deep Space Gateway）载人空间站、月面有人探索、载人火星探索的全球探索路线图。美国历届政府均把载人探索作为太空计划的首要任务。2017 年 6 月 30 日，特朗普政府签署行政令重设国家航天委员会，提出 2033 年载人登陆火星的目标，主导月球轨道深空门户站的建设。2017 年 12 月 11 日，特朗普政府宣布将重返月球，并最终前往火星。2016 年 8 月 5 日，美国政府批准私营公司开展登月活动，月球开发将向民间企业开放，月球被纳入地球经济圈。美国太空探索技术公司（SpaceX）、洛克希德 · 马丁空间系统公司等航天公司也提出了雄心勃勃的载人火星探索计划。载人深空探索已成为当前国际太空活动的前沿热点。正如航海时代创造的奇迹一样，载人深空探索将会创造人类发展史的下一个奇迹。面对一个遥远的、存在众多未知因素的星球，载人深空探索将会面临一系列前所未有的挑战。

航天强国不仅是中国传统文化的“飞天”梦想，更是新时期建设社会主义现代化强国的重要目标，是实施科教兴国战略的重要领

域。习近平多次强调，发展航天事业、建设航天强国，是我们不懈追求的“航天梦”。党的十九大报告也进一步明确提出建设航天强国的战略目标。我国在载人和探月重大科技专项工程的基础上，正计划开展载人登月、月球基地建设等重大任务。

在载人深空探索活动中，地外生存是人类实现长期太空飞行（地球和月球轨道任务、地球和火星长期飞行任务）、地外长期居住（月球和火星基地）、地外移民的基本能力。从地球上携带资源来开展载人深空探索，任务成本代价极高，技术上也难以实现，因此，必须对飞行器废弃物原位资源和地外天体原位资源加以有效利用，才能大大减少从地球上携带的物资量，使载人深空探索任务具备可行性。

借助太阳能实现地外人工光合成，是支撑未来航天强国的必要手段。地外人工光合成是模拟地球上绿色植物的自然光合作用，通过光电催化，可控地将二氧化碳转化成氧气和含碳燃料的化学过程，是太空探索的核心能力。它不仅可通过人工光合成将人类呼吸产生的二氧化碳转化为氧气，实现密闭空间的废弃原位资源再生循环，大大降低载人空间站、载人深空飞船的物资供应需求，而且可利用火星等地外大气环境丰富的二氧化碳和水等原位资源生产氧气与燃料，实现人类在其他行星的地外生存，支撑可承受、可持续的载人深空探索任务。因此，通过地面实验、空间实验，分阶段、分步骤研究、开发和验证地外人工光合成技术，将有力支撑载人航天的后续发展。

目前国际空间站采用电解水的方式为宇航员补充氧气，针对空间站和火星二氧化碳的原位资源利用（*in-situ* resource utilization，ISRU），正在研发高温热还原或电解技术，将二氧化碳还原成氧气，但运行条件苛刻（高温、高压条件），能耗高。地外人工光合成技术是现阶段唯一可能取得国际首创的研究成果、解决长期载人航天的关键难题、引领未来载人深空探测发展方向的新技术。

（二）自然光合成的模拟和高效光化学能的转换

自然界的许多生命过程受太阳光主导或依赖于太阳光，因此研究自然光合成是生物学、物理学、化学和材料学交叉学科研究中的一个重要研究领域。在众多研究方向中，植物和藻类对太阳光的利用一直是生命科学领域的重大基础理论课题，包括植物对光的感知、吸收、传递、转化和利用等一系列生物学过程的本质和规律，以及光合作用和光能利用、生物质能及高效光化学能转换的前沿科学技术。它与当今人类面临的能源与环境等国际重大问题密切相关，世界发达国家高度重视太阳能光化学转化和利用，推出了一系列的重大研究计划，如欧盟的“太阳能-氢”计划、“人工树叶”计划，并成立相关研究联盟，如美国的人工光合作用联合中心、阿贡-西北太阳能研究中心，瑞典的人工光合作用联盟及澳大利亚的人工光合作用网络等。各国科学家正在通过生物学、材料学、物理学、化学等学科，试图揭示太阳能光化学转化的机制，并通过人工模拟光合作用机理产生可再生的清洁能源，构建可持续的新能源体系。《国家中长期科学和技术发展规划纲要（2006—2020 年）》中面向国家重大战略需求的基础研究部分，明确提出了光能高效利用的机理研究。同时，光合作用机理研究也是《创新 2050：科学技术与中国的未来》中提出的 22 个科技战略问题之一。

自然光合成提供了一个相对完整的环境，以更复杂、更有效的方式来实现太阳能的利用。植物等通过数十种（或更多种）酶协同的生物级联催化过程实现了连续和选择性地用最简单的单元（如 CO_2、N_2 和 H_2O）构建复杂的大分子。生物体系可使用活性炭作为反应性结构单元，以促进 C—C 偶联，从而避免解吸附反应物再活化产生的能量损失。生物系统的自我修复和生物繁殖速度很快，使其具有潜在的可扩展过程，并减轻了对敏感组件潜在的不稳定性对系统的影响。但是生物光合系统缺点明显，尤其是无法最大限度地实现太阳能的转化。大部分植物的光合作用效率为 0.1%，最高也不超过 6%。除此之外，当太阳光强太大时（超过 20% 太阳光强），会导致光损伤，从而降低转化效率。

相比于自然光合作用，人工光合系统中半导体纳米材料具有更宽光谱的光吸收能力，可以设计具有串联结构的互补光吸收器。基于人工合成半导体材料的光合作用器件通过掺杂和异质结，可以直接控制电荷分离，系统的相对简单性使得它们更易于以模块化方式进行修改和改进。但是人工光合系统的不足之处在于其高的转化效率通常以牺牲昂贵的高纯度半导体为代价，在电解质溶液的长期浸泡过程中，这种昂贵的材料容易降解，且不能自修复。

利用人工系统的功能组件（电极、纳米材料、光吸收剂等）与自然机制（酶的形式或细胞内的整个代谢途径）有机结合，通过“半”人工光合作用可实现太阳能燃料和太阳能化学品的高效生产，同时优化界面电荷转移，克服自然和人工光合作用的局限性，将二者的优势各自发挥到极致，实现自修复、低成本的高转化效率。“半”人工光合成包含模仿或基于光合系统Ⅱ（photosystem Ⅱ，PS Ⅱ）的“半”人工光氧化和“半”人工光还原系统。

（三）水伏发电材料体系的构建

近年来，以我国学者为代表的研究者发现，水与低维材料通过表界面相互作用可以直接输出电能。类比于光伏（photovoltaic）技术，这类生电现象被称为水伏（hydrovoltaic）效应。水伏效应使得人们能够用纳米材料从水运动和循环过程中直接捕获能量，为水能的利用提供了全新的方式。

研究者发现，通过碳纳米管（carbon nanotube，CNT）、石墨烯等低维碳材料与水直接相互作用，可以将水中更丰富的热能和机械能转化为有用的电能。例如，石墨烯可通过双电层的边界运动将拖动和下落水滴的能量直接转化为电能（拖曳势），把波动能转化为电能（波动势）。最近更是发现廉价的炭黑等纳米结构材料可通过大气环境下无所不在的水的自然蒸发持续产生伏级的电能。蒸发发电带来的最大优势是它不需要任何机械能输入。在环境蒸发条件下，1 cm 大小的炭黑片可稳定输出1V 的电压。类比于光伏、压电等能量转化效应，这类通过材料与水作用直接将水能转化为电能的现象被称为水伏效应。2018 年 12 月的《自然-纳米技术》以封面亮点标题的形式提出了“水伏学”（hydrovoltaics）一词，并指出水伏是从水中获取电能的全新途径。水伏效应的理论与技术目前仍处于初期研究阶段，但其所展示的发展潜力和独特应用前景已透出水伏科学技术的曙光。

五、小结

新能源材料与技术走向实用化的关键是性能（包括转化效率和使用寿命）与开发成本。新能源材料领域经过多年的发展和积累，正孕育着重大突破，处于迈向大规模应用的关键时期，国际竞争激烈。本书针对重大科学问题展开研究，以太阳能为主体，氢能、锂电等其他新能源为补充，涵盖能量转换材料，储能材料，新型节能材料，以及新能源材料发展的新概念、新应用、新机遇等四个新能源技术领域。要实现新能源供给和高效利用，需要在以下重点领域形成颠覆性创新技术：①高适应性、全光谱、高效率的太阳能发电创新技术体系；②高通量的锂电储能、电解水产氢或光催化分解海水创新技术体系，以满足电、氢的规模转化需求；③基于地外生存的人工光合成材料原位制备与体系构建，以满足载人航天中亟待解决的关键性资源需求；④以高丰度元素为基础的燃料电池发电创新技术体系；⑤面向未来的突破性新能源材料与技术。

Abstract

The 21st century is an era of energy revolution. Facing the constraints of resources and environment all over the world, many countries have treated the low carbonization as the first strategic position of energy structure, and committed themselves to establishing the strategic goals of a large proportion of renewable energy in the future, striving to build a new comprehensive sustainable energy system based on solar energy, including energy conversion systems based on solar cells, carbon dioxide conversion, hydrogen fuel cells, and energy storage systems based on hydrogen energy and lithium batteries. The development of new energy is an important and urgent strategic task related to the overall situation of China's socialist modernization, the great rejuvenation and long-term development of the Chinese nation, and it is also an important pillar for China to build an innovative and powerful country in science and technology. Scientific research on new energy materials is the cornerstone for the development of new energy industry. And its development is the fundamental driving force to enhance the independent innovation capability of China's energy science and technology.

New energy materials are an important part of materials science, including energy conversion materials, energy storage materials, new energy-saving materials, and new concepts and applications of new energy materials developed in recent years. The development of new energy materials determines the world's energy-driven trend and the progress of energy science and technology. It is an important guarantee for China to achieve the peak carbon emissions target before 2030, to promote clean and sustainable development and smart energy technology. And it is also an indispensable key support for high-tech fields such as information, transportation, aerospace, biomedicine, and national defense construction. The new energy supply mode supporting the future industrial transformation must meet several conditions: ① The energy utilization system can obtain energy from solar energy so as to ensure the continuous supply of energy; ② The energy carrier must circulate in a closed loop without consuming other resources on a large scale, so as to realize the sustainable development of environment and energy; ③ The energy carrier must be the earth's high abundance element, and its calorific value is significantly higher than the existing petroleum to support the sharply rising energy demand; ④ A fast, efficient, high-density and safe energy storage system must be developed. This book will divide the branches of new energy materials based on specific scientific problems, including 4 branches: energy conversion materials, energy storage materials, new energy-saving materials, new concepts, new applications and opportunities for the development of new energy materials, and each sub-report is self-contained, so as to sort out and grasp the specific scientific problem, research status, existing problems and future development direction of each discipline design. The situation of each branch is briefly summarized as follows.

1. Materials and technologies for hydrogen energy conversion and applications

The energy recycling system based on hydrogen energy is one of the energy utilization systems that can meet the above conditions. As a zero-carbon energy resource, hydrogen energy has the characteristics of high energy density, high conversion efficiency, abundant reserves and wide application range, which can realize zero emission and zero pollution in the whole process from development to utilization, and is the most promising and efficient alternative energy. Hydrogen energy is a secondary energy, and fuel cell is the best tool for clean utilization of hydrogen energy. The utilization of hydrogen energy spans three fields: electricity, heat supply and fuel power. Therefore, the technological development of hydrogen energy and fuel cell will play an important role in the future global energy structure reform.

In the process of energy conservation and emission reduction, China has also promoted the utilization of hydrogen energy to the national strategic level. In April 2016, the National Development and Reform Commission and the National Energy Administration issued the Energy Technology Innovation Action Plan in (2016—2030), and at the same time issued the Action Roadmap for Key Innovation in Energy Technology Revolution, which listed "hydrogen energy and fuel cell technology innovation" as one of the 15 key tasks. On August 8, 2018, the National Science and Technology Innovation Plan of the 13th Five-Year Plan cleared the development of hydrogen energy and fuel cells as the "disruptive technology that leads industrial changes". At the same time, local governments began to actively lay out hydrogen energy and fuel cell industries. In September 2017, Shanghai took the lead in issuing the Development Plan of Shanghai Fuel Cell Vehicles. It is planned that by 2020, the hydrogen fuel cell industry chain will create an output value of ¥15 billion in Shanghai, 5—10 hydrogen refueling stations, and two passenger car demonstration zones will be built, with the operation scale reaching 3,000 vehicles. Subsequently, Xi'an, Guangzhou, Wuhan, Foshan and other places successively released the development plans of hydrogen fuel cell industry. Shenhua, SINOPEC and other energy and chemical groups have announced the layout of hydrogen refueling stations and other infrastructure construction industries throughout the country, and their participation has accelerated the development of hydrogen energy industry. At the beginning of 2018, under the guidance of the Ministry of Science and Technology, the Ministry of Industry and Information Technology, the Ministry of Finance, the Ministry of Transport, the State-owned Assets Supervision and Administration Commission of the State Council, the Energy Bureau, the China Development Bank, the Chinese Academy of Sciences, the Chinese Academy of Engineering, the China Association for Science and Technology, and the National Standardization Administration Committee, the interdisciplinary, cross-industry and inter-departmental "China Hydrogen Energy and Fuel Cell Industry Innovation Strategic Alliance" was formally established. Its establishment also indicates that China's hydrogen energy and fuel cell industry has begun to enter standardization and acceleration.

With the relatively complete hydrogen efficient utilization technology, "Promoting the construction of charging and hydrogenation facilities" was first written in the Report on the Work of the Government 2019. On the basis of hydrogen production from fossil energy, accelerating the development of clean production technology of hydrogen fuel is directly related to the deployment of the national energy strategic plan, and is the key to achieving national sustainable economic development in the future. Focusing on the direct decomposition of seawater to produce hydrogen by photocatalytic materials, it can not only effectively realize the integration of efficient conversion and storage of solar energy, but also use the richest seawater resources on the earth as the material source of energy, which greatly reduces the demand for clean fresh water resources in the process of hydrogen energy production. It has become the key direction of clean energy utilization, green energy industry construction, and sustainable development to build a hydrogen production device with solar energy photocatalytic decomposition of seawater with an efficiency of more than 15% by improving the solar energy conversion and reaction rate of photocatalytic materials.

2. Efficient conversion and utilization of solar energy

Solar energy is one of the important clean energy sources. China is rich in solar energy resources, with about 1.6×10^{15}W every year, which is the basis of realizing sustainable development of energy. In the long run, the effective development and utilization of solar energy is of great significance to optimize China's energy structure. Solar energy will not pollute the environment during its use, and it has attracted wide attention because of its inexhaustible and remarkable characteristics. At present, the main way of solar energy utilization is solar cell, also known as "photovoltaic cell". Solar cells convert solar energy into electric energy by means of light-electricity or light-heat-electricity. However, at present, most solar cells have some disadvantages, such as high manufacturing cost, waste pollution, low energy conversion efficiency and poor stability, which lead to their great limitations in practical applications. Therefore, developing economical and environmentally-friendly materials to prepare efficient and stable solar cells has become the focus of research.

At the same time, solar energy and other new renewable energy sources have problems such as low energy density, uneven distribution, large circadian/seasonal changes, and are difficult to store. As a result, the problem of grid-connected consumption of clean energy has become increasingly serious. Among them, water, wind and solar power abandonment are focused and difficult issues in the development of clean energy power, which seriously restrict the sustainable development of renewable energy power generation. Renewable energy generation is often uncoordinated with power plant network, and it takes a lot of manpower and material resources to set up power system and transmission channel. Therefore, developing a new and reliable electrochemical energy conversion and storage system to convert renewable energy into secondary energy which is convenient for storage, transportation and reuse is an urgent task to solve the bottleneck problem of efficient conversion and utilization of solar energy.

By using solar photocatalytic technology to convert solar energy into secondary energy, such as hydrogen energy, the problem of "three power abandonment" in renewable energy power generation can be effectively solved. At present, the utilization technology of hydrogen energy is gradually maturing. And fuel cells have been practically used, and some "green" products such as hydrogen-powered automobiles and hydrogen-powered steam turbines have commercialized. The maturity of high-efficiency hydrogen energy utilization technology raises the requirement of rapid development of hydrogen production technology. The development of low-cost, large-scale clean hydrogen production technology by high-efficiency solar energy conversion will surely become a breakthrough technology to meet the urgent needs of the "hydrogen economy" era.

3. New energy storage materials and systems

Energy storage materials and technologies are important measures to achieve the targets of distributed power generation, energy saving, and emission reduction. They are also an effective supplement to centralized power generation. Energy storage technology is widely used in all aspects of power system transmission and distribution, and has attracted great attention from domestic and foreign industry. In the first year of the "Thirteenth Five-Year Plan", the National Development and Reform Commission and the Energy Bureau also successively promulgated the Notice of Electric Energy Storage Participating in Power Grid Peak-shaving Auxiliary Services in the "Three-North Region", Energy Technology Innovation Action Plan (2016—2030), Notice on Actively Promoting the Cooperation Model between Government and Social Capital in the Energy Field. And other policies have greatly stimulated the development of China's energy storage industrialization, and the

100 MW energy storage power station has risen to the national strategic level.

Under the guidance of strong policies, the main types of chemical energy storage technologies have developed rapidly. Megawatt chemical energy storage technology has been variously demonstrated. And lithium ion battery energy storage mode has started commercial operation in China. After Zhangbei National Wind and Solar Energy Storage and Transmission Demonstration Project, a number of representative demonstration energy storage power stations and integrated solar energy-storage power stations have emerged one after another, such as Woniushi 5 MW energy storage power station in Liaoning, Jiuquan MW-class energy storage power station in Gansu, Qinghai 15 MW energy storage power station, etc. On the user side, Baoqing MW-class energy storage power station, and Xinwangda residential park MW-class mobile energy storage system in Shenzhen have also been demonstrated and applied. At present, energy storage technology is gradually moving from experimental demonstration to commercial promotion. The wide application of large-scale energy storage system needs to break through the key technologies of energy storage equipment such as high safety, low cost, long life and high energy conversion efficiency.

4. New concepts, applications, and opportunities for the development of new energy materials

The development of new energy materials based on solar energy is increasingly characterized by diversification, low carbonization, intellectualization, and distribution. The discovery of many novel phenomena, the presentation of new concepts, and the breakthrough of traditional materials in new fields have brought great challenges and vitality to the development of new energy materials. In the future, the application of new energy materials in the aerospace field, the research on the simulated natural photosynthesis, and the construction of hydrovoltaic power generation material system are all likely to achieve important breakthroughs, trigger a new revolution in energy science and technology, and promote profound changes in human lifestyles and modes of production.

(1) Extraterrestrial artificial photosynthetic materials and technologies

With the expansion of human exploration territory, challenging space missions such as returning to the moon, and manned landing on Mars have become feasible. "Expanding human existence in the solar system and better understanding our position in the universe" has become the common goal of human beings. In January, 2018, the International Space Exploration Coordination Group (ISECG), composed of the 14 national space agencies, put forward step by step the global exploration roadmap for implementing the Deep Space Gateway in lunar orbit, the manned exploration on the lunar surface, and the manned Mars exploration. Successive American governments have regarded manned exploration as the primary task of the space program. On June 30, 2017, the Trump administration signed an executive order to re-establish the National Space Commission, put forward the goal of manned landing on Mars in 2033, and led the construction of deep space portal station in lunar orbit. On December 11, 2017, Trump administration announced that they would return to the moon and eventually go to Mars. On August 5, 2016, the U.S. government approved private companies to launch lunar landing activities. The development of the moon will be open to private enterprises, and the moon will be included in the earth's economic circle. SpaceX, Lockheed Martin Space System Company and other space companies have also put forward ambitious manned Mars exploration plans. Manned deep space exploration has become the forefront of international space activities. Just like the miracle created in the maritime era, manned deep space exploration will create the next miracle in the history of human development. Facing a distant planet with many unknown factors, manned deep space exploration will face a series of unprecedented challenges.

Becoming an aerospace power is not only the dream of Chinese traditional culture, but also an important goal of building a socialist modernization power in the new period and an important field of implementing the strategy of rejuvenating the country through science and education. President Xi Jinping has repeatedly stressed that "developing the aerospace industry and building a space power is our unremitting pursuit of the space dream". The report of the 19th National Congress of the Communist Party of China further clearly put forward the strategic goal of building a space power. On the basis of major scientific and technological special projects for manned and lunar exploration, China is planning to carry out major tasks such as manned landing on the moon and building a lunar base.

In manned deep space exploration activities, extraterrestrial survival is the basic ability of human beings to realize long-term space flight (Earth and Moon orbit mission, long-term ground fire mission), long-term residence (Moon and Mars base) and extraterrestrial migration. Carrying resources from the earth to carry out manned deep space exploration is extremely costly and technically difficult to achieve. Therefore, the *in-situ* resources of aircraft wastes and extraterrestrial celestial bodies must be effectively utilized in order to greatly reduce the materials carried from the Earth and make the manned deep space exploration mission feasible.

Solar energy is a necessary means to realize artificial photosynthesis outside the earth and support the future "space power". artificial photosynthesis is a chemical process that simulates the natural photosynthesis of green plants on the earth, and can control the conversion of carbon dioxide into oxygen and carbonaceous fuel through photoelectrocatalysis. It is the core capability of space exploration. It can not only convert carbon dioxide generated by human respiration into oxygen through artificial photosynthesis, and realize the recycling of abandoned *in-situ* resources in confined space, which greatly reduces the material supply demand of manned space station and manned deep space spacecraft, but also produce oxygen and fuel by using the rich *in-situ* resources of carbon dioxide and water in the atmospheric environment outside Mars, so as to realize human survival outside other planets and support the affordable and sustainable manned deep space exploration mission. Therefore, the research, development and verification of extraterrestrial artificial photosynthesis technology will strongly support the subsequent development of manned spaceflight.

At present, the International Space Station uses electrolyzed water to supplement oxygen for astronauts. Aiming at the *in-situ* utilization of CO_2 resources in space station and Mars, high temperature thermal reduction or electrolysis technology is being developed to reduce CO_2 into oxygen, but the operating conditions are harsh (high temperature and high pressure) and the energy consumption is high. Extraterrestrial artificial photosynthesis technology is the only new technology that can achieve the first international research results, solve the key problems of long-term manned spaceflight and lead the future development direction of manned deep space exploration.

(2) Simulation of natural photosynthesis and efficient conversion of photochemical energy

Many life processes in nature are dominated by or depend on sunlight, so the study of natural photosynthesis is an important research field in the interdisciplinary research of biology, physics, chemistry and materials science. Among many research directions, the utilization of sunlight by plants and algae has always been a major basic theoretical problem in the field of life sciences, including the nature and laws of a series of biological processes such as the perception, absorption, transmission, transformation and utilization of light by plants, and the cutting-edge science and technology of photosynthesis and light energy utilization, biomass energy and efficient photochemical energy conversion. It is closely related to major international issues such as energy and environment. Developed countries in the world attach great importance to the photochemical conversion and utilization of solar energy, and have launched a series of major research programs, such as the "Solar-Hydrogen" project and the "Artificial Leaf " project of the European Union, and established related research alliances, such as the "Joint Center for Artificial Photosynthesis" of the United States, "Argonne-Northwestern Solar Energy Research Center", Swedish Consortium for Artificial Photosynthesis and Australia scientists from all over the world are trying to reveal the mechanism of photochemical conversion of solar energy through biology, materials science, physics, chemistry and other disciplines, and to generate renewable clean energy and build a sustainable new energy system through artificial simulation of photosynthesis mechanism. In the Outline of National Medium- and Long-Term Program for Science and Technology Development (2006—2020), the research on the mechanism of efficient utilization of light energy is clearly put forward in the basic research section facing the major strategic needs of the country. At the same time, the study of photosynthesis mechanism is also one of the 22 strategic issues of science and technology put forward in Innovation 2050: Science and Technology and China's Future.

Natural photosynthesis provides a relatively complete environment and realizes the utilization of solar energy in a more complex and effective way. Plants and algae have realized the continuous and selective construction of complex macromolecules from the simplest units (such as CO_2, N_2 and H_2O) through the cooperative biological cascade catalysis of dozens (or more) enzymes. Activated carbon can be used as reactive structural unit in biological system to promote C—C coupling and avoid energy loss caused by reactivation of desorption reactants. Self-repair and reproduction of biological system are very fast, which makes it have potential expansion process and reduces the worry about the instability of sensitive components. However, the shortcomings of biological photosynthesis system are obvious, especially the inability to maximize the conversion of solar energy. The photosynthetic efficiency of most plants is 0.1%, and the highest is no more than 6%. In addition, when the solar intensity is too high (more than 20% of the solar intensity), it will also lead to light damage, and the repair of light damage also needs energy, thus reducing the conversion efficiency.

Compared with natural photosynthesis, semiconductor nanomaterials in artificial photosynthesis system have wider spectral light absorption capacity, so complementary light absorbers with series configuration can be designed. Photosynthetic devices based on synthetic semiconductor materials can directly control charge separation through doping and heterojunction. The relative simplicity of the system makes them easier to modify and improve in a modular way. However, the disadvantage of artificial photosynthesis system is that high conversion efficiency usually comes at the expense of expensive high-purity semiconductors. During the long-term immersion in electrolyte solution, this expensive material is easy to degrade and cannot repair itself.

The functional components of artificial system (electrodes, nanomaterials, light absorbers, etc.) are organically combined with natural mechanisms (the form of enzymes or the whole metabolic pathway in cells), and efficient production of solar fuel and solar chemicals can be realized through semi-artificial photosynthesis. Learn the experience of interface charge transfer, equilibrium electron flux, catalytic reactivity and reaction mechanism. The aim is to overcome the limitations of natural and artificial photosynthesis, and bring their respective advantages to the extreme, so as to achieve self-repair, low cost and high conversion efficiency. Semi-artificial photosynthesis includes semi-artificial photo-oxidation imitating or based on PS Ⅱ system, and semi-artificial photo-reduction system.

(3) Construction of material system for hydrovoltaic power

generation

In recent years, researchers, represented by Chinese scholars, found in low-dimensional materials such as graphene and carbon nano-films that water and low-dimensional materials can directly output electrical energy through surface/interface interaction. Compared with photovoltaic technology, this kind of electricity generation phenomenon is called hydrovoltaic effect. Hydrovoltaic effect enables people to use nano-materials to directly capture energy from water flow, water waves, raindrops, evaporation, humidity and other water movement and circulation processes, which provides a new way for water energy utilization.

In recent years, researchers, represented by Chinese scholars, have found that through the direct interaction of low-dimensional carbon materials such as carbon nanotubes and graphene with water, more abundant thermal energy and mechanical energy in water can be converted into useful electrical energy for output. For example, graphene can directly convert the energy of dragging and falling water droplets into electric energy (drag potential) and wave energy into electric energy (wave potential) through the boundary movement of electric double layer. Recently, it has been found that cheap nano-structured materials such as carbon black can continuously generate volt-level electric energy through the natural evaporation of ubiquitous water in atmospheric environment. The biggest advantage of evaporative power generation is that it does not need any mechanical energy input. Under the condition of environmental evaporation, a carbon black sheet with a size of one centimeter can stably output a voltage of 1 V. Compared with photovoltaic, piezoelectric and other energy conversion effects, this phenomenon of directly converting water energy into electric energy through the interaction between materials and water is called "hydrovoltaic effect". In December 2018, *Nature Nanotechnology* put forward the word "hydrovoltaics" in the form of the cover highlighted title, and pointed out that it is a brand-new way to obtain electric energy from water. The theoretical and technical research on the underwater effect is still in its infancy, but its development potential and unique application prospect have already revealed the dawn of underwater science and technology.

5. Conclusion

The key to the practical application of new energy materials and technologies is performance (including conversion efficiency and service life) and development cost. After years of development and accumulation, the field of new energy materials is gestating a major breakthrough, which is in a critical period of large-scale application, and the international competition is fierce. This book focuses on major scientific issues, with solar energy as the main body, hydrogen energy, lithium battery and other new energy sources as supplements, covering four new energy technology fields, such as new concepts, new applications and new opportunities in the development of energy conversion materials, energy storage materials, new energy-saving materials and new energy materials. To achieve total energy supply and efficient utilization, disruptive innovative technologies need to be formed in the following key areas: ① Innovative technology system of solar power generation with high adaptability, full spectrum and high efficiency; ② The innovative technical system of high-throughput lithium battery energy storage, hydrogen production by electrolysis of water or photocatalytic decomposition of seawater can meet the scale conversion requirements of electricity and hydrogen; ③ *In-situ* preparation and system construction of artificial photosynthetic materials based on extraterrestrial survival to meet the urgent need of key resources in manned spaceflight; ④ Innovative technology system of fuel cell power generation based on high abundance elements; ⑤ Breakthrough new energy materials and technologies facing the future.

目　录

第一章 能量转换材料

第一节　热催化产氢材料

一、热催化产氢材料研究概述

氢气是一种重要的化工原料，大量的氢气被用于精炼石油、合成氨、制备甲醇及费-托反应等。此外，氢气还因具有能量密度高、使用终端清洁等特点，被广泛认为是一种极具应用前景的燃料及储能介质。然而氢在自然界中（与 O 结合以 H_2O 的形式存在）并无单质的沉积矿储，因此产氢成为一个重要的科学问题，深刻影响着人们的生活生产以及未来能源结构转型。

工业上由热催化（蒸汽重整）天然气（主要成分是甲烷）、石油和煤制成的氢气，分别占世界氢产量的 48%、30% 和 18%（Santhanam et al., 2017）。其中，甲醇-水蒸气重整是目前最便宜的氢气来源，也是工业上成熟的生产过程。在高温蒸汽（700～1000 ℃）的存在下，镍催化剂催化甲烷裂解产生氢气和一氧化碳（$CH_4 + H_2O \longrightarrow 3H_2 + CO$）。现已有几乎零能量损失转化甲烷制氢的报道（Malerod-Fjeld et al., 2017）。但所得产物是纯度不佳的混合气体，不能被直接用于能源工业。燃料电池中由于贵金属电极的使用，对氢气纯度有很高的要求，特别是在低温下工作的燃料电池。对在 80 ℃下工作的质子交换膜燃料电池（proton exchange membrane fuel cell，PEMFC）而言，氢气中的一氧化碳和二氧化硫（SO_2）的含量要求在 ppm① （μg/g）级水

① $1ppm=1\times10^{-6}$。

平。石油冶炼由于使用了各式各样的催化剂，对氢气的纯度要求达到99.99%以上（毛宗强和毛志明，2015）。水煤气反应（$CO+H_2O \longrightarrow CO_2 + H_2$；$\Delta H = -41.1$ kJ/mol）可以用来降低一氧化碳浓度，同时产氢，是另一个重要的热催化反应。该反应是一个放热过程，因此低温有利于反应平衡。但低温环境阻碍了动力学过程，对反应速率不利。

美国页岩气革命正在深刻影响世界能源格局，将提供大量的页岩气（主要成分是甲烷）。加上成熟的蒸汽-甲烷重整制氢工艺，一条经济的工业产氢链渐渐初显。但在应用端，尤其是在原位产氢应用中，低温高效净化氢气仍是研究的热点和难点，为此我们需要寻找更加稳定的水煤气反应催化剂。

其他的热催化产氢方法还包括热化学循环、部分氧化及离子重整等。

二、热催化产氢材料研究的历史及现状

热催化产氢技术是利用不同的能源通过热处理释放出氢气的，主要包括蒸汽重整、热化学循环制氢、部分氧化、等离子重整、生物质热化学制氢等技术。从天然气中生产氢气是目前最经济的氢气来源，天然气重组技术也是短期内向“氢经济”发展的一种主要制氢办法（毛宗强和毛志明，2015）。

（一）蒸汽重整

1. 蒸汽甲烷重整

蒸汽甲烷重整（steam methane reforming，SMR）制氢（图1-1）的研究始于20世纪20年代后期，30年代在美国建立了以天然气为原料的蒸汽转化炉，到70年代，英国帝国化学工业有限公司开发了弱碱催化剂用于天然气蒸汽重整制氢，该工艺至今仍被广泛应用（毛宗强和毛志明，2015）。在高温（700～1000 ℃）下，蒸汽甲烷重整主要反应为

$$CH_4 + H_2O = CO + 3H_2 \qquad \Delta H = +49\ \text{kcal}^{①}/\text{mol}$$

第二阶段在约360 ℃下进行低温、放热的水煤气反应，产生额外的氢气：

$$CO + H_2O = CO_2 + H_2 \qquad \Delta H = -10\ \text{kcal/mol}$$

随着反应的进行，CO_2 在此处取代 H_2O 成为反应物，因此会发生下面的反应：

$$CH_4 + CO_2 = 2CO + 2H_2 \qquad \Delta H = +59\ \text{kcal/mol}$$

上述反应均需催化剂的存在，最常用的催化剂是Ni。蒸汽甲烷重整反应

① 1 cal=4.1868 J。

是强吸热反应，具有能耗高的缺点，其燃料成本占总成本的52%～68%。该过程的主要副产品是CO、CO_2和其他温室气体。

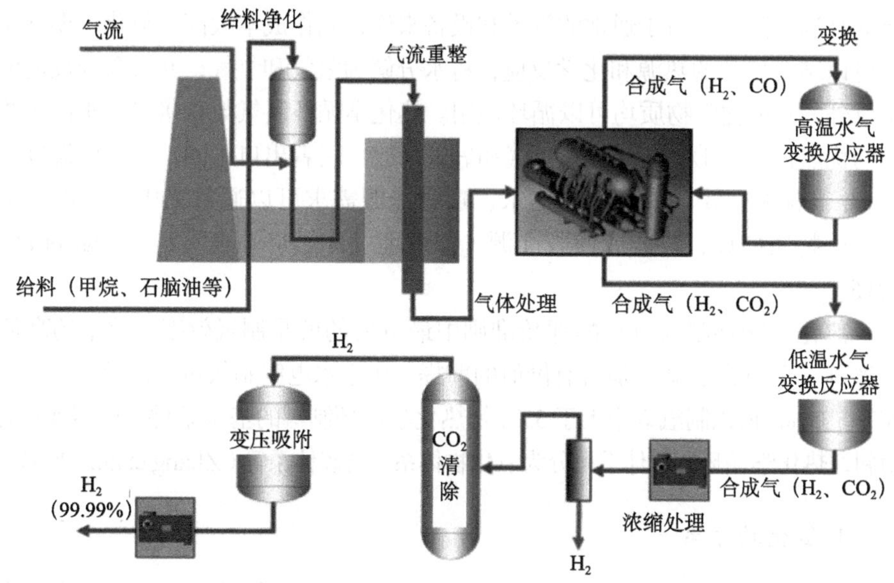

图1-1　蒸汽甲烷重整制氢示意图

2. 煤气化制氢

煤气化是指煤与催化剂在一定的温度、压力等条件下发生化学反应而转化为煤气的工艺过程。煤气化过程中使用蒸汽和精心控制的气体浓度来破坏煤中的分子键，形成氢气和一氧化碳的气态混合物。煤气化制氢曾经是主要的制氢方法，随着石油工业的兴起，特别是蒸汽甲烷重整制氢技术的出现，煤气化制氢技术呈现逐步减缓发展的态势。

我国的煤炭资源丰富，是世界上少数以煤炭为主的国家之一，1997年我国的煤炭消费占一次能源的73.5%。在中国，煤炭将长期是一次能源消费的主要来源。煤炭的价格相对低廉，而天然气价格较高，资源储量并不大，因此对我国大规模制氢并减排CO_2而言，煤气化制氢是一个重要的途径。

3. 石焦油

与煤类似，石焦油也可以通过煤气化在富氢合成气中转化。在这种情况下，合成气主要由氢气、一氧化碳和硫化氢组成，这取决于焦炭进料的硫含量（Gemayel et al.，2014）。

（二）热化学循环制氢

水在约 2700 K 时自发解离为氢气和氧气，但是若要满足这种热解离需要的高温和高压，对于通常的管道和设备来说，制作成本太高。热化学循环制氢可以结合单独的热源和化学反应，将水分解为氢气和氧气，并且参与反应的水以外的其他化学物质均可以循环使用。热化学循环制氢与直接热解水制氢相比的优点在于：①不需要分离氢气和氧气的膜，两者出口不同；②反应温度在 1000～1200 K；③零或低电能需求，这种能源需求可以通过集中太阳能、地热、生物质燃烧、核能或回收能源（即填埋气体燃烧）来满足（Acar et al., 2016）。

热化学循环制氢过程的评价准则中最重要的就是制氢效率，效率的高低是评判一个热化学循环是否有价值的前提。由于水电解制氢过程的总体效率为 26%～35%，所以制氢效率大于 35% 是热化学循环制氢的基本条件。按照涉及的材料，热化学循环制氢体系可分为氧化物体系、含硫体系等（Zhang et al., 2005）。

1. 氧化物体系

基于金属氧化物氧化还原的热化学循环如图 1-2 所示。第一步发生在太阳能反应器中，金属氧化物在吸热反应中分解产生金属。第二步是在相对较低的温度 800～1100 K 下发生放热反应。在该反应中，金属水解产生金属氧化物和氢气。金属氧化物被循环到太阳能反应器，同时氢气被收集以供进一步使用。该方法的优点在于过程步骤简单，氢气和氧气在不同步骤下生成，不存在高温气体分离等困难的分离问题。

2. 含硫体系

如图 1-3（a）所示，混合硫循环中二氧化硫被水氧化产生硫酸和氢气。在这种电解中，二氧化硫和水形成的硫酸可用于阳极去极化，产生硫酸的同时会有两个质子和两个电子，这两个质子和两个电子穿过电解质隔板，并分别通过外部回路传导到阴极，在那里它们重新结合形成氢气。这需要电力，其理论电池电位仅为 0.17 V，小于常规水电解的 1.23 V 的理论电压。

热化学硫碘（SI）循环被认为是一种有前景的大规模高效制氢的方法。如图 1-3（b）所示，它可以由太阳能或核能驱动，表现出高热效率和低环境污染。用 800～1200 ℃的热将水分解产生氢气，其过程可以连续操作，闭路循环，只需要加入水，其他物料循环使用，预期效率可以达到约 52%，联合过程（制氢与发电）效率可达 60%。

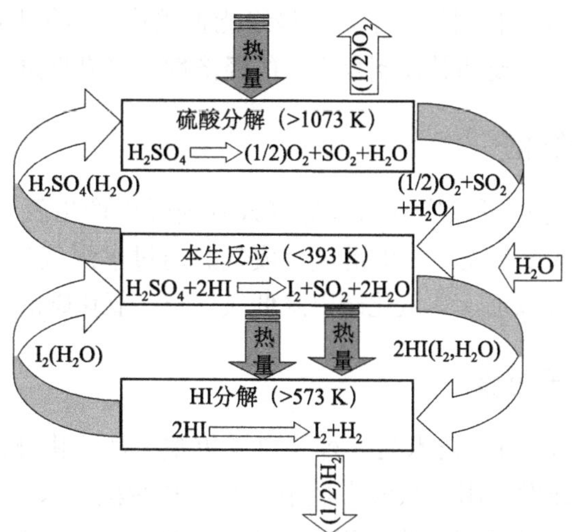

图 1-2 基于两步金属氧化物氧化还原对的热化学循环示意图（Yadav & Banerjee，2016）

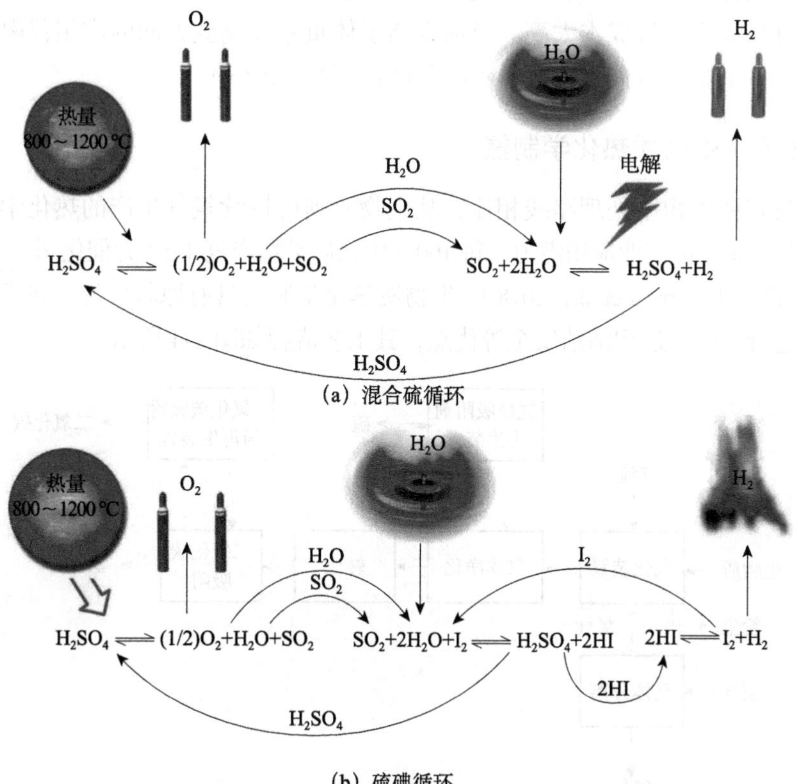

图 1-3 两个基于硫的循环的反应顺序示意图（Sattler et al.，2017）

目前热化学循环制氢技术仍处于研究阶段，距离商业化还很遥远，其成本无法与蒸汽甲烷重整制氢竞争。其最终能否成功，不仅取决于热化学循环制氢本身的技术，还要和其他制氢方法的经济性、可靠性进行比较。

（三）部分氧化

通过部分氧化也可以实现从天然气或其他烃中产生氢。燃料-空气或燃料-氧气混合物部分燃烧，产生富氢合成气。通过水煤气反应获得氢气和二氧化碳，此过程可以加入二氧化碳以降低氢气与一氧化碳的比率。

（四）等离子重整

克维乐（Kværner）技术是一种等离子重整方法，由挪威克维乐公司于20世纪80年代开发，用于从液态烃中生产氢和炭黑。采用等离子弧废物处理技术，在等离子体转换器中可以利用甲烷来生产氢气、热能等。等离子体转化碳氢化合物制氢具有反应速率快、温度低、装置体积小等优点，但是其击穿电场太高导致成本太高。目前等离子体重整在制氢方面的应用仅限于科学研究，工业应用实例很少（毛宗强和毛志明，2015）。

（五）生物质热化学制氢

与其他生物质处理路线相比，从生物质到可持续氢气生产的热化学处理路线具有很大的工业应用潜力。每100 g生物质平均产氢4 g（无催化剂）和7 g（有催化剂）（Arregi et al.，2018）。生物质热化学制氢具有原料广泛、在常温常压下进行、生产过程清洁安全等优点，其工艺流程如图1-4所示。

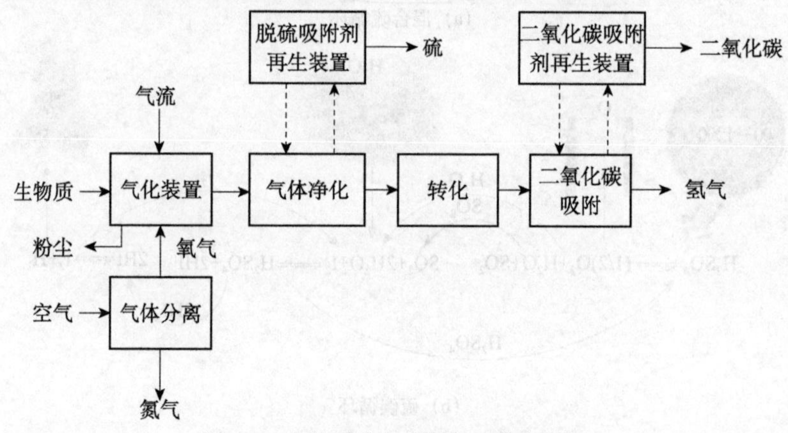

图1-4　生物质热化学制氢工艺流程图

三、热催化产氢材料研究目前面临的重大科学问题及前沿方向

在终端燃料电池应用中，氢气因清洁与高效而成为最具潜力和前景的燃料。但其活泼的化学性质，尤其是极宽的爆炸极限（4%～76%），在其运输储存补给环节存在严重的安全隐患。此外，氢气的体积能量密度很低。这些无法回避的问题决定了氢气本身不是理想的能量载体，特别是在包括汽车在内的移动体系中。解决上述矛盾的一种合理思路是将氢气储存在以甲醇为代表的稳定液体中，再原位释放以供给燃料电池（Lin & Ma，2018）。其中，甲醇液相重整制氢（$CH_3OH+H_2O \longrightarrow CO_2+3H_2$）是关键环节。

在甲醇液相重整制氢过程中，水裂解除了提供氢源，还产生含氧物种（主要认为是 OH^*），该物种与 CO 反应从而缓解其对催化剂的毒化。这一过程可认为是水煤气反应（$CO+H_2O \longrightarrow CO_2+H_2$）。在实现低温高效稳定的制氢中，水的活化是重要研究内容。事实上，在 Cu（111）面上水煤气转换的低效基准反应（benchmark reaction）中，H_2O 在金属 Cu 上的缓慢裂解被认为是速率控制步骤（rate-determining step）（Gokhale et al.，2008）。据此，贵金属-载体（noble metal-support）的双功能催化结构被广泛研究，以期在贵金属和载体上分别实现含碳物种与水的活化。

在贵金属中，催化的最优活性常见于 Pt 或 Au，而水活化载体的选择则更为广泛。早先，一类以含氧空位的可还原金属氧化物为载体的结构，如 Pt-CeO_2（Fu et al.，2003）被报道可以有效地催化水煤气反应，其性能普遍优于其他不可还原金属氧化物载体。随后实验分析表明（Shekhar et al.，2012），在可还原氧化物载体 TiO_2 上，H_2O 的表观反应级数（apparent H_2O order）显著低于其在不可还原氧化物载体 Al_2O_3 上的，这意味着可还原金属氧化物载体的活性正是得益于其可以更有效地吸附水分子物种（O^*、OH^* 及 H_2O）。碱金属离子协同的贵金属-惰性载体（如 Al_2O_3 和 SiO_2），可被归为另一类低温高效的水煤气催化剂。其催化机制的理论解释在于（Yang 等，2015），碱金属离子稳定的原子级别分散的 Pt 氧化物种 $PtK_xO_y(OH)_z$ 提供的活性位点在热力学上更利于反应进行，其中包括促进水的分解。在这两种体系中，均引入了更能促进水分解产生 OH^* 的组分，从而加速反应进行。同样的思路还被应用于电催化电极材料设计中。引入与 OH^* 结合更强的其他金属，Pt 电极在实验中氧化甲醇的耐用性得到显著的提升（Huang et al.，2018）。

燃料电池可以支持移动能源（包括汽车和船）和分布式能源战略，但其

对氢气纯度提出了新的更高的要求。实现低温高效净化氢气是热催化产氢的一个重要目标。

四、研究总结与展望

值得注意的是，在之前研究的水煤气反应体系中，对于 H_2O 参与反应的讨论基本基于朗缪尔-欣谢尔伍德（Langmuir-Hinshelwood，LH）机理，即 H_2O 先在表面分解产生 OH^*，而后被吸附在表面的 OH^* 继续参与下一步反应。在此机理下，反应活性会在原理上受到限制。表面对 OH^* 的吸附弱则不利于 H_2O 裂解产生 OH^*，对 OH^* 吸附强则不利于 OH^* 脱附继续下一步反应。即使选取吸附强度适中的表面，根据火山图拟合关系，此时整体反应速率出现火山口。此极限即为反应活性在理论上的上限，难以突破。另外一个问题是，在双功能或更多组分的催化体系中，OH^* 与其要反应的物种往往不在同一个位点。理论计算预测，在 Pt-Ru 合金中，如果 Ru 上产生的 OH^* 与吸附在 Pt 上的 CO 距离在 4 Å 以上，两者将不会反应。一个良好的催化剂对 OH^* 与含碳氧化物都有良好的吸附作用，因此它们通过扩散而产生的相互反应将被抑制。这些问题将会在动力学与热力学上阻碍 H_2O 参与反应以及限制产氢效率的提升。寻找 LH 机理之外的水裂解产氢途径十分必要。

五、学科发展政策建议与措施

国际氢能协会预计，到 2050 年，全球环境 20% 的二氧化碳的减排要靠氢气来完成。氢能汽车将占到全球车辆的 20%～25%，并承担 18% 以上的能源需求，主导脱碳社会。对此，我们需要抢占能源转型先机，加快研发体系建设，完善产业发展环境，提升关键基础材料的产品质量稳定性，降低生产成本，增强产业支撑能力。同时，加强配套政策支持，推进人才队伍建设（Tu et al.，2017）。

首先，通过热力学计算和理论可行性论证来寻找关键的化学反应材料，如低温高效水煤气催化剂；其次，用实验证实其可行性并对动力学过程进行评价（Zhang et al.，2005），为进行流程与仪器设计，可以对相关的物理性质和热力学数据进行测量，对于过程中的关键反应步骤，需要材料来验证；最后，进行经济性评价。

田　浩（南方科技大学）、唐叔贤（南方科技大学）

本节参考文献

毛宗强，毛志明 . 2015. 氢气生产及热化学作用 . 北京：化学工业出版社 .

Acar C, Dincer I, Naterer G F. 2016. Review of photocatalytic water-splitting methods for sustainable hydrogen production. International Journal of Energy Research, 40: 1449-1473.

Arregi A, Amutio M, Lopez G, et al. 2018. Evaluation of thermochemical routes for hydrogen production from biomass: a review. Energy Conversion and Management, 165: 696-719.

Fu Q, Saltsburg H, Flytzani-Stephanopoulos M. 2003. Active nonmetallic Au and Pt species on ceria-based water-gas shift catalysts. Science, 301: 935-938.

Gemayel J E, Macchi A, Hughes R, et al. 2014. Simulation of the integration of a bitumen upgrading facility and an IGCC process with carbon capture. Fuel, 117: 1288-1297.

Gokhale A A, Dumesic J A, Mavrikakis M. 2008. On the mechanism of low-temperature water gas shift reaction on copper. Journal of the American Chemical Society, 130: 1402-1414.

Huang L, Zhang X P, Wang Q Q, et al. 2018. Shape-control of Pt-Ru nanocrystals: tuning surface structure for enhanced electrocatalytic methanol oxidation. Journal of the American Chemical Society, 140: 1142-1147.

Lin L L, Ma D. 2018. Low-temperature hydrogen production from water and methanol using Pt/α-MoC catalysts. Abstracts of Papers of the American Chemical Society, 1: 255 .

Malerod-Fjeld H, Clark D, Yuste-Tirados I, et al. 2017. Thermo-electrochemical production of compressed hydrogen from methane with near-zero energy loss. Nature Energy, 2: 923-931.

Santhanam K S V, Press R J, Miri M J, et al. 2017. Introduction to Hydrogen Technology. 2nd edition. New York: John Wiley & Sons.

Sattler C, Roeb M, Agrafiotis C, et al. 2017. Solar hydrogen production via sulphur based thermochemical water-splitting. Solar Energy, 156: 30-47.

Shekhar M, Wang J, Lee W S, et al. 2012. Size and support effects for the water-gas shift catalysis over gold nanoparticles supported on model Al_2O_3 and TiO_2. Journal of the American Chemical Society, 134: 4700-4708.

Tu H L, Li F T, Ma F. 2017. The Development status and prospect of China's critical basic materials. Strategic Study of Chinese Academy of Engineering, 19(3): 125-135.

Yadav D, Banerjee R. 2016. A review of solar thermochemical processes. Renewable and Sustainable Energy Reviews, 54: 497-532.

Yang M, Liu J L, Lee S, et al. 2015. A common single-site Pt(Ⅱ)-$O(OH)_x$- species stabilized by sodium on "active" and "inert" supports catalyzes the water-gas shift reaction. Journal of the American Chemical Society, 137: 3470-3473.

Zhang P, Yu B, Chen J, et al. 2005. Study on the hydrogen production by thermochemical water splitting. Progress in Chemistry, 17: 643-650.

第二节 电催化产氢材料

一、电催化产氢材料研究概述

（一）电催化产氢材料研究的意义

能源发展是迭代发展，出于能源安全和环保压力的考虑，世界各国都在大力发展新能源。氢能作为零碳绿色的新能源，具有能量密度大、转化效率高、储量丰富和适用范围广等特点，可实现从开发到利用全过程的零碳排放、零污染，是最具发展潜力的高效替代能源。氢能是二次能源，氢能的利用横跨燃料动力、电力、供热三个领域（图 1-5），因此氢能与燃烧电池的技术发展在未来全球能源结构变革中居于重要的地位，具有重要的作用。

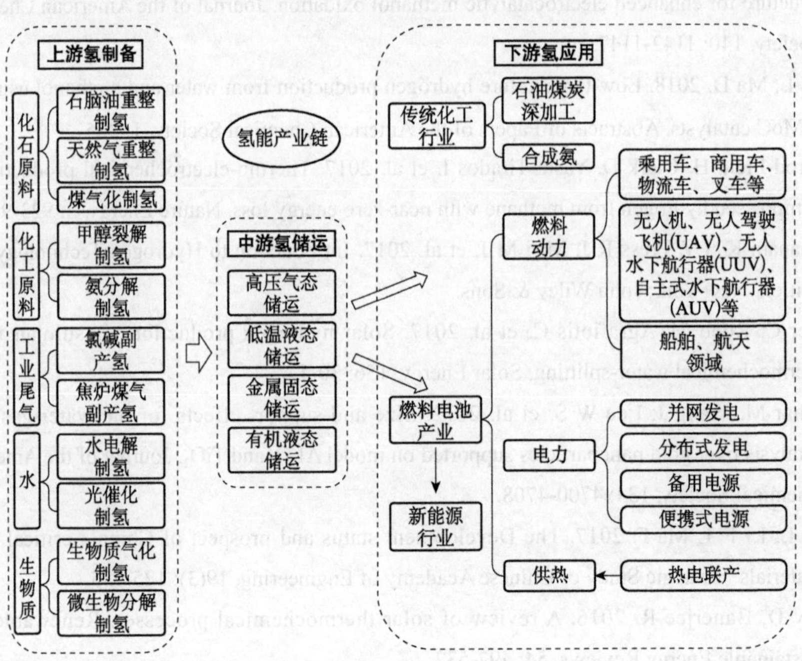

图 1-5 氢能产业链示意图

电解水产氢技术作为相对成熟的清洁产氢技术，具有为“氢经济”持续供应氢气的潜在应用前景（图 1-6）（Zou & Zhang，2015），但该技术因高昂的成本发展受限。尽管电解水技术已有较长的发展历史，但亟待持续的技术进步和材料创新，持续推动电解水的产物性能不断提升、成本不断降低、稳定性不断提高。因此，开发高效的非贵金属电催化产氢材料，是制备低成本氢气的先决条件（Artero et al.，2011；Marshall et al.，2007；Wang et al.，2012；Thoi et al.，2013；Chen et al.，2011）。

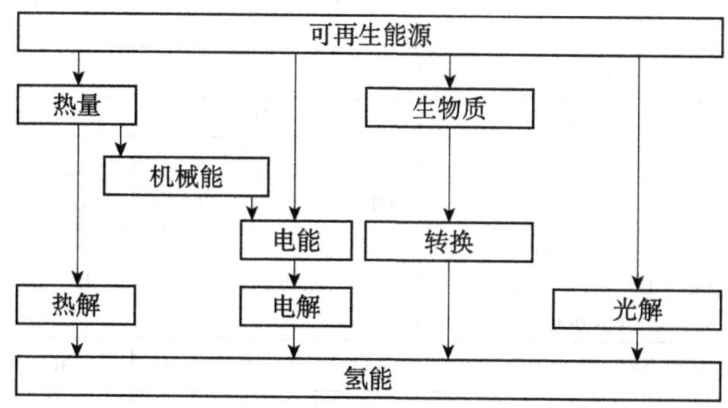

图 1-6　可再生能源制氢过程

（二）电催化产氢材料的基本原理

自 1800 年尼科尔森（Nicolson）等发现了电解水产氢现象以来，电催化产氢技术已经历了较长时期的发展。目前流行的中低温电催化技术按电解质性质可分为碱性液态电解水、（固态）碱性膜电解水和（固态）酸性质子交换膜（proton exchange membrane，PEM）电解水三类。这三类电解水产氢技术及其优缺点总结如表 1-1 所示（Sapountzi et al.，2017）。其中，碱性液态电解水是最成熟的电解产氢技术，已初步实现商用。国际知名的相关企业包括德国 Nara 电气（De Nora S.A.P）、挪威水电（Norsk Hydro）、美国 Electrolyzer 公 司（Electrolyzer Corp. Ltd）、美 国 Teledyne 能 源 系 统（Teledyne Energy Systems）公司、美国通用电气（General Electric）等。这些制造企业及其产品路线如表 1-2 所示（Zeng & Zhang，2010）。然而碱性电解的缺点仍然显著，存在电化学转换效率低、气体渗透严重等问题，需要从材料和系统两方面综合分析，优化碱性液态电解水产氢系统，降低系统能耗，提高催化反应速率和氢能转换效率。

表 1-1　主要电催化产氢技术

项目	碱性液态电解水	（固态）碱性膜电解水	（固态）酸性质子交换膜电解水
酸碱性	碱性		酸性
电解质类别	液态电解质	高分子电解质	高分子电解质
反应装置示意图	O_2 e^- e^- H_2	e^- e^- H_2O H_2O OH^- O_2 H_2	e^- e^- H_2O H^+ O_2 H_2
传输离子	OH^-	OH^-	H^+
温度 /℃	20～80	20～200	20～200
电解质形态	液态	固态（高分子）	固态（高分子）
阳极析氧反应	$4OH^- \longrightarrow 2H_2O+O_2+4e^-$	$4OH^- \longrightarrow 2H_2O+O_2+4e^-$	$2H_2O \longrightarrow 4H^++4e^-+O_2$
阳极材料	Ni>Co>Fe（氧化物）、钙钛矿（$Ba_{0.5}Sr_{0.5}Co_{0.8}Fe_{0.2}O_{3-\delta}$）、$LaCoO_3$	Ni 基材料	IrO_2、RuO_2、$Ir_xRu_{1-x}O_2$
阴极析氢反应	$2H_2O+4e^- \longrightarrow 2OH^-+H_2$	$2H_2O+4e^- \longrightarrow 2OH^-+H_2$	$4H^++4e^- \longrightarrow 2H_2$
阴极材料	Ni 合金	Ni、Ni-Fe、$NiFe_2O_4$	Pt/C、MoS_2
能量转换效率	59%～70%	65%～82%	65%～82%
实用性	商用	实验室阶段	接近商用
优点	成本低廉，材料稳定，技术成熟	将碱性和酸性固体电解质相结合	集成化设计，快速响应，低温启动，高纯氢气
缺点	电解质腐蚀性强，气体易渗透，低动力学速率	低 OH^- 离子迁移速率	电解质和催化材料成本高
发展方向	稳定性、可靠性有待提高，析氧反应效率有待提高	电解质性能和稳定性有待提高	贵金属材料的用量有待下降

表 1-2　主要相关企业电催化产氢技术一览表

参数	德国 Nora 公司	挪威水电	美国 Electrolyzer 公司	美国 Teledyne 能源系统	美国通用电气
阳极材料	Ni/ 软钢衬底	Ni/ 钢衬底	Ni/ 钢衬底	Ni 片	贵金属 / 聚四氟乙烯（PTFE）
阴极材料	Ni/ 钢衬底	Ni/ 钢衬底	钢	Ni 片	贵金属 /PTFE
气压 /MPa	无加压	无加压	无加压	0.2	0.4
温度 /℃	80	80	72	82	80
电解质	29% KOH	25% KOH	28% KOH	35% KOH	全氟磺酸（Nafion）
电流密度 /（A/m²）	1500	1750	1340	2000	5000
电压 /V	1.85	1.75	1.9	1.9	1.7
法拉第效率 /%	98.5	98.5	>99.9		
氧气纯度 /%	99.6	99.3～99.7	99.7	>98.0	>98.0
氢气纯度 /%	99.9	98.9～99.9	99.9	99.99	>99.0

电解水装置通常包含三个关键部件，即电解质、阴极、阳极。其中析氢及析氧催化材料分别涂覆在阴阳极侧，用来提高电解水速率。当提供外电压时，水分子在阴阳极分别分解出氢气和氧气，因此可分为两个半反应，即水氧化反应［析氧反应（oxygen evolution reaction，OER）］和水还原反应［析氢反应（hydrogen evolution reaction，HER）］。根据反应介质（电解质）pH变化，半反应方程各有不同。

全反应方程式：$2H_2O \longrightarrow 2H_2 + O_2$　（1-1）

酸性介质：

阴极：$2H^+ + 2e^- \longrightarrow H_2$　（1-2）

阳极：$2H_2O \longrightarrow 4H^+ + O_2 + 4e^-$　（1-3）

碱性或中性介质：

阴极：$2H_2O + 2e^- \longrightarrow H_2 + 2OH^-$　（1-4）

阳极：$4OH^- \longrightarrow 2H_2O + O_2 + 4e^-$　（1-5）

在两种电解质环境中，电解水的热力学电位在标准状态下［25 ℃、1个标准大气压（standard atmospheric pressure，atm）］均为1.23 V，并可通过提高温度显著降低电解水的热力学电位。但是在实际操作中，电解水施加电位往往远高于热力学电位，高于热力学电位的部分称为过电位（或过电势，η），包含阴阳极突破材料本征动力学势垒所需电位（η_c、η_a），以及克服其他电阻所需的电位（η_{other}）。因此，实际工作电位（E_{op}）可描述为（Cook et al.，2010）

$$E_{op} = 1.23\ \text{V} + \eta_a + \eta_c + \eta_{other} \tag{1-6}$$

从式（1-6）可知，降低过电位或操作电位是提高电解水能量效率的核心。通过优化电解池，可有效降低电阻及η_{other}，而通过选取高活性的析氧或析氢电催化材料，可有效降低η_c和η_a。此外，优化电催化材料、有效改善材料结构、增加电极的有效活性面积也是降低电极电位的重要因素。

酸性介质中，HER主要包含三种可能的步骤。福尔默（Volmer）步骤$H^+ + e^- \longrightarrow H_{ads}$。电子和质子反应，在电极表面形成吸附氢原子。随后发生塔费尔（Tafel）步骤$2H_{ads} \longrightarrow H_2$，或海罗夫斯基（Heyrovsky）步骤$H_{ads} + H^+ + e^- \longrightarrow H_2$，或同时发生Tafel步骤及Heyrovsky步骤。无论通过哪种路径发生HER，H_{ads}均会生成。因此氢吸附自由能（ΔG_H）均可作为HER的有效描述符。例如ΔG_H在Pt表面约为0，因此Pt是目前最优的析氢反应催化材料。如果$\Delta G_H >> 0$，则H_{ads}强吸附于催化材料表面，加强初始的Volmer步骤，而减慢随后的Tafel步骤或Heyrovsky步骤。反之，H_{ads}不易吸附于催化材料表面，则减慢Volmer步骤，从而限制整体HER的发生。因此，理想的HER

催化材料应具有适当的表面化学和结构性质，使表面 ΔG_H 接近于 0。

有效标定 HER 催化材料的活性的参数主要包括材料活性、塔费尔斜率、材料稳定性、法拉第效率（Faraday efficiency，FE）、周转频率（turnover frequency，TOF）等。

二、电催化产氢材料研究的历史及现状

（一）电解水产氢的历史

电解水产氢已经历了两个多世纪的发展，其发展历程可参见图 1-7。1789 年，杨·鲁道夫·德曼（Jan Rudolph Deiman）和阿德里安·派斯·范·特鲁斯维克（Adriaan Paets van Troostwijk）利用静电发电原理，采用金电极第一次把莱顿瓶中的自然水电解成氢气和氧气（Millet，1996）。1800 年，亚历山德罗·伏打（Alessandro Volta）发明了伏打电堆，随后被威廉·尼克尔森（William Nicolson）和安东尼·卡莱尔（Anthony Carlisle）用于直流电电解水，并首次证明电解水可产生氢气和氧气，其应用潜能得到关注。1833 年，法拉第总结出法拉第定律，电解水概念获得全面的科学认知和关注。1869 年，格拉姆（Gramme）发明直流发电机后，电解水产氢技术研究得到显著发展。1888 年，德米特里·拉奇诺夫（Dmitry Lachinov）首次实现工业化电解水产氢。20 世纪 20 年代，大量不同设计的电解水产氢装置获得开发，并于同期在全球范围内建造了多个 100 MW 级电解水设施。然而，第二次世界大战中，由于石油等碳氢燃料和水电的推广与盛行，以及铵类化肥厂的建造，电解水产氢一度停滞。直至 20 世纪 70 年代，随着石油危机的爆发，电解水产氢作为能源危机的解决途径重新获得重视。在当前能源危机及环境问题大背景下，电解水制氢技术成为一种解决环境与能源危机，实现可持续发展的最有前景的技术方法（Carmo et al.，2013）。近些年来，电解水制氢技术有了进一步长足的发展。

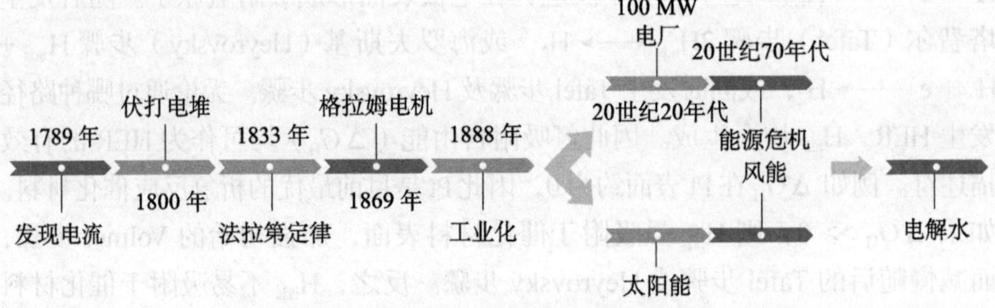

图 1-7 电解水产氢发展历程（Zou & Zhang，2015）

（二）电催化产氢材料国内外发展现状

图 1-8 显示了目前主要用于构筑电催化产氢材料的元素。根据其物理和化学性质的不同，这些元素可分为 3 类。Pt 基析氢催化剂；非贵金属催化剂，主要包括铁（Fe）、钴（Co）、镍（Ni）、铜（Cu）、钼（Mo）、钨（W）；非金属催化剂，主要包括硼（B）、碳（C）、氮（N）、磷（P）、硫（S）、硒（Se）等。

元素周期表（节选）

1	2	3	4	5	6	7	8	9	10	11	12	13	14	15	16	17	18
H																	He
Li	Be											B	C	N	O	F	Ne
Na	Mg											Al	Si	P	S	Cl	Ar
K	Ca	Sc	Ti	V	Cr	Mn	Fe	Co	Ni	Cu	Zn	Ga	Ge	As	Se	Br	Kr
Rb	Sr	Y	Zr	Nb	Mo	Tc	Ru	Rh	Pd	Ag	Cd	In	Sn	Sb	Te	I	Xe
Cs	Ba	La~Lu	Hf	Ta	W	Re	Os	Ir	Pt	Au	Hg	Tl	Pb	Bi	Po	At	Rn

■ Pt基析氢催化剂
▨ 非贵金属催化剂
■ 非金属催化剂

图 1-8 构筑 HER 催化剂的主要元素（Zou & Zhang，2015）

图 1-9 显示了各金属元素的地壳丰度。通过比较，可获得以下结论：①作为理想的电催化产氢材料，Pt 的丰度仅为 $3.7\times10^{-6}\%$，远低于其他元素；②其他 6 种金属元素的丰度排序依次为 W = Mo < Co < Cu < Ni << Fe。因此在设计和构筑新型催化材料时，应关注金属丰度和潜在成本。由于 Fe、Ni 的地壳丰度最高、售价最低，因此应首选设计制备铁基或镍基电催化材料。

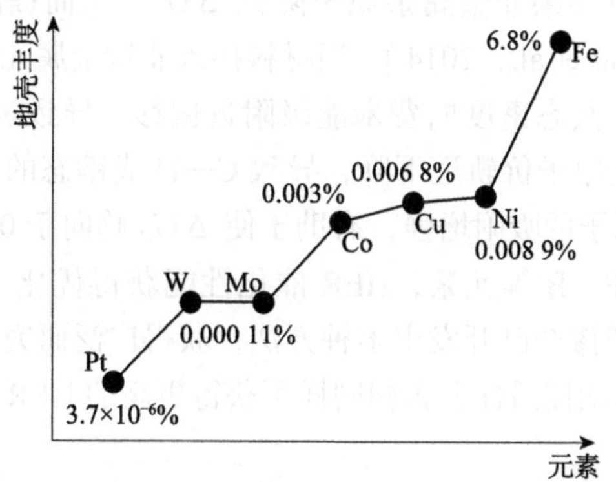

图 1-9 各元素的地壳丰度对比（Zou & Zhang，2015）

1. 酸性介质中的 HER 电催化材料

质子交换膜电解池及光伏器件可在酸性条件下运行，并具有高效率、低重量等特点，可广泛应用于电催化产氢。因此，开发酸性介质中高效稳定的 HER 电催化材料具有更大的商业价值（Hambourger et al.，2008；Le Goff et al.，2009）。同时基于 HER 的反应机理，酸性电解质中具有更多质子，更利于产氢。对于酸性介质中的 HER 电催化材料，目前已有大量报道。然而由于酸性条件的强腐蚀作用，电催化材料的稳定性差。其中，碳材料因在酸性条件下的抗腐蚀能力强而受到广泛关注。目前已有大量稳定高效的碳基非贵金属电催化产氢材料的报道。这些材料主要可分为以下几类：①负载型电催化材料；②包覆型电催化材料；③单原子电催化材料；④自支撑型电催化材料。

1）负载型电催化材料

负载型电催化材料的制备方法主要包括电沉积、水热法或低温热解法等。负载型电催化材料的金属相均暴露在表面，可以有效提高金属与电解液的接触面积。目前已有大量酸性环境下负载型电催化材料的报道，主要包括非贵金属硫化物、磷化物、硒化物、氮化物、碳化物等，主要关注材料的催化性能及稳定性（Vrubel & Hu，2012；Chen，1996）。其中碳基非贵金属硫化物、磷化物目前是 HER 性能最优的材料。

目前负载型电催化材料的性能优化主要依靠改变碳载体和活性材料的结构与电子性质实现。其中碳基材料主要包括碳纳米管、石墨烯、氧化石墨烯（graphene oxide，GO）、多孔碳材料等。这些碳材料因其可控的分子结构、优越的导电性、孔结构与化学稳定性，成为催化材料的优异负载材料（Liu et al.，2015）。碳载体和活性材料的结合，可有效提高电化学活性材料的导电性和分散性。密度泛函理论（density functional theory，DFT）计算表明，纳米碳材料可通过掺杂非金属杂原子降低 ΔG_H，从而提高 HER 活性（Jiao et al.，2016；Zheng et al.，2014）。当材料掺入非贵金属元素，在杂原子附近形成碳活性中心，其态密度向费米能级附近偏移，导致氢吸附能显著增加。同时具有活性的碳原子价轨道下降，导致 C—H 成键态的电子填充增强，从而使碳原子与氢原子的吸附增强，有助于使 ΔG_H 趋向于 0。实验证明，当碳骨架掺杂 N、S、P、B 等元素，HER 催化性能获得优化（Ito et al.，2015）。尤其是针对 N 原子掺杂已开发出多种方法，获得广泛研究和关注。此外，通过掺杂非金属元素调控活性金属相同样可获得更高的 HER 性能（Duan et al.，2016）。

（1）碳载非贵金属硫化物催化材料。每种层状金属硫化物均具有六方晶格结构。硫化物块体材料依靠层间范德瓦耳斯力，由单层材料堆叠构成。同时层状硫化物可生成其他各种低维材料，如纳米管、纳米带、纳米团簇或纳米粒子（nanoparticle，NP）。低维材料的表面缺陷，如边缘、拐角等处是 HER 的主要活性位点。在已有的报道中，基于硫化钼（MoS_x）（Firmiano et al.，2012；Yang et al.，2016；Liao et al.，2013）、WS_2、CoS_2 的金属硫化物受到广泛关注（Yang et al.，2013；Shifa et al.，2015）。其中 MoS_x 的研究最为广泛，这种材料的活性中心集中在边缘与缺陷处，而无缺陷的二维平面无催化性能（Gao et al.，2013）。因此在纳米尺度上对 MoS_x 材料进行调控，使其暴露更多的边界，可获得更高的 HER 催化性能。得益于金属前驱体和碳载体之间的强化学键，MoS_x 纳米粒子可具有低尺度和高分散性。此外，碳载体和 MoS_x 之间紧密的电子相互作用将促进碳载体和 MoS_x 之间的快速电荷转移，从而加速氢的生成。

（2）碳载非贵金属磷化物催化材料。非贵金属磷化物是一类重要的 HER 催化材料。磷原子可掺入金属原子的晶格间，磷原子的引入使金属原子之间的距离略有增加，并且随着金属的 d 带带隙收缩，导致费米能级附近的金属态密度增强，使金属磷化物具有贵金属的特性。此外，非贵金属磷化物在化学成分上模拟生物氢酶活性位点，有可能替代铂基电催化剂。然而，目前大多数非贵金属磷化物导电性差，活性位点不足。用碳纳米材料担载非贵金属金属磷化物可有效解决导电性和材料结构性质的限制。目前已经建立了许多非贵金属磷化物基复合材料，如 CoP@C（Zhang et al.，2015a；Pan et al.，2016；Wang et al.，2015）、FeP@C（Zhu et al.，2016；Zhang et al.，2015b）、NiP@C（Pan et al.，2015）。所报道的非贵金属磷化物基复合材料的催化机理类似于氢化酶的催化机理。金属位点和磷原子分别作为氢化物受体中心与质子受体中心（Popczun et al.，2013）。金属位点在整个系统中充当活性中心（Artero et al.，2011）。

（3）碳载非贵金属碳化物催化材料。非贵金属碳化物是由填充入非贵金属晶格空隙中的碳原子与金属形成的金属型碳化物。非贵金属碳化物的表面性质和催化性能类似于贵金属。在已发表的碳载非贵金属碳化物催化材料中，碳化钼因其独特的 d 带电子结构而显示出良好的 HER 性能（Levy & Boudart，1973）。例如，阿齐克（Adzic）等通过钼酸铵原位掺碳合成了修饰在碳纳米管上的 β-Mo_2C（Mo_2C/CNT）。Mo_2C/CNT 具有优异的活性，过电位仅为 152 mV，塔费尔斜率仅为 55.2 mV。其中钼原子充当氢吸附位点，碳化物配体的引入改变了 Mo_2C 的 d 电子构型，提供了相对适中的 Mo—H 结合强度（Chen et al.，2013）。同样，在碳纳米管-石墨烯复合载体上担载的 Mo_2C 已被证明是高效的 HER 电催化剂（Youn et al.，2014）。碳纳米管-石墨烯复合载体构建三维大孔结构，导致快速的电子和电解质转移。另外，分散良好的 Mo_2C 纳米粒子提供了更多的催化位点。此外，Mo_2C 与石墨烯片的结合（Mo_2C-G）（He & Tao，2015）或与石墨碳片的结合（Mo_2C-GCS）被证实为高效且稳定的电催化剂（Cui et al.，2014）。Mo_2C 基复合材料的高活性源自 Mo_2C 和碳载体之间的共价结合，降低了 HER 中的氢结合能。

（4）碳载非贵金属氮化物催化材料。与非贵金属碳化物相似，非贵金属氮化物由填充入非贵金属晶格空隙中的氮原子形成。近年来，一些团队利用非贵金属氮化物来研究 HER，例如还原氧化石墨烯（reduced graphene oxide，rGO）担载磷改性的氮化钨（P-WN/rGO）（Yan et al.，2015）。P-WN/rGO 复合材料 10 mA/cm^2 处过电位仅为 85 mV。磷改性可以极大地改善 HER 在酸性溶液中的催化性能和耐久性。掺杂磷后，WN/rGO 界面发生明显的极化，导致 WN/rGO 表面发生大量的电负性变化，形成的负电荷表面有利于 HER。

2）包覆型电催化材料

尽管酸性介质中负载型电催化剂的研究取得了很大进展，但由于活性组分易受酸腐蚀，其稳定性并不够高。在石墨烯壳层或碳纳米管中封装金属纳米粒子，特别是对于易受酸腐蚀和氧化的金属相来说，这是保持催化材料反应活性和稳定性的有效方法（Zhou et al.，2016a；Lu et al.，2016）。目前相关的研究主要集中在 Fe、Ni、Co 及其衍生物（Zhou et al.，2015a，2016b；Fan et al.，2015；Deng et al.，2014，2015），W 氧化物（Wu et al.，2015）和 Mo_2C 基复合材料的设计上（Ma et al.，2015）。研究证明，包覆型电催化材料是 HER 优异的电催化材料，石墨烯外壳封装未导致内部金属纳米粒子催化活性的降低。目前氮掺杂碳包覆材料越来越受到人们的关注。由于氮原子独特的孤电子对，其电导率和催化能力均可调控。由于电负性的差异，氮掺杂可以激活相邻的碳原子，并进一步增加活性位点的密度。结合实验结果和密度泛函理论计算，包覆型电催化材料优异的催化活性源于金属核与掺杂剂之间的协同效应。根据涂层金属芯的类型，我们将包覆型电催化材料分为纳米金属或合金内核型、金属化合物内核型。

3）单原子电催化材料

一般认为，小尺寸金属纳米粒子具有优异的催化性能或选择性。理论上，负载型催化剂的分散极限是金属以独立的单原子形式存在。为了最大限度地发挥金属的催化效率，降低制造成本，单原子电催化材料的合成是一个重要的突破。单原子催化材料的研究是一个非常年轻的领域。迄今，应用于电化学 HER 的单原子电催化材料很少，主要集中在 Co 或 Ni 金属。

然而，单原子电催化材料仍不完美。当金属颗粒减少到单原子水平时，比表面积急剧增加，导致金属表面自由能急剧增加。金属颗粒在制备或反应过程中容易聚集形成团簇，这会不可避免地导致催化剂的失活。尽管报道的单原子电催化材料显示出潜在的活性，但是由于低负载，单原子电催化材料的性能仍然远远没有达到实际应用水平。此外，目前的制备方法需要使用大量的酸来洗掉大部分的金属，这是不经济和不环保的。发展高效的单原子电催化材料制备策略迫在眉睫。

4）自支撑型电催化材料

上述催化剂大部分必须使用聚合物黏合剂（全氟磺酸或 PTFE）黏附在电极表面，费时且消耗成本。一方面，聚合物黏合剂不可避免地增加了串联电阻；另一方面，它可能遮挡活性位点并抑制气体的渗透性，从而降低材料的催化效率（Zhu et al.，2015）。直接在集电体上生长活性物质可使 HER 更加直接和平稳。自支撑型电极的制备不仅可以大大简化电极制备过程，还可以有效提高活性材料的负载质量。通常自支撑型电极的制备通过以下两种方法获得：①活性材料在导电基底［如碳布（CC）］表面生长；②直接将电极材料制备成薄膜。参照合成方法，自支撑型电极也可分为两类，即碳布电极和薄膜电极。

（1）碳布电极。目前碳布电极的研究主要集中于非贵金属磷化物（Yang et al.，2015）、硫化物（Yan et al.，2015）、硒化物（Wang et al.，2015）、碳化物（Fan et al.，2015；Zhou et al.，2015b）等。特别是非贵金属磷化物和硫化物由于其高固有活性而被广泛应用于制备碳布电极。对于非贵金属磷化物/碳布电极（MP/CC），制备过程通常通过两个步骤进行：①在碳布上生长金属氧化物、氢氧化物或纯金属；②用亚硫酸氢钠低温磷化。与负载型电催化剂电致发光的原因相似，其优异的性能可以归结为：①碳布的高导电性及其与 MP 纳米粒子之间的强电子耦合；②独特的三维（3D）结构推动了 H_2 的释放；③金属和 P 原子分别作为氢化物受体与质子受体中心，同时 P 进一步促进了金属氢化物的形成。

（2）薄膜电极。除在碳布或碳纤维上合成催化材料之外，自支撑型电极的制备还可以通过自支撑膜来实现。通常，薄膜的形成往往需要利用石墨烯、氧化石墨烯、碳纳米管等碳材料的力学特性、柔韧性和分子间作用力。

令人印象深刻的是，罗摩克里希纳（Ramakrishna）团队基于 CoS_2、rGO 和 CNT 设计，并采用水热法和真空过滤相结合的方法建立了自支撑型三维电极（CoS_2/rGO-CNT）（Peng et al.，2014）。连续的三维分级结构、复合纳米粒子与 CNT 之间的紧密接触，以及 CoS_2、rGO 和 CNT 的协同耦合共同促成了高 HER 活性。这项工作为先进的自支撑型电极设计提供了一个新的方向。

自支撑型 3D 电极的应用不仅避免了串联电阻，而且能有效防止测试过程中活性物质的剥离。更重要的是，在集电器上集成活性材料可以大大增加质量负载，从而实现 H_2 的规模生产。然而，目前报道的大多数催化剂由于其低比表面积和孔结构而受到应用限制。此外，大部分材料是通过电沉积或水热工艺制备的，需要进一步采用磷化（NaH_2PO_2）或硫化（S 粉）工艺，在此过程中形成有毒的 PH_3 或 H_2S。因此，开发简易和环境友好的方法来合成 3D 多孔电极是非常重要的。

2. 碱性介质中的 HER 电催化材料

尽管目前已开发了大量适应酸性介质的 HER 电催化材料，并且显示出优异的活性，但是通常在高温下操作的酸性质子交换膜电解池不可避免地产生电解液蒸汽或酸雾，不仅污染氢气，而且容易腐蚀装置。相反，碱性电解槽通常在较高的温度、较低的蒸汽压下工作，能够产生高纯度的氢气。在碱性介质中，非贵金属电催化剂可保持稳定性，避免腐蚀溶解，因此相对于酸性质子交换膜电解池，碱性电解也是一种较优的选择。然而，基于不同电解质下的水电解机理，碱性电解质下的 HER 反应比酸性电解质下的 HER 反应更难。碱性溶液中氢的生成需要附加的水分解步骤，这可能引入额外的势垒，影响总反应速率，因此，碱性介质中 HER 高性能催化材料的研究相对较少。到目前为止，3d 过渡金属（如 Fe、Co、Ni）已被广泛用作碱性介质中的阴极材料（Danilovic et al.，2012）。在这些材料中，Ni 基催化材料被认为是最有前途的电极材料。然而，Ni 活性较差。在过去的几十年里，人们的兴趣已经转移到 Ni 基合金和高比表面积雷尼型电极的设计上（Safizadeh et al.，2015）。目前，杂原子掺杂的石墨碳材料也已经应用于碱性 HER（Jiao et al.，2015）。由于掺杂原子（如 N、B、S、P）的大表面积、良好的导电性和活化效果，掺杂碳材料具有潜在的 HER 活性（Duan et al.，2015）。因此非贵金属基碳复合材料在水分解中的应用也可以扩展到碱性电解质，但活性相对较弱。然而，一般廉价的 3d 过渡金属、合金、氧化物和氢氧化物催化材料，由于稳定性差，在酸性电解质中性能和稳定性较差，可被应用于碱性介质中，并表现出对 HER 的优异催化性能。

3. 中性介质中的 HER 电催化材料

大多数关于电化学 HER 的研究都是在酸性和碱性条件下进行的。然而，强碱和酸性条件对水分解装置提出了更高的要求。相反，中性电解液环境友好、腐蚀性小，对电解装置要求低，降低了水分解系统的成本。此外，电极材料在反应过程中较少受到电解质的影响，因此几乎所有种类的电催化剂都可以应用于中性溶液。然而中性介质欧姆损耗较大且质子浓度较低，与碱性和酸性条件相比，在中性介质中进行的 HER 动力学较慢（Zou et al.，2014）。为了在中性介质中达到相当高的反应速率，水分子必须充当反应物，因此需要使用高得多的外加电压来达到中性溶液中与酸碱溶液相同水平的电流密度。近年来，在中性条件下报道了一些关于非贵金属（如 Ni、Fe、Cu、W、Mo、Co）电催化材料的研究（Jiang et al.，2014；Zhao et al.，2015；Mitov et al.，2012；Shi & Hu，2015；Du et al.，2015）。在已知的高地球丰度的电催化剂中，FeP、CoP 和 CoS 显示出相对优越的活性（Zhang et al.，2016）。然而在中性条件下，非贵金属碳基复合材料和无金属碳材料在 HER 中的应用很少。

所有这些工作都有助于为开发高效稳定的中性介质中水还原电催化剂铺平道路，然而中性溶液中催化剂表面质子还原的动力学过程和分子表面原位变化等方面的工作还有待进一步研究。此外，中性缓冲溶液的制备相对复杂，探索在硫酸钠溶液等电解质中具有优异析氢活性的催化材料极具挑战性。

4. 双功能电催化材料

尽管目前已开发的非贵金属碳基复合材料主要用于高效制氢，但是用于全水分解（同时将水分解成氢和氧）的双功能催化剂的研究也有一定的吸引力（Duan et al.，2016；Chen et al.，2016）。两电极体系不仅更接近实际应用的电解池，而且可以避免使用昂贵的铂作为对电极。然而，几乎所有非贵金属 OER 催化剂仅在中性或碱性介质中保持一定的稳定性。同时，除了少数能在中性和碱性介质中工作，大多数最近开发的 HER 催化剂仅在酸性介质中保持高效及高稳定性。因此，很难找到合适的 HER 和 OER 双功能催化剂，它们能在相同的酸碱度范围内很好地工作。因此，探索全水分解电催化剂是一个具有挑战性和紧迫性的问题。

三、电催化产氢材料研究目前面临的重大科学问题及前沿方向

未来的“氢经济”是一个利用氢储存和运输能源的系统。在氢循环过程中有三个主要过程（图 1-10）：收集可再生能源（如太阳能和风能），利用可再生能源将 H_2O 分解成 H_2 和 O_2，以及通过 H_2 与 O_2 反应来重新释放能量。高效 HER 电催化材料设计开发的主要目标之一是利用廉价且地壳丰度高的非贵金属电催化材料取代昂贵且稀有的贵金属电催化材料。

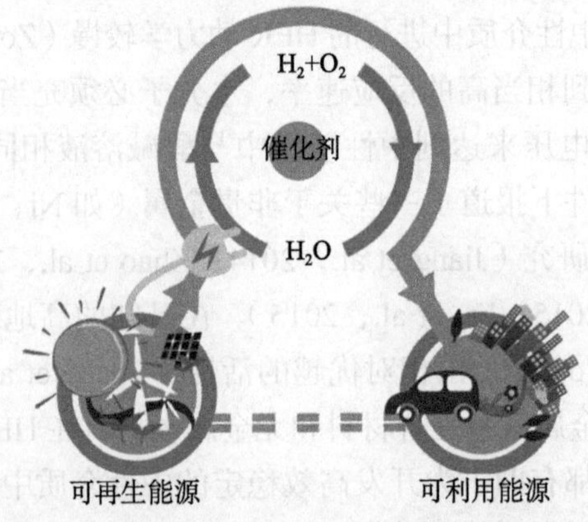

图 1-10　未来氢循环路径（Zou & Zhang，2015）

过去的几十年中，非贵金属碳基电催化材料作为 Pt 基材料的替代品，在更高效的 HER 的构建方面不断取得突破，主要集中在高地壳丰度的元素，如 Mo、Fe、Co、Ni、W 等，并与石墨烯、rGO、CNT、泡沫碳、碳气凝胶、多孔碳材料、功能化碳材料、碳布、碳纸等导电纳米碳材料复合。碳材料的引入大大改善了催化体系的导电性和微纳结构。通过碳和金属离子之间的紧密结合，与块状金属材料相比，金属纳米颗粒具有更小的尺度和更高的分散性，有利于暴露更多的活性位点。此外，碳载体和活性金属的协同效应进一步促进了更好的电荷转移，增强了催化活性。

核壳结构的设计可有效防止金属腐蚀，保持材料稳定性。然而实现核壳结构厚度和成分可控仍然是一大挑战。此外，利用 N、S、P、B、O 等元素对碳壳进行化学改性，调控碳壳电负性，可进一步改善材料的催化性能。采用多原子掺杂的催化材料使组分更加复杂，因而难以识别实际的活性位点。尽管密度泛函理论计算可在一定程度上预测反应中间物种和活性中心，但理论模型并不能完全反映真实的催化体系，因此对多组分体系的催化机理仍有

待进一步研究。

大多数非贵金属碳基电催化材料适用于酸性电解质。金属氧化物和氢氧化物对酸不稳定，通常适用于碱性电解质，且活性相对较差。许多碱性稳定性差的金属化合物，如 WS_2、WC、MoS_2、Mo_2C 等，更适用于酸性体系。

尽管 HER 电催化材料已经取得了巨大的研究进展，但在寻找性能优异的 HER 电催化材料方面进展仍相对缓慢。主要原因之一是，材料选择主要通过实验筛选。对于材料结构与催化性能之间的构效关系及机理研究仍很匮乏。其中的关键性科学问题总结如下。

1. 机理分析

对 HER 电催化材料的机理研究不仅具有科学意义，而且可以为材料的性能优化提供理论指导。对大多数已知的 HER 电催化材料，尤其是复合催化材料的研究仍然缺乏原子水平的机理分析。此外，碱性 HER 电催化材料催化机理较模糊。因此，需要将理论模拟和原位表征技术相结合，以阐明催化机理。

2. 标准化测试

建立标准化测试有助于比较不同材料，并筛选最佳 HER 电催化材料与合成方法。由于材料质量负载、电极制备方法与反应电解质等存在差异，很难直接比较各种催化材料。例如在大多数研究中，测得电流仅对电极几何面积归一化，但催化材料的负载量常被忽略。由于总测量电流往往与催化材料的质量或物质的量有关，所以仅通过电极面积归一化无法实现不同材料之间性能评估的真实性。因此，研究人员应该提供尽可能多的电极表征信息，以便进行不同材料间的比较。其中重要的性能参数主要包括塔费尔斜率、交换电流密度、催化材料负载量、按质量或电极面积归一化判定的催化活性、法拉第效率、材料稳定性等。此外，应大力推动 HER 的标准化研究。在相同条件下，使用标准方法对电催化材料性能进行客观比较，有助于评估现有电催化材料，并为新催化体系的开发提供参考。在这方面，加州理工学院的麦克罗里（McCrory）等做了一些非常有意义的研究（Zou et al.，2013；Kanan & Nocera，2008）。

同时，亟须开发原位谱学技术探测反应过程中催化材料的活性位点和界面结构。尽管已经进行了一些光谱测量，如原位拉曼光谱、X 射线吸收光谱，但是目前的原位光谱测量技术仍非常有限。

3. 探索新材料

探索新型 HER 电催化材料将是未来几年的核心研究目标之一。一种理想的非铂 HER 电催化材料应该满足几个标准：①与 Pt 相比拟的高产氢效率；②至少几年的耐久性；③在较宽的 pH 范围内，甚至在所有 pH 下，具有高化学 / 催化稳定性；④低成本，确保廉价的氢气生产；⑤规模化的可扩展性，以确保广泛的产业化应用。然而迄今没有一种已知的非贵金属 HER 电催化材料具有上述优点，因此一个可行的办法是为特定的应用选择合适的 HER 催化材料。过渡金属 Fe 和 Ni 化合物由于 Fe 和 Ni 的天然丰度及化合物的潜在催化活性，将是未来 HER 电催化材料的首选。此外，应积极开发耐海水的 HER 催化材料，使海水的直接利用成为可能。

4. HER 与 OER 电催化材料的结合

在电催化材料和半导体光催化材料上，电化学水分解的效率不仅由 HER 和 OER 电催化材料本身决定，而且由它们的相容性决定。同时 HER 电催化材料和半导体材料的适应性对光催化或光电化学水分解的效率产生重要影响。因此，HER 电催化材料各方面的综合评估对将其用于真正的电化学或光电化学水分解系统来说是必要的。

四、研究总结与展望

回顾电催化产氢材料的发展历程，特别是最近十余年的研究成果，可见其取得了长足的进步。在能源与环境问题强大需求的推动下，国际上电催化产氢领域的研究已由基础理论研究逐步转向电催化材料的应用基础研究，由电催化材料制备逐步转向电催化材料体系设计。在研究手段方面，已能够从分子、原子水平上揭示电催化材料的基本物理性质以及电催化材料的构效关系，从飞秒尺度上研究 HER 反应过程和机理。包括第一性原理与分子动力学模拟在内的现代科学计算方法逐渐在电催化产氢材料特性与电催化基本原理研究方面起到重要作用。以电化学、纳米材料科学和化学热力学与动力学为基础的较为完整的电催化基础理论体系已经初步建立。电催化产氢已发展成物理学、化学、能源、环境等多学科交叉领域，成为热门研究领域之一。电催化产氢的最新研究进展与发展趋势主要体现在对电催化产氢转化效率的限制因素认识与效率提升、电催化机理认识与表征手段、基于新物理机制的电催化材料设计、改善电催化效率的方法、电催化反应中材料构效关系等方面。

五、学科发展政策建议与措施

电催化产氢材料学科的发展经历了两百多年的历史，依然保持着旺盛的发展动力，这不仅是因为该学科的发展有重要的应用前景，还因为这一学科自身的挑战性，其对相关学科的发展具有极大的促进作用。基于我国电催化产氢材料学科的发展呈现良好的上升势头，有必要整合现有的研究队伍，以项目群形式组织人才队伍，以发挥各自优势，攻克该领域重大基础问题。

第一，优先发展高效电催化产氢材料设计理论、方法，提出可用于指导高效电催化产氢材料开发的物理化学参数指标。

第二，优先发展电化学能量转化反应的动力学原位检测手段，揭示多界面电荷传输机制。

第三，优先发展先进材料制备科学技术、先进薄膜材料制备技术，阐明材料构效关系，为电催化产氢技术工程化提供材料制备方面的科学支撑。

第四，引入新奇物理化学效应，注重电催化产氢材料学科与其他学科的交叉融合研究。

姚颖方（南京大学）

本节参考文献

Artero V, Chavarot-Kerlidou M, Fontecave M. 2011. Splitting water with cobalt. Angewandte Chemie International Edition, 50: 7238-7266.

Carmo M, Fritz D L, Merge J, et al. 2013. A comprehensive review on PEM water electrolysis. International Journal of Hydrogen Energy, 38: 4901-4934.

Chen G F, Ma T Y, Liu Z Q, et al. 2016. Efficient and stable bifunctional electrocatalysts Ni/Ni_xM_y（M = P, S）for overall water splitting. Advanced Functional Materials, 26: 3314-3323.

Chen J G G. 1996. Carbide and nitride overlayers on early transition metal surfaces: preparation, characterization, and reactivities. Chemical Reviews, 96: 1477-1498.

Chen W F, Wang C H, Sasaki K, et al. 2013. Highly active and durable nanostructured molybdenum carbide electrocatalysts for hydrogen production. Energy & Environmental Science, 6: 943-951.

Chen Z W, Higgins D, Yu A P, et al. 2011. A review on non-precious metal electrocatalysts for

PEM fuel cells. Energy & Environmental Science, 4: 3167-3192.

Cook T R, Dogutan D K, Reece S Y, et al. 2010. Solar energy supply and storage for the legacy and nonlegacy worlds. Chemical Reviews, 110: 6474-6502.

Cui W, Cheng N Y, Liu Q, et al. 2014. Mo_2C nanoparticles decorated graphitic carbon sheets: biopolymer-derived solid-state synthesis and application as an efficient electrocatalyst for hydrogen generation. ACS Catalysis, 4: 2658-2661.

Danilovic N, Subbaraman R, Strmcnik D, et al. 2012. Enhancing the alkaline hydrogen evolution reaction activity through the bifunctionality of Ni $(OH)_2$/metal catalysts. Angewandte Chemie International Edition, 51: 12495-12498.

Deng J, Ren P J, Deng D H, et al. 2014. Highly active and durable non-precious-metal catalysts encapsulated in carbon nanotubes for hydrogen evolution reaction. Energy & Environmental Science, 7: 1919-1923.

Deng J, Ren P J, Deng D H, et al. 2015. Enhanced electron penetration through an ultrathin graphene layer for highly efficient catalysis of the hydrogen evolution reaction. Angewandte Chemie International Edition, 54: 2100-2104.

Du H F, Gu S, Liu R W, et al. 2015. Tungsten diphosphide nanorods as an efficient catalyst for electrochemical hydrogen evolution. Journal of Power Sources, 278: 540-545.

Duan J J, Chen S, Jaroniec M, et al. 2015. Heteroatom-doped graphene-based materials for energy-relevant electrocatalytic processes. ACS Catalysis, 5: 5207-5234.

Duan J J, Chen S, Vasileff A, et al. 2016. Anion and cation modulation in metal compounds for bifunctional overall water splitting. ACS Nano, 10: 8738-8745.

Fan M H, Chen H, Wu Y Y, et al. 2015. Growth of molybdenum carbide micro-islands on carbon cloth toward binder-free cathodes for efficient hydrogen evolution reaction. Journal of Materials Chemistry A, 3: 16320-16326.

Fan X J, Peng Z W, Ye R Q, et al. 2015. M_3C (M: Fe, Co, Ni) nanocrystals encased in graphene nanoribbons: an active and stable bifunctional electrocatalyst for oxygen reduction and hydrogen evolution reactions. ACS Nano, 9: 7407-7418.

Firmiano E G S, Cordeiro M A L, Rabelo A C, et al. 2012. Graphene oxide as a highly selective substrate to synthesize a layered MoS_2 hybrid electrocatalyst. Chemical Communications, 48: 7687-7689.

Gao M R, Xu Y F, Jiang J, et al. 2013. Nanostructured metal chalcogenides: synthesis, modification, and applications in energy conversion and storage devices. Chemical Society Reviews, 42: 2986-3017.

Hambourger M, Gervaldo M, Svedruzic D, et al. 2008. [FeFe]-hydrogenase-catalyzed H_2 production in a photoelectrochemical biofuel cell. Journal of the American Chemical Society,

130: 2015-2022.

He C Y, Tao J Z. 2015. Synthesis of nanostructured clean surface molybdenum carbides on graphene sheets as efficient and stable hydrogen evolution reaction catalysts. Chemical Communications, 51: 8323-8325.

Ito Y, Cong W T, Fujita T, et al. 2015. High catalytic activity of nitrogen and sulfur co-doped nanoporous graphene in the hydrogen evolution reaction. Angewandte Chemie International Edition, 54: 2131-2136 .

Jiang N, Bogoev L, Popova M, et al. 2014. Electrodeposited nickel-sulfide films as competent hydrogen evolution catalysts in neutral water. Journal of Materials Chemistry A, 2: 19407-19414.

Jiao Y, Zheng Y, Davey K, et al. 2016. Activity origin and catalyst design principles for electrocatalytic hydrogen evolution on heteroatom-doped graphene. Nature Energy, 1: 16130.

Jiao Y, Zheng Y, Jaroniec M, et al. 2015. Design of electrocatalysts for oxygen- and hydrogen-involving energy conversion reactions. Chemical Society Reviews, 44: 2060-2086.

Kanan M W, Nocera D G. 2008. *In situ* formation of an oxygen-evolving catalyst in neutral water containing phosphate and Co^{2+}. Science, 321: 1072-1075.

Le Goff A, Artero V, Jousselme B, et al. 2009. From hydrogenases to noble metal-free catalytic nanomaterials for H_2 production and uptake. Science, 326: 1384-1387 .

Levy R B, Boudart M. 1973. Platinum-like behavior of tungsten carbide in surface catalysis. Science, 181: 547-549.

Liao L, Zhu J, Bian X, et al. 2013. MoS_2 formed on mesoporous graphene as a highly active catalyst for hydrogen evolution. Advanced Functional Materials, 23: 5326-5333.

Liu L, Zhu Y P, Su M, et al. 2015. Metal-free carbonaceous materials as promising heterogeneous catalysts. ChemCatChem, 7: 2765-2787.

Lu J, Zhou W J, Wang L K, et al. 2016. Core-shell nanocomposites based on gold nanoparticle@ zinc-iron-embedded porous carbons derived from metal-organic frameworks as efficient dual catalysts for oxygen reduction and hydrogen evolution reactions. ACS Catalysis, 6: 1045-1053.

Ma R G, Zhou Y, Chen Y F, et al. 2015. Ultrafine molybdenum carbide nanoparticles composited with carbon as a highly active hydrogen-evolution electrocatalyst. Angewandte Chemie International Edition, 54: 14723-14727.

Marshall A, Borresen B, Hagen G, et al. 2007. Hydrogen production by advanced proton exchange membrane (PEM) water electrolysers-reduced energy consumption by improved electrocatalysis. Energy, 32: 431-436.

Millet P, Andolfatto F, Durand R. 1996. Design and performance of a solid polymer electrolyte water electrolyzer. International Journal of Hydrogen Energy, 21: 87-93.

Mitov M, Chorbadzhiyska E, Rashkov R, et al. 2012. Novel nanostructured electrocatalysts for hydrogen evolution reaction in neutral and weak acidic solutions. International Journal of Hydrogen Energy, 37: 16522-16526.

Pan Y, Lin Y, Chen Y J, et al. 2016. Cobalt phosphide-based electrocatalysts: synthesis and phase catalytic activity comparison for hydrogen evolution. Journal of Materials Chemistry A, 4: 4745-4754.

Pan Y, Yang N, Chen Y J, et al. 2015. Nickel phosphide nanoparticles-nitrogen-doped graphene hybrid as an efficient catalyst for enhanced hydrogen evolution activity. Journal of Power Sources, 297: 45-52.

Peng S J, Li L L, Han X P, et al. 2014. Cobalt sulfide nanosheet/graphene/carbon nanotube nanocomposites as flexible electrodes for hydrogen evolution. Angewandte Chemie International Edition, 53: 12594-12599.

Popczun E J, McKone J R, Read C G, et al. 2013. Nanostructured nickel phosphide as an electrocatalyst for the hydrogen evolution reaction. Journal of the American Chemical Society, 135: 9267-9270.

Safizadeh F, Ghali E, Houlachi G. 2015. Electrocatalysis developments for hydrogen evolution reaction in alkaline solutions—a review. International Journal of Hydrogen Energy, 40: 256-274.

Sapountzi F M, Gracia J M, Weststrate C J, et al. 2017. Electrocatalysts for the generation of hydrogen, oxygen and synthesis gas. Progress in Energy and Combustion Science, 58: 1-35.

Shi J L, Hu J M. 2015. Molybdenum sulfide nanosheet arrays supported on Ti plate: an efficient hydrogen-evolving cathode over the whole pH range. Electrochimica Acta, 168: 256-260.

Shifa T A, Wang F M, Cheng Z Z, et al. 2015. A vertical-oriented WS_2 nanosheet sensitized by graphene: an advanced electrocatalyst for hydrogen evolution reaction. Nanoscale, 7: 14760-14765.

Thoi V S, Sun Y J, Long J R, et al. 2013. Complexes of earth-abundant metals for catalytic electrochemical hydrogen generation under aqueous conditions. Chemical Society Reviews, 42: 2388-2400.

Vrubel H, Hu X L. 2012. Molybdenum boride and carbide catalyze hydrogen evolution in both acidic and basic solutions. Angewandte Chemie International Edition, 51: 12703-12706.

Wang C D, Jiang J, Zhou X L, et al. 2015. Alternative synthesis of cobalt monophosphide@C core-shell nanocables for electrochemical hydrogen production. Journal of Power Sources, 286: 464-469.

Wang K, Xi D, Zhou C J, et al. 2015. $CoSe_2$ necklace-like nanowires supported by carbon fiber paper: a 3D integrated electrode for the hydrogen evolution reaction. Journal of Materials Chemistry A, 3: 9415-9420 .

Wang M, Chen L, Sun L C. 2012. Recent progress in electrochemical hydrogen production with earth-abundant metal complexes as catalysts. Energy & Environmental Science, 5: 6763-6778.

Wu R, Zhang J F, Shi Y M, et al. 2015. Metallic WO_2-carbon mesoporous nanowires as highly efficient electrocatalysts for hydrogen evolution reaction. Journal of the American Chemical Society, 137: 6983-6986.

Yan H J, Tian C G, Wang L, et al. 2015. Phosphorus-modified tungsten nitride/reduced graphene oxide as a high-performance, non-noble-metal electrocatalyst for the hydrogen evolution reaction. Angewandte Chemie International Edition, 54: 6325-6329.

Yan Y, Xia B Y, Li N, et al. 2015. Vertically oriented MoS_2 and WS_2 nanosheets directly grown on carbon cloth as efficient and stable 3-dimensional hydrogen-evolving cathodes. Journal of Materials Chemistry A, 3: 131-135.

Yang J, Voiry D, Ahn S J, et al. 2013. Two-dimensional hybrid nanosheets of tungsten disulfide and reduced graphene oxide as catalysts for enhanced hydrogen evolution. Angewandte Chemie International Edition, 52: 13751-13754.

Yang L J, Zhou W J, Lu J, et al. 2016. Hierarchical spheres constructed by defect-rich MoS_2/carbon nanosheets for efficient electrocatalytic hydrogen evolution. Nano Energy, 22: 490-498.

Yang X L, Lu A Y, Zhu Y H, et al. 2015. Rugae-like FeP nanocrystal assembly on a carbon cloth: an exceptionally efficient and stable cathode for hydrogen evolution. Nanoscale, 7: 10974-10981.

Youn D H, Han S, Kim J Y, et al. 2014. Highly active and stable hydrogen evolution electrocatalysts based on molybdenum compounds on carbon nanotube-graphene hybrid support. ACS Nano, 8: 5164-5173.

Zeng K, Zhang D K. 2010. Recent progress in alkaline water electrolysis for hydrogen production and applications. Progress in Energy and Combustion Science, 36: 307-326.

Zhang P L, Wang M, Chen H, et al. 2016. A Cu-based nanoparticulate film as super-active and robust catalyst surpasses Pt for electrochemical H_2 production from neutral and weak acidic aqueous solutions. Advanced Energy Materials, 6: 1502319.

Zhang Z, Hao J H, Yang W S, et al. 2015. Defect-rich CoP/Nitrogen-doped carbon composites derived from a metal-organic framework: high-performance electrocatalysts for the hydrogen evolution reaction. ChemCatChem, 7: 1920-1925.

Zhang Z, Hao J H, Yang W S, et al. 2015. Modifying candle soot with FeP nanoparticles into high-performance and cost-effective catalysts for the electrocatalytic hydrogen evolution reaction. Nanoscale, 7: 4400-4405.

Zhao J, Tran P D, Chen Y, et al. 2015. Achieving high electrocatalytic efficiency on copper: a low-cost alternative to platinum for hydrogen generation in water. ACS Catalysis, 5: 4115-4120.

Zheng Y, Jiao Y, Li L H, et al. 2014. Toward design of synergistically active carbon-based catalysts for electrocatalytic hydrogen evolution. ACS Nano, 8: 5290-5296.

Zhou W J, Lu J, Zhou K, et al. 2016. $CoSe_2$ nanoparticles embedded defective carbon nanotubes derived from MOFs as efficient electrocatalyst for hydrogen evolution reaction. Nano Energy, 28: 143-150.

Zhou W J, Xiong T L, Shi C H, et al. 2016. Bioreduction of precious metals by microorganism: efficient gold@N-doped carbon electrocatalysts for the hydrogen evolution reaction. Angewandte Chemie International Edition, 55: 8416-8420.

Zhou W J, Zhou J, Zhou Y C, et al. 2015. N-doped carbon-wrapped cobalt nanoparticles on N-doped graphene nanosheets for high-efficiency hydrogen production. Chemistry of Materials, 27: 2026-2032.

Zhou W J, Zhou Y C, Yang L J, et al. 2015. N-doped carbon-coated cobalt nanorod arrays supported on a titanium mesh as highly active electrocatalysts for the hydrogen evolution reaction. Journal of Materials Chemistry A, 3: 1915-1919.

Zhu X H, Liu M J, Liu Y, et al. 2016. Carbon-coated hollow mesoporous FeP microcubes: an efficient and stable electrocatalyst for hydrogen evolution. Journal of Materials Chemistry A, 4: 8974-8977.

Zhu Y P, Liu Y P, Ren T Z, et al. 2015. Self-supported cobalt phosphide mesoporous nanorod arrays: a flexible and bifunctional electrode for highly active electrocatalytic water reduction and oxidation. Advanced Functional Materials, 25: 7337-7347.

Zou X X, Goswami A, Asefa T, 2013. Efficient noble metal-free (electro) catalysis of water and alcohol oxidations by zinc-cobalt layered double hydroxide. Journal of the American Chemical Society, 135: 17242-17245.

Zou X X, Huang X X, Goswami A, et al. 2014. Cobalt-embedded nitrogen-rich carbon nanotubes efficiently catalyze hydrogen evolution reaction at all pH values. Angewandte Chemie International Edition, 53: 4372-4376.

Zou X X, Zhang Y. 2015. Noble metal-free hydrogen evolution catalysts for water splitting. Chemical Society Reviews, 44: 5148-5180.

第三节　电催化 CO_2 还原材料

一、电催化 CO_2 还原材料研究概述

（一）电催化 CO_2 还原材料研究的意义

1800～2000 年，化石能源的广泛使用导致大气中的 CO_2 浓度从 280 ppm 升高到 380 ppm（图 1-11），这将会导致全球气候发生变化，并造成海洋酸化，破坏整个地球的生态系统平衡，减少 CO_2 排放对人类社会与地球的可持续发展具有重要的意义（Zhu et al.，2016；Wu et al.，2017）。

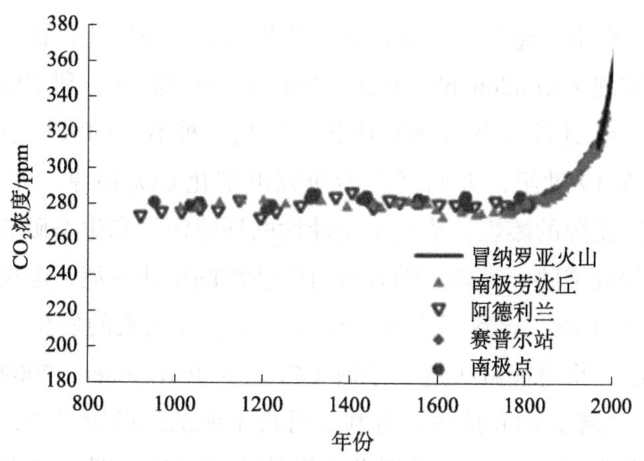

图 1-11　800～2000 年 CO_2 浓度

目前 CO_2 最主要来自发电厂的排放，减少空气中的 CO_2 含量主要有两种途径：减少化石能源的使用，以及将产生的 CO_2 转化为其他物质来进行资源化重新利用。如图 1-12 所示，目前常见的几种 CO_2 转化方法包括化学法、光化学法、电化学法、生物化学法等，可以将 CO_2 转化为碳酸盐、CO、CH_4、HCOOH 以及其他更为复杂的有机分子。其中，大自然的光合作用就是采用光化学法将 CO_2 转化为糖类的（El-Khouly et al.，2017）。这种方法效率较低，无法处理人类排放的大量 CO_2。因此，突破自然的限制，采用人工的方法大规模处理 CO_2 是必不可少的一种途径。CO_2 转换成碳酸盐只是酸碱中和反应，

C 原子的价态没有发生改变，因此不需要更多的能量，但需要消耗大量的原材料。

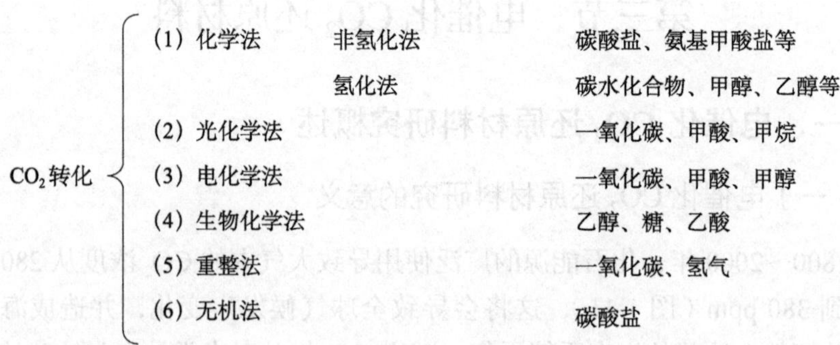

图 1-12 常见 CO_2 转化方法（Mikkelsen et al.，2010）

CO_2 分子稳定，化学反应活性差，要想进行还原转化，往往需要高温高压活化才能实现（Kondratenko et al.，2013）。在室温下，利用电化学法或光电化学法将 CO_2 还原成各种燃料或化学品是一种潜在的大规模减少碳排放的方法。早在 19 世纪，人们就开始研究电催化 CO_2 还原，尤其是最近 30 年，大量矿物能源的燃烧，造成了全球性的环境污染和生态破坏，对人类的生存和发展构成威胁。另外，随着可再生能源的迅速发展，电网无法消纳大规模的可再生能源，导致了弃光、弃风、弃水等现象的发生。利用可再生能源还原 CO_2，将可能解决这一矛盾（Centi & Perathoner，2009；Kalamaras et al.，2018）。减少 CO_2 排放，有利于可再生能源的健康发展，因此近年来利用可再生能源还原 CO_2 又受到学术界的广泛关注。图 1-13 是最近 100 多年来关于电催化 CO_2 还原的 SCI 论文发表情况。分为几个阶段：1976 年以前，电催化CO_2还原领域每年发表论文1～2篇，经常全年没有1篇论文发表；1976～1990 年，几乎每年都有论文发表，但每年数量不到 10 篇；1991 年发表论文数量突然增长到 50 余篇，然后缓慢增长到 2010 年的 157 篇；2010 年之后发表论文数量出现爆发式增长，到 2018 年已经达到年发表 1353 篇 SCI 论文。这说明近 10 年来，CO_2 电催化还原已经成为学术界的研究热点，然而依然存在能量转换效率不高、反应产物的选择性较差，以及反应速率慢等难题需要克服。

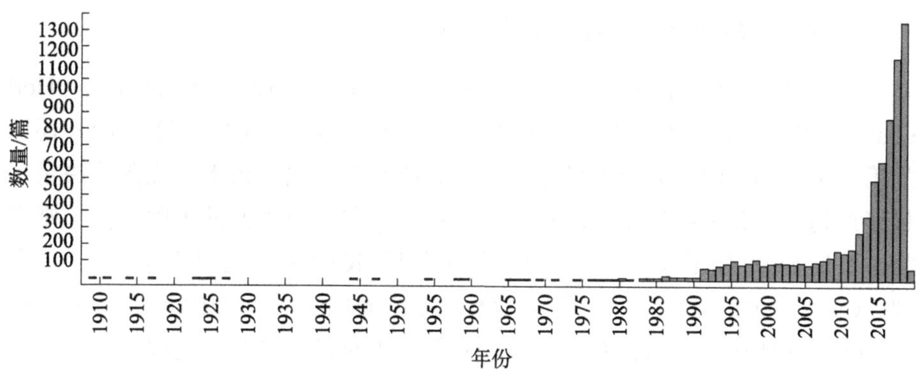

图 1-13　近 100 多年关于电催化 CO_2 还原发表的论文数量统计（截至 2019 年 3 月）

（二）CO_2 还原机理

1. CO_2 还原热力学要求

如表 1-3 所示，在 CO_2 还原半反应中，不同电子转移过程的电位不同，过电位也不同。虽然热力学的电子转移数越少过电位越负，但是动力学过电位相对较低，热力学与动力学的共同效应导致大部分的 CO_2 还原产物是两电子转移过程的产物 CO 和 HCOOH，所以目前的 CO_2 还原反应中法拉第效率最高的是 CO 和 HCOOH。目前，实现多电子产物的高法拉第效率是一个主要挑战。另外，CO_2 还原的平衡电位与 H^+ 的还原电位接近，因此经常会有析氢副反应的发生，降低了 CO_2 还原反应的法拉第效率。所以，目前法拉第效率比较高的电催化材料析氢过电位都比较高（Qiao et al.，2014；Lee et al.，2019）。

表 1-3　CO_2 还原半反应方程

CO_2 还原半反应	标准电极电位 /（V *vs.* RHE，pH=7）
$CO_2(g) + 2H^+ + 2e^- \longrightarrow CO(g) + H_2O(l)$	−0.106
$CO_2(g) + 2H^+ + 2e^- \longrightarrow HCOOH(l)$	−0.250
$CO_2(g) + 4H^+ + 4e^- \longrightarrow HCHO(l) + H_2O(l)$	−0.070
$CO_2(g) + 6H^+ + 6e^- \longrightarrow CH_3OH(l) + H_2O(l)$	0.016
$CO_2(g) + 8H^+ + 8e^- \longrightarrow CH_4(g) + 2H_2O(l)$	0.169
$2CO_2(g) + 12H^+ + 12e^- \longrightarrow C_2H_4(g) + 4H_2O(l)$	0.064
$2CO_2(g) + 12H^+ + 12e^- \longrightarrow C_2H_5OH(l) + 3H_2O(l)$	0.084

资料来源：Zhao 等（2019）

2. 最简单的两电子还原动力学过程

1993 年霍里（Hori）等发现了选择性产 CO 和 HCOOH 的方法（Hori et al.，1994）。如图 1-14 所示，这两个两电子转移过程不同，这是 CO_2 还原的第一步产物，$CO_2^{\cdot-}$ 是重要的中间产物，$CO_2^{\cdot-}$ 在金属电极上的吸附能决定了最终产物到底是 CO 还是 HCOOH。如果 $CO_2^{\cdot-}$ 在金属表面的吸附能高，则产物主要是 CO；如果吸附能低，则产物主要是 HCOOH。他们还有一个重要发现，Cu 是一种最为特殊的电催化 CO_2 还原的催化剂，产物中常见的 HCOOH 和 CO 都不是其主产物，而 CH_4、C_2H_4 等烃类是其主产物。这些烃类产物都是多电子过程，其过电位一般会很高，但是在 Cu 电催化上，则有很多不同的反应。后来进一步的研究表明，其还原产物的种类有十余种，这是 Cu 最为奇特的性质（Gao et al.，2019）。

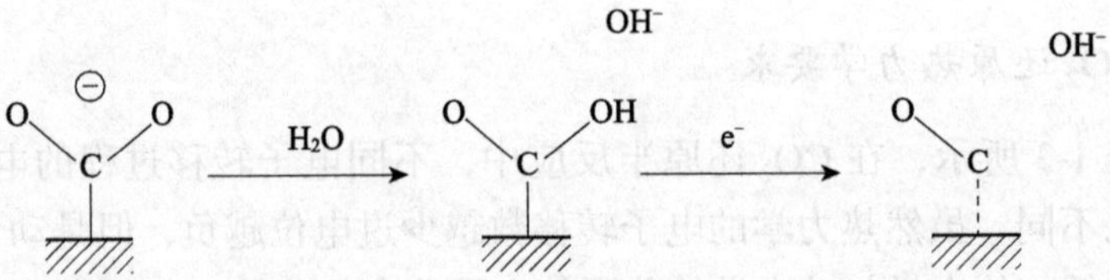

（a）水溶液中金属（Au、Ag、Cu、Zn）表面吸附 $CO_2^{\cdot-}$

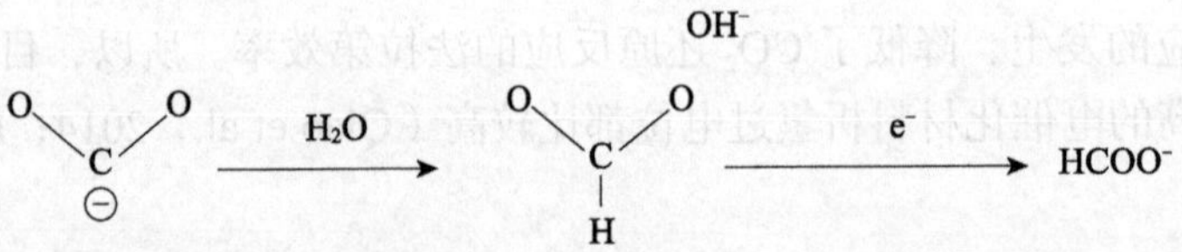

（b）水溶液中未在金属（Cd、Sn、In、Pb、Tl、Hg）表面吸附 $CO_2^{\cdot-}$

图 1-14　电催化 CO_2 还原中两电子产物的基本反应原理

（三）CO_2 还原电催化材料

1. 单一金属电催化材料

金属、金属氧化物、金属硫化物、金属有机框架材料（metal organic frameworks，MOFs）等不同的电催化剂已广泛用于电催化 CO_2 还原。其中，金属由于具有导电性高、催化活性好、成分可调以及制备简单等优点，研究也最为广泛（Low et al.，2019；Lei et al.，2016；Zhang et al.，2018；Kuhl et al.，2014）。

1954 年，美国俄勒冈大学的杜鲁门·蒂特（Truman Teeter）等发现，在 Hg 电极上 CO_2 被还原为 HCOOH 的法拉第效率为 100%（Teeter & Van Rysselberghe，1954）。

1985 年，Hori 等对金属电极进行全产物分析，包括气相产物和液相产物（Hori et al.，1985）（表 1-4），系统研究了 Zn、Cd、In、Sn、Pb、Ag、Au、Cu、Ni、Fe 等金属（图 1-15），发现 Cd、In、Sn、Pb 电催化还原的主要产物是 HCOOH，Ag 和 Au 的主要还原产物是 CO，Cu 上主要是产生 CH_4，Ni、Fe 的主要还原产物是 H_2。图 1-15（Zhang et al.，2018）根据 Hori 等的研究绘制而成，能清楚地看到各种金属电极在水溶液体系中电催化还原 CO_2 的主产物。Ag、Au、Pd 主要产 CO；Cd、In、Sn、Hg、Tl、Pb、Bi 等的主要产物是 HCOOH；而 Zn 的还原产物重复性差，有时是 CO，有时是 HCOOH；Cu 最为特殊，主要产物除了两电子转移过程的 CO 和 HCOOH，还包括 CH_4、C_2H_4、醇类等多电子过程还原产物，多达 20 余种。

表 1-4　各种金属的 CO_2 还原产物

电极	电极电位 /（V *vs.* SHE）	法拉第效率 /%				
		$HCOO^-$	CO	CH_4	H_2	合计
Cd	−1.66 ± 0.02	65.3/67.2	6.2/11.1	0.2	14.9/22.2	93/100
Sn	−1.40 ± 0.04	65.5/79.5	2.4/4.1	0.1/0.2	13.4/40.8	94/110
Pb	−1.62 ± 0.03	72.5/88.8	0.3/0.6	0.1/0.2	3.8/30.9	94/100
In	−1.51 ± 0.05	92.7/97.6	0.9/2.2	0.0	1.6/4.5	93/102
Zn	−1.56 ± 0.08	17.6/85.0	3.3/63.3	0.0	2.2/17.6	90/98
Cu	−1.39 ± 0.02	15.4/16.5	1.5/3.1	37.1/40.0	32.8/33.0	87/92
Ag	−1.45 ± 0.02	1.6/4.6	61.4/89.9	0.0	10.4/35.3	99/106
Au	−1.14 ± 0.01	0.4/1.0	81.2/93.0	0.0	6.7/23.2	100/105
Ni	−1.39	0.3	0.0	1.2	96.3	98
Fe	−1.42	2.1	1.4	0.0	97.5	101

资料来源：Hori 等（1985）

注：表中法拉第效率两个数值的为最低效率 / 最高效率。

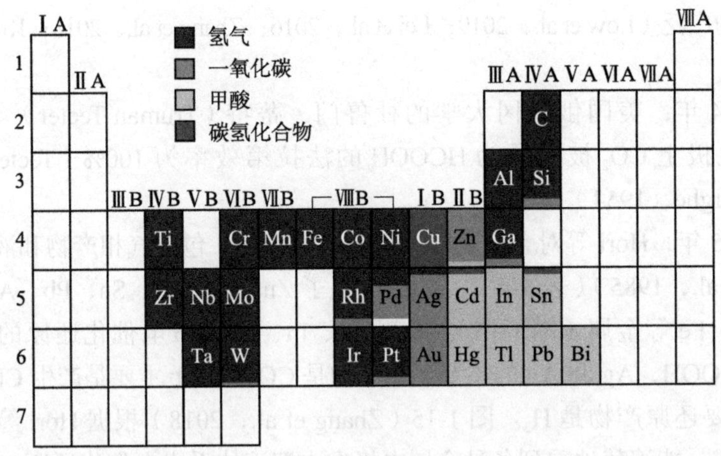

图 1-15　不同金属的 CO_2 还原产物分布图（Zhang et al.，2018）
金属元素所在格颜色有变化代表产生多种产物

在 CO_2 电催化还原领域中，Cu 作为一种特殊的催化剂，是当前研究最热门的材料（Klingan et al.，2018；Bevilacqua et al.，2016；Luc et al.，2019）。关于氧化预处理后的 Cu（OD-Cu）对电催化 CO_2 还原活性的影响受到许多科学家的关注（Mandal et al.，2018；Cavalca et al.，2017），相关研究得到了一致的结果：OD-Cu 提高了对 C2 产物的选择性，抑制了 C1 产物的生成（Lum & Ager，2019）。

马修 · 卡南（Matthew W. Kanan）课题组（Li & Kanan，2012）研究了 Cu 通过高温退火处理在表面形成的一层 Cu_2O 对 CO_2 还原反应的电催化活性的影响（图 1-16）。将高纯铜箔（>99.9999%）在 85% 的 H_3PO_4 中处理后，然后在空气中高温退火，会在 Cu 电极表面形成一层 Cu_2O 薄膜，在 0.5 mol/L 的 CO_2 饱和 $NaHCO_3$ 溶液中进行电催化还原反应。通过对工作电极反应前后的表征发现：经过高温退火处理以后，Cu 箔氧化形成 Cu_2O 和 Cu(Ⅱ)（CuO），而且氧化以后的 Cu 箔中氧含量达到 51%。然后在 −0.5 V *vs.* RHE 电位下经过 CO_2 催化还原反应之后，X 射线衍射（X-ray diffraction，XRD）数据中只有金属 Cu 的特征峰，X 射线光电子能谱（X-ray photoelectron spectroscopy，XPS）数据中可以看到催化剂表面的氧含量还高达 39%。

将铜箔在 500 ℃温度下退火处理 12 h 后与多晶铜的电化学性能进行对比，实验结果表明，CO 和甲酸的过电位明显降低：退火处理后的铜箔，−0.5～−0.3 V *vs.* RHE 的电位范围内，CO 达到最高法拉第效率 45%，过电位为 0.19～0.39 V。甲酸在 −0.65～−0.45 V *vs.* RHE 的电位范围内达到最高法拉第

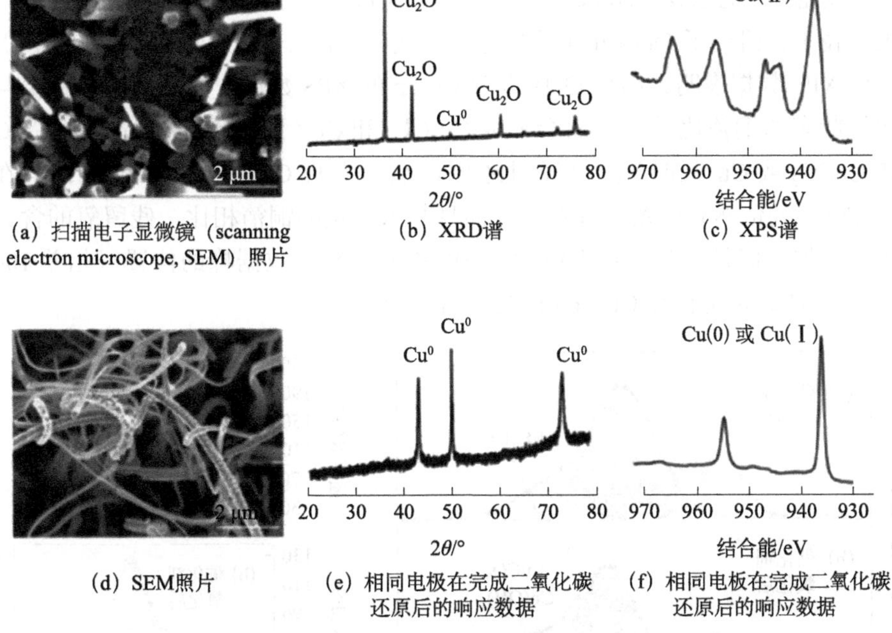

（a）扫描电子显微镜（scanning electron microscope, SEM）照片　（b）XRD谱　（c）XPS谱

（d）SEM照片　（e）相同电极在完成二氧化碳还原后的响应数据　（f）相同电板在完成二氧化碳还原后的响应数据

图 1-16　500℃高温退火后 OD-Cu 的表征数据（Li & Kanan，2012）

效率 33%，过电位为 0.25～0.45 V；多晶铜的 CO 最高法拉第效率的电位在 −0.8 V *vs.* RHE，过电位为 0.69 V。甲酸在 −0.9～−0.7 V *vs.* RHE 的范围内达到最高法拉第效率，最高为 20% 左右，过电位达到 0.5～0.7 V（图 1-17）。

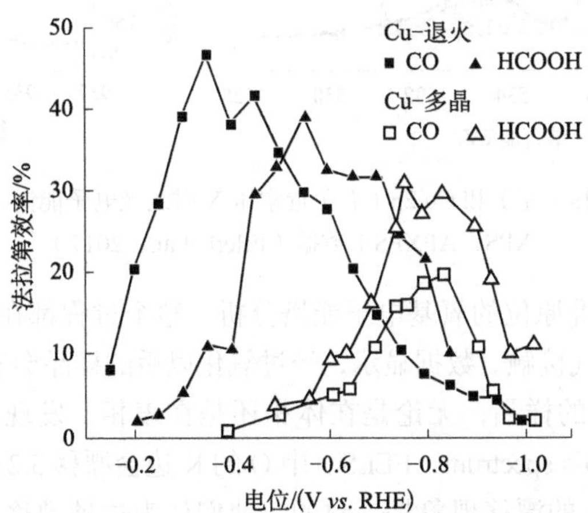

图 1-17　500 ℃处理后的 OD-Cu 与单晶铜的 CO 和 HCOOH 法拉第效率图（Li & Kanan，2012）

安德斯·尼尔森（Anders Nilsson）等通过电化学氧化的方法，结合原位表征技术，研究了 OD-Cu 在催化过程中残留氧的作用（Eilert et al.，2017）。原位 XPS 数据表明，结合 O 1s 以及 Cu 2p 的 XPS 数据，氧化以后的铜箔与初始状态的铜箔比较，表面会形成 Cu（Ⅰ）和 Cu（Ⅱ），分别对应的是 Cu_2O 和 $CuCO_3$ 或 $Cu(OH)_2$。再经过还原以后，发现 Cu（Ⅱ）的峰消失，但在 531.7 eV 处依然有残留氧峰存在，而且与未氧化的铜箔相比，残留氧的含量明显增加。但是因为 Cu（Ⅰ）和 Cu (0) 之间的 XPS 数据峰的位置只相差 100 mV，无法确定是否有 Cu（Ⅰ）存在（图 1-18）。

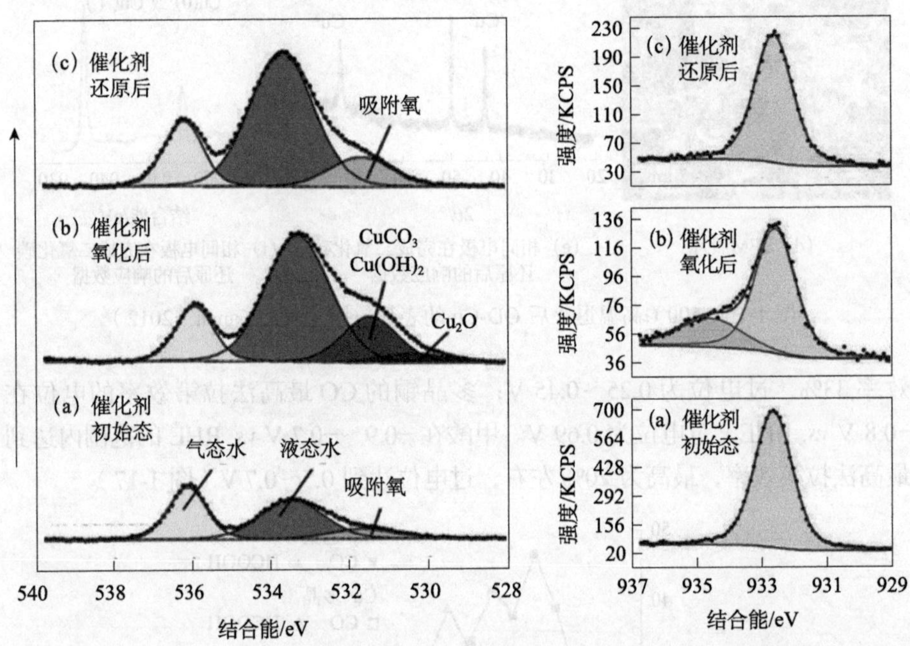

图 1-18　原位 O 1s（左）和 Cu 2p（右）近常压 X 射线光电子能谱（ambient-pressure XPS，APXPS）数据（Eilert et al.，2017）

此外，结合准原位的氧基电子能谱分析，整个过程都在保护气的氛围中进行，避免与空气接触。数据显示，经过氧化以后的样品会有 Cu_2O 存在，但是经过还原以后的样品，无论是在体相还是在表相，发现电子能量损失谱（electron energy loss spectrum，EELS）中 O 的 K 边会漂移 3.2 eV 和 3.5 eV，这种现象与 APXPS 的漂移现象是一致的，他们认为这种漂移是 O 的周围化学环境变化造成的。再结合 EELS 中 Cu 的 L 边数据，他们认为经过还原以后的样品中的残留 O 并不会与 Cu 形成 Cu_2O，而是以一种金属 Cu 和残留氧共存的形式存在。

Beatriz Roldan Cuenya 课题组结合一些新的技术手段研究了 OD-Cu 在二氧化碳催化还原过程中的化学状态以及催化活性（Mistry et al.，2016）。他们利用氧等离子体处理的手段对铜箔进行激活处理，分别用 20 W 氧等离子体处理 2 min，用 100 W 氧等离子体处理 2 min，用 100 W 氧等离子体处理 10 min，得到一种多孔的 CuO、CuO_2 以及 CuO/CuO_2 共存材料（图 1-19）。利用氧等离子体处理后的 Cu 片，在表面以 CuO 的形式存在，在亚表面以 CuO_2 的形式存在，而在体相以金属形式的 Cu 存在。然后对处理后的 Cu 进行原位还原处理以及表征发现，在 −0.91 V *vs.* RHE 的电位下反应 1 h 之后，虽然氧化态的 Cu 有部分被还原为金属态的铜，但在这种材料表面依旧有残留的氧化态的 Cu 存在，残留的 O 与 Cu 结合形成 CuO_2 保留在催化剂的表面。在富氧区域，氧原子占比达到 3%～28%。

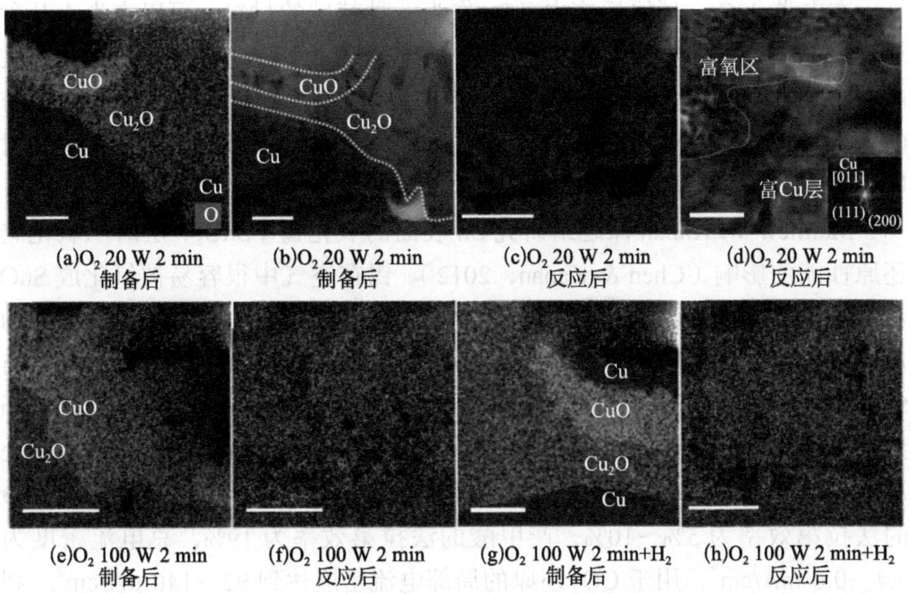

(a)O_2 20 W 2 min 制备后　(b)O_2 20 W 2 min 制备后　(c)O_2 20 W 2 min 反应后　(d)O_2 20 W 2 min 反应后

(e)O_2 100 W 2 min 制备后　(f)O_2 100 W 2 min 反应后　(g)O_2 100 W 2 min+H_2 制备后　(h)O_2 100 W 2 min+H_2 反应后

图 1-19 OD-Cu 的透射电子显微镜（transmission electron microscope，TEM）数据

对催化剂的二氧化碳还原催化性能测试发现，利用 20 W 的氧等离子体处理的 Cu 片对 C_2H_4 的选择性最高，在 −0.9 V *vs.* RHE 的电位下，乙烯的法拉第效率可以达到 60% 左右，乙烯的法拉第电流可以达到约 7.2 mA/cm^2。在长达 5 h 的测量下，一直可以保持很好的稳定性（图 1-20）。

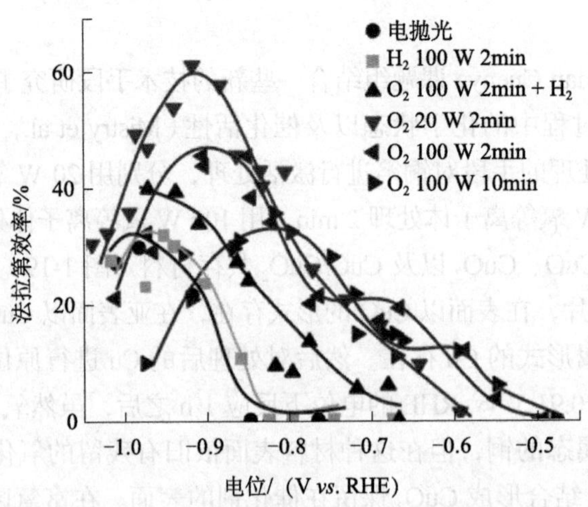

图 1-20　等离子体处理后的 OD-Cu 对乙烯的法拉第效率

在电催化 CO_2 还原反应中，Cu 作为一种特殊的材料，可以产生十几种产物。但是有一些金属（如 Sn、Bi 等），在电催化 CO_2 还原反应中，对甲酸的选择性比较好。由于这个特性，有很多科学家对其进行了研究，发现了与 Cu 很相似的现象：Sn、Bi 这些材料的氧化物在电催化 CO_2 还原过程中，对甲酸的产生有很大的影响（Zhao et al.，2019；Liu et al.，2019）。

Matthew W. Kanan 课题组研究 Sn 表面的氧化锡（SnO_x）层对二氧化碳还原性能的影响（Chen & Kanan，2012）。锡在空气中很容易被氧化成 SnO 和 SnO_2，他们用直接的锡片进行 XPS 分析，发现锡片表面 SnO_x 与 Sn 的原子比 SnO_x : Sn = 95 : 5。在 90℃的 HBr 处理 10 min 后，锡片表面 SnO_x 与 Sn 的原子比 SnO_x : Sn = 78 : 83，把样品在 −0.7 V *vs*. RHE 的电位下反应 12 h 后，SnO_x : Sn = 89 : 11，说明在反应过程中 SnO_x 可以稳定存在。然后对处理前后的样品进行 CO_2 催化测量，发现未处理的锡片，CO_2 被还原成 CO 的法拉第效率为 5%～10%，产甲酸的法拉第效率为 19%，总电流密度为 0.4～0.6 mA/cm²，用于 CO_2 还原的局部电流密度达到 92～140 μA/cm²。利用酸处理过后的锡片测量，总电流密度提高到 3～4 mA/cm²，但是 CO_2 被还原成 CO 的法拉第效率只有 0.5%，产甲酸的法拉第效率只有 0.3%，用于 CO_2 还原的局部电流密度有 24～32 μA/cm²。对比说明，将 Sn 表面的 SnO_x 层去掉会抑制 CO_2 的还原反应，99% 的电流用于产氢。他们利用电沉积法在钛片上电沉积 Sn/SnO_x，经过分析，SnO_x : Sn = 93 : 7，然后进行 CO_2 还原反应测试，发现 90% 以上的电流是用于 CO_2 还原反应的（图 1-21）。

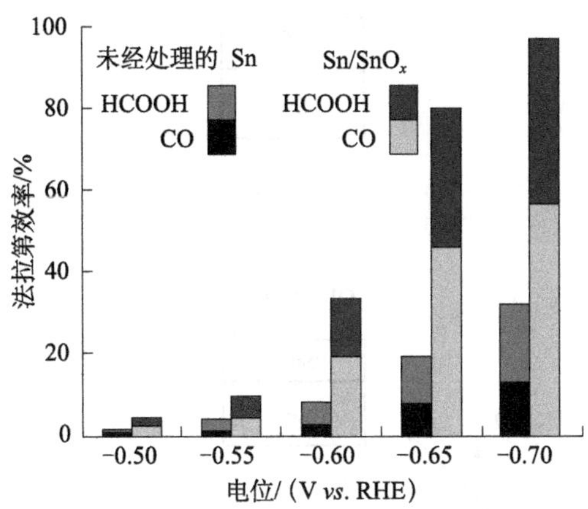

图 1-21 未经处理的锡片和电镀在钛片上的 Sn/SnO$_x$ 的法拉第效率
（Chen & Kanan，2012）

安德鲁·博卡斯利（Andrew B. Bocarsly）课题组利用原位衰减全反射红外光谱（ATR-IR）研究在电催化 CO_2 还原过程中 Sn 表面的 Sn $(OH)_x$ 的作用（Baruch et al.，2015）。他们确认了 Sn $(OH)_x$ 在电催化过程中可以稳定存在，并且认为表面 Sn $(OH)_2$ 的形成对电催化 CO_2 还原生成 HCOOH 有着重要的作用（图 1-22）。

利用离子束溅射的方法在 ZnSe 晶体衬底上沉积 Sn 薄膜。结合 Sn 3d 以及 O 1s 的 XPS 数据，可以判断制备沉积出来的 Sn 在表面以金属 Sn、氧化物 SnO$_x$ 以及 Sn $(OH)_x$ 的形式存在。在 Ar 气氛下 0.1 mol/L 的 K_2SO_4 电解液中，在不同的电位下对 Sn 进行原位表征，发现在 −1.4 V *vs*. RHE 的电位下，表面的 SnO$_x$ 虽然会发生还原，但是依然会有残留的 SnO$_x$ 存在；而在电位跃迁到 −1.6 V *vs*. RHE 下进行反应时，SnO$_x$ 会被进一步还原，但依旧会有一层稳定厚度的 SnO$_x$ 存在。在 CO_2 的气氛 0.1 mol/L 的 K_2SO_4 电解液中，在 −1.8 V *vs*. Ag/AgCl 的电位下，前 30 min 通入 CO_2，后 30 min 停止通入 CO_2，会发现有一个关于表面 $SnCO_3$ 的峰在停止通入 CO_2 后逐渐消失，因此确定表面 $SnCO_3$ 的形成对电催化 CO_2 还原有很重要的作用。然后比较了 SnO_2 和 Sn_6O_4 $(OH)_4$ 的催化活性，发现在 −1.4 V *vs*. Ag/AgCl 的电位下，SnO_2 的 HCOOH 法拉第效率低于 Sn_6O_4 $(OH)_4$ 的效率，因此他们认为，在电催化 CO_2 还原生成 HCOOH 的过程中，表面的 Sn $(OH)_2$ 才是真正具有发挥催化活性的物质。

雷立旭课题组对比了不同温度下煅烧得到的表面 SnO$_x$ 以及自然氧化的

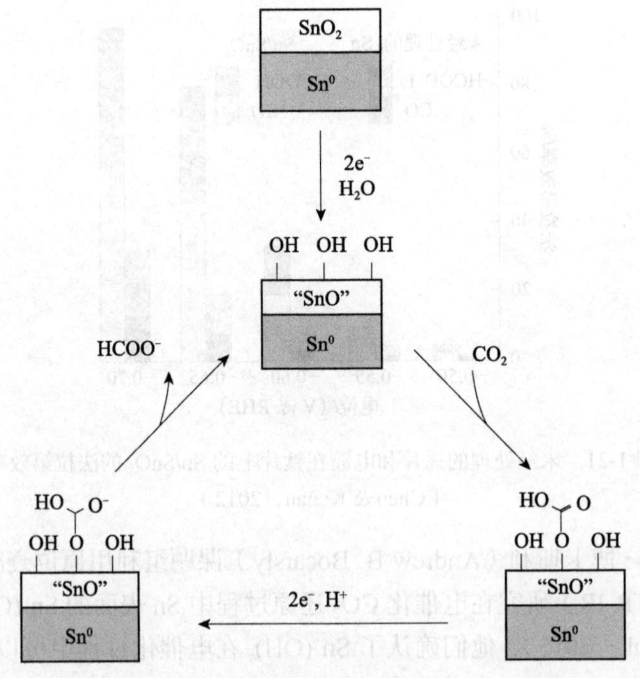

图 1-22 在 SnO_2 表面的 CO_2 还原反应机理图（Baruch et al.，2015）

"SnO"：二价 Sn 的氧化物

SnO_x 对电催化 CO_2 还原的影响（Zhang et al.，2015）。周晓冬课题组研究了 Sn 催化剂表面的 SnO_x 厚度对电催化 CO_2 还原的影响（Wu et al.，2014）。当表面 SnO_x 的厚度为 3.5 nm 时，甲酸法拉第效率可以达到 64%；当表面 SnO_x 厚度达到 7 nm 时，CO 的选择性最好，最高法拉第效率可以达到 35% 左右。

韩国首尔大学 Ki Tae Nam 课题组研究了 BiO_x/C 电催化 CO_2 还原的催化活性（Lee et al.，2018）。他们利用水热法在碳上担载 BiO_x 纳米颗粒催化剂。如图 1-23 所示，结合扫描透射电子显微术（scanning transmission electron microscopy，STEM）、XPS 以及 X 射线吸收光谱术（X-ray absorption spectroscopy，XAS）数据，可以确定制备出来的样品是一种碳担载 BiO_x 的催化剂。然后在−1.15 V *vs.* RHE 电位下，进行原位 XAS（*in situ* XAS）表征，发现在催化过程中，虽然有部分的 BiO_x 被还原，但依然还是有部分 BiO_x 残留下来。

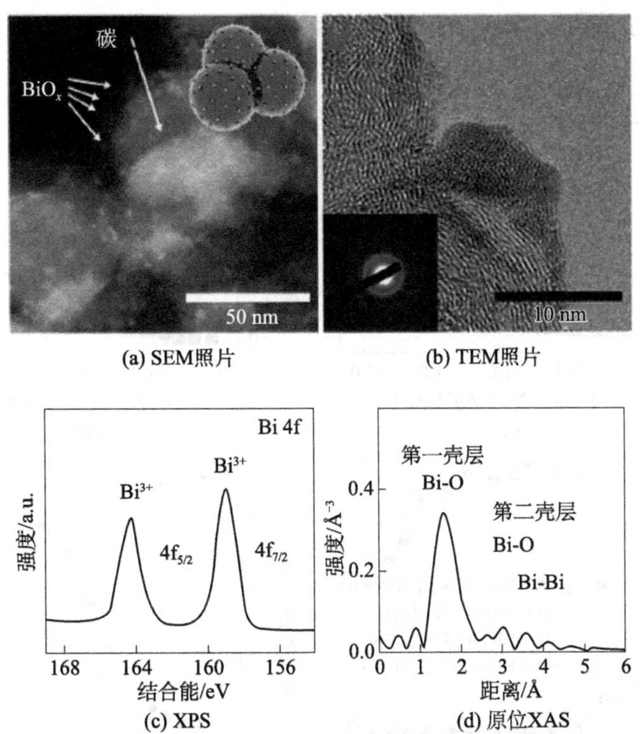

(a) SEM照片　(b) TEM照片

(c) XPS　(d) 原位XAS

图 1-23　BiO_x/C 的基本表征（Lee et al.，2018）

在 0.5 mol/L $KHCO_3$/0.5 mol/L $NaClO_4$ 电解液中进行电催化 CO_2 还原反应，结果显示，在 −1.37 V *vs.* RHE 的电位下，HCOOH 的法拉第效率可以达到 92.1% 左右，而且可以保持在很宽的电位条件下；在 −1.7 V *vs.* RHE 的电位下，HCOOH 的法拉第效率依然可以维持在 93.4% 左右（图 1-24）。

除 Cu、Sn、Bi 这些容易被氧化的金属之外，研究发现，Au、Ag 这种不易被氧化的金属被氧化处理后，它们的催化活性也有很大提升。

威尔逊·史密斯（Wilson A. Smith）课题组研究了氧化预处理后的 Ag（OD-Ag）的电催化 CO_2 活性（Ma et al.，2016）。在 0.2 mol/L NaOH 电解液中，他们利用电化学氧化的方法对 Ag 箔进行阳极氧化处理，在 Ag 箔的表面得到 Ag_2O，然后在 0.1 mol/L $KHCO_3$ 溶液中进行电催化 CO_2 还原反应。与多晶 Ag 相比，他们发现，OD-Ag 的 CO 法拉第效率得到了提高，最高可以达到 90% 以上，而且相比于多晶 Ag，OD-Ag 产 CO 的过电位也明显减小。

Matthew W. Kanan 课题组也研究了 Au 在经过电化学氧化后的电催化 CO_2 还原活性（Chen et al.，2012）。如图 1-25 所示，CO 法拉第效率有了很大的提升，最高可以达到 96% 左右，而且过电位明显降低，只有 140 mV。

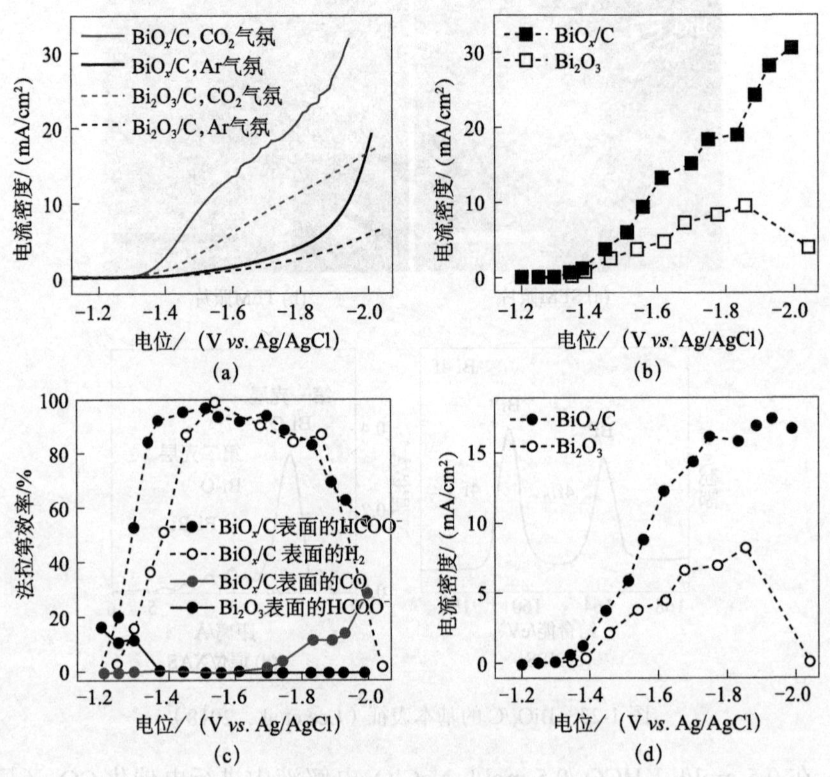

图 1-24　BiO$_x$/C 的催化活性图（Lee et al.，2018）

(a) 在二氧化碳和氩气饱和条件下，在 0.5 mol/L 碳酸氢钠和 0.5 mol/L 氯化钠中，在 50 mV/s 下对 BiO$_x$/C 和商用 Bi_2O_3 进行线性伏安扫描；(b) 总电流密度与电位的关系；(c) 不同材料表面各产物的法拉第效率；(d) HCOOH 的电流密度

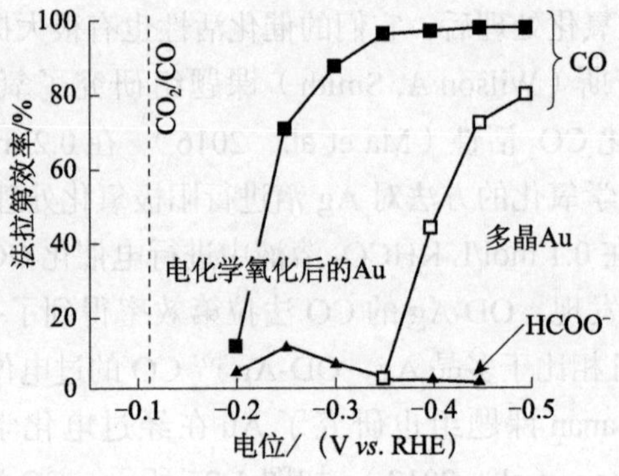

图 1-25　电化学氧化后的 Au 与多晶 Au 的催化活性对比（Chen et al.，2012）

2. 二元合金电催化材料

参考电化学防腐蚀的原理以及不同金属的标准氧化还原电位（表 1-5），我们发现，由不同金属组成的合金成分中，两种金属中必定有一个更容易发生氧化的金属与氧结合在表面，形成一层表面氧化物（图 1-26）。

表 1-5　常见金属的标准氧化还原电位

金属-金属离子平衡	电极电位 /（V *vs.* RHE）
Au/Au^{+}	1.850
Pd/Pd^{2+}	0.915
Ag/Ag^{+}	0.799
Cu/Cu^{+}	0.520
Bi/Bi^{3+}	0.308
H/H^{+}	0
Sn/Sn^{2+}	−0.130
Zn/Zn^{2+}	−0.760

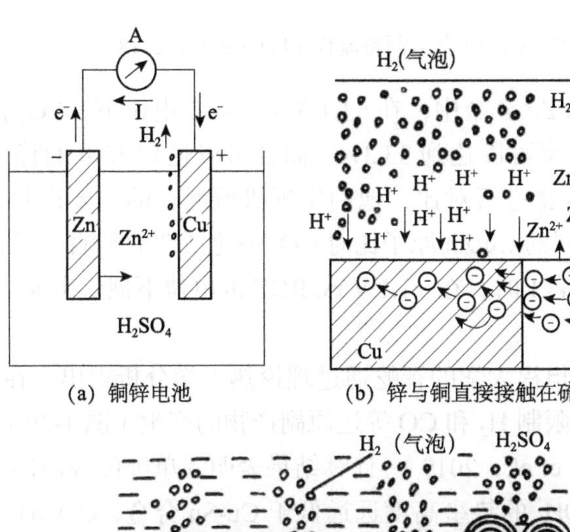

(a) 铜锌电池

(b) 锌与铜直接接触在硫酸溶液中溶解

(c) 含杂质的工业用锌在硫酸溶液中溶解

图 1-26　电化学防腐蚀原理图

铜作为一种特殊的催化剂，除金属 Cu 之外，Cu 基合金也被许多课题组进行了大量研究。天津大学杜希文等 2018 年在 *Langmuir* 上发表了关于 Cu-Zn 合金催化 CO_2 还原生成乙烯（C_2H_4）的文章（Feng et al.，2018），利用纳秒脉冲激光对浸入水中的 Cu、Zn 进行烧蚀，合成 Cu-ZnO 纳米颗粒，电还原生成 Cu-Zn 纳米合金（图 1-27）。

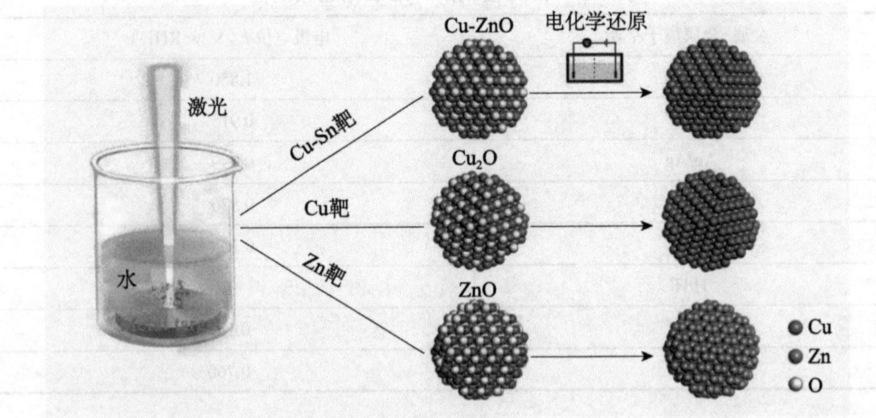

图 1-27　Cu-Zn 合金制备流程（Feng et al.，2018）

经过测量，在 Cu-Zn 合金中，在 −1.1 V *vs*. RHE 电位下，CO_2 催化生成 C_2H_4 的最高法拉第效率可以达到 33.3%，而且可以达到 15 h 的稳定性。与 CuO_2/ZnO 的机械混合物进行对比，他们发现机械混合的主要产物是 CO 和 H_2。他们也研究了不同的 Cu/Zn 原子比对 CO_2 催化活性的影响，当 Cu-Zn 合金的 Cu/Zn 原子比为 4∶1 时，在 −1.1 V *vs*. RHE 的电位下催化生成 C_2H_4 的活性最高（图 1-28）。

2018 年，美国斯坦福大学的崔屹通过理论热力学分析得出，在 Cu 电催化剂中掺入 Sn，可以限制 H_2 和 CO 等还原副产物的产生（图 1-29），倾向于产生 HCOOH（Zheng et al.，2019）。计算结果表明，单纯的 Sn 和 Cu，其表面吸附中间物种 $COOH^*$ 的势垒高度远远低于 Cu-Sn 合金，$COOH^*$ 是产 CO 的中间物种，如果势垒高，则能避免 CO 的产生。因此，Cu-Sn 合金产 CO 得到抑制。相似地，H^* 在 Cu、Sn 上的自由能远低于 CuSn 和 $CuSn_3$，H^* 是产氢的中间物种，因此，Cu、Sn 电催化剂产氢活性将高于 CuSn 和 $CuSn_3$。更进一步的细节表明，CuSn 产氢的过电位远低于 $CuSn_3$，因此，$CuSn_3$ 比 CuSn 产氢更少，产 HCOOH 的选择性更高。与理论分析结果一致，通过电沉积法制备的 $CuSn_3$ 催化剂在 −0.5 V *vs*. RHE 电位下显示了 95% 的产 HCOOH 法拉第效率

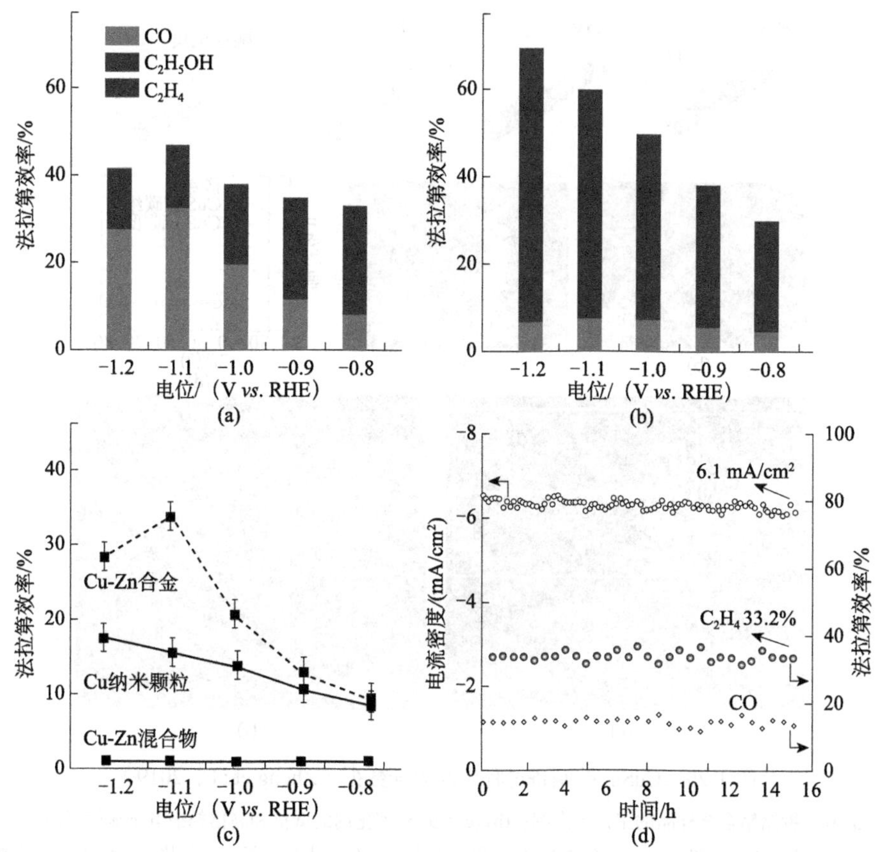

图 1-28　Cu-Zn 合金以及 Cu-Zn 混合物的催化活性图（Feng et al., 2018）

(a) Cu-Zn 合金法拉第效率；(b) Cu-Zn 混合物法拉第效率；(c) Cu-Zn 合金、Cu-Zn 混合物和 Cu 纳米颗粒催化产乙烯的法拉第效率；(d) 电解还原 15 h 稳定性测试

（图 1-30）。这是目前 CO_2 还原成 HCOOH 的最高性能。催化剂在 50 h 测试的过程中，活性没有明显衰减。原位 X 射线吸收近边结构（X-ray absorption near edge structure，XANES）结果表明，电子从 Sn 到 Cu 转移，使 Sn 在电催化测试条件下显示了更正的氧化态。电沉积制备的 $CuSn_3$ 是小的纳米颗粒，镶嵌在非晶结构中。−0.5 V *vs.* RHE 的 HCOOH 还原电流密度为 31 mA/cm^2。

通过对反应前后的 $CuSn_3$ 合金的原位电子结构和 XPS 表征发现，在反应后，合金中的 Cu 和 Sn 以一种氧化物与金属的形式共同存在。Sn L_3 边原位表征数据进一步表明，在反应过程中，Sn 以金属 Sn 和氧化物 SnO 的形式存在；Cu K 边原位表征数据显示 Cu 在电催化还原 CO_2 过程中主要以金属 Cu 的形式存在。

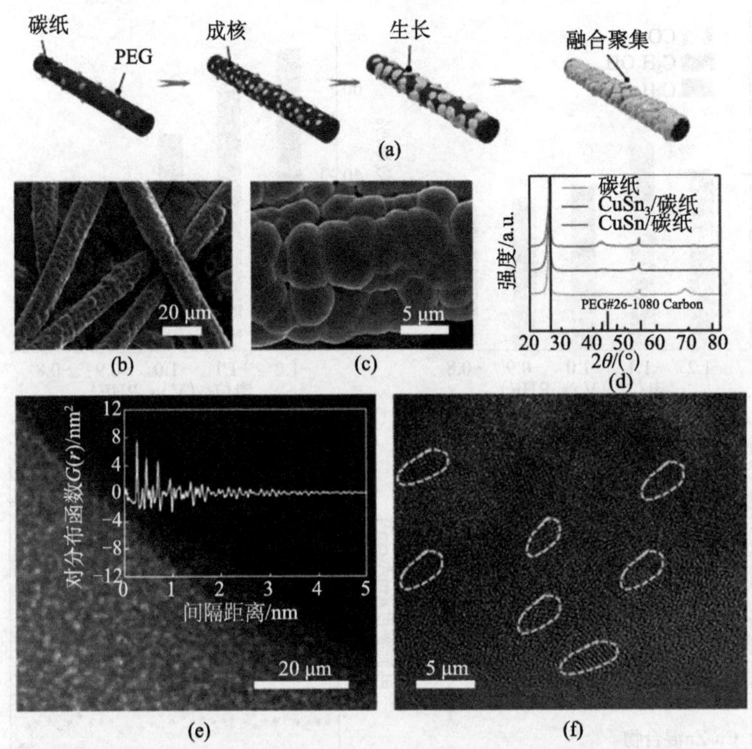

图 1-29　$CuSn_3$ 的制备过程以及表征数据（Zheng et al.，2019）

(a) 制作铜锆钴合金样品的工艺示意图；(b) (c) $CuSn_3$ 催化剂的低倍和高倍扫描电子显微镜图像；(d) 碳纸基材、CuSn 和 $CuSn_3$ 电极的 XRD 结果，其中仅观察到来自碳纸的布拉格峰；(e) $CuSn_3$ 电极的高角环形暗场扫描透射（high-angle annular dark-field-STEM，HAADF-STEM）[图 (e) 中因为没有原子对距离超过粒子直径，$CuSn_3$ 的对分布函数峰在 5 nm 处减小]；(f) 高分辨率 TEM 图像

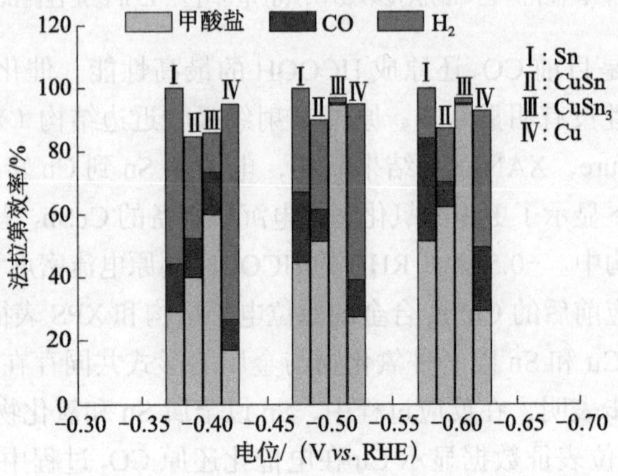

图 1-30　Sn、CuSn、$CuSn_3$、Cu 的催化活性对比（Zheng et al.，2019）

安德鲁·格沃思（Andrew A. Gewirth）课题组研究了 Cu-Ag 合金在 1 mol/L KOH 的电解液中对 CO_2 的催化活性（Hoang et al.，2018）。他们利用电沉积的方法在碳纸上得到 Cu-Ag 合金，然后通过加入差热分析（differential thermal analysis，DTA）来控制合金的形貌，得到 Cu 纳米线、Cu-Ag 大尺寸颗粒（Cu-Ag poly）以及 Cu-Ag 纳米线三种样品。通过 XRD 表征发现，在 Cu-Ag 纳米线样品中含有 Cu_2O，而通过 XPS 表征发现，Cu-Ag 大尺寸颗粒的合金中存在一些 CuO 的卫星峰，表明在 Cu-Ag 合金表面存在一些铜的氧化物 CuO。结合 X 射线吸收精细结构（X-ray absorption fine structure，XAFS），Ag 原子周围主要与 Cu 以及 Ag 原子结合，而 Cu 原子除与 Ag 以及 Cu 原子结合外，还会与 O 原子结合（图 1-31）。结合所有数据表明，Cu-Ag 合金是一种体相为金属 Cu、表相为 Cu_2O 的催化剂材料。

此外，在 1 mol/L KOH 的电解液中，在 −0.7 V *vs*. RHE 的电位下进行反应过程中，利用原位拉曼光谱对材料进行表征，在 526 cm^{-1} 的位置有一个峰，代表的是 Cu—O 峰，证明 Cu-Ag 纳米线在催化过程中依然会有 Cu_2O 存在，是 CuO_2 和金属 Cu、Ag 共存的状态。催化剂表面的 Cu_2O 以及内部的 Ag 共同作用，可以促进电催化 CO_2 还原生成 C_2 产物。

对 Cu-Ag 合金进行催化活性测试，发现当 Ag 的含量为 6% 的时候，CO_2 还原催化活性最高，在 −0.7 V *vs*. RHE 的电位下，C_2H_4 的法拉第效率可以达到 60%，C_2H_5OH 的法拉第效率可以达到 25% 左右，而且在这个电位下，总电流密度可以达到 300 mA/cm^2（图 1-32）。

2017 年，美国伊利诺伊大学厄巴纳-香槟分校的凯尼斯（Kenis）等制备了有序相、无序相、偏析相三种 Cu-Pd 合金（Ma et al.，2017）。从 XRD 数据分析中可以看到，CuPd 偏析相中主要由金属 Cu、氧化物 Cu_2O 以及金属 Pd 组成。XPS 对 Cu-Pd 合金的分析发现，在 Cu-Pd 合金中，有序相合金表面的 Cu 以金属态的 Cu 存在；而偏析相合金表面的 Cu 主要以 CuO 的形式存在；偏析相合金表面主要以 Cu_2O 的形式存在。XPS 数据显示，无论是在有序相、无序相还是在偏析相中，Pd 都是以一种金属态的形式存在的。

如图 1-33 所示，从 HAADF-STEM 图中可以观察到，无论是有序相、无序相还是偏析相，在合金的表面都存在一层很薄的 Cu，只是偏析相的 Cu 层最厚，最薄的地方都有约 3 nm，而无序相表面的 Cu 层只有约 1 nm，有序相表面的 Cu 层甚至不到 1 nm。

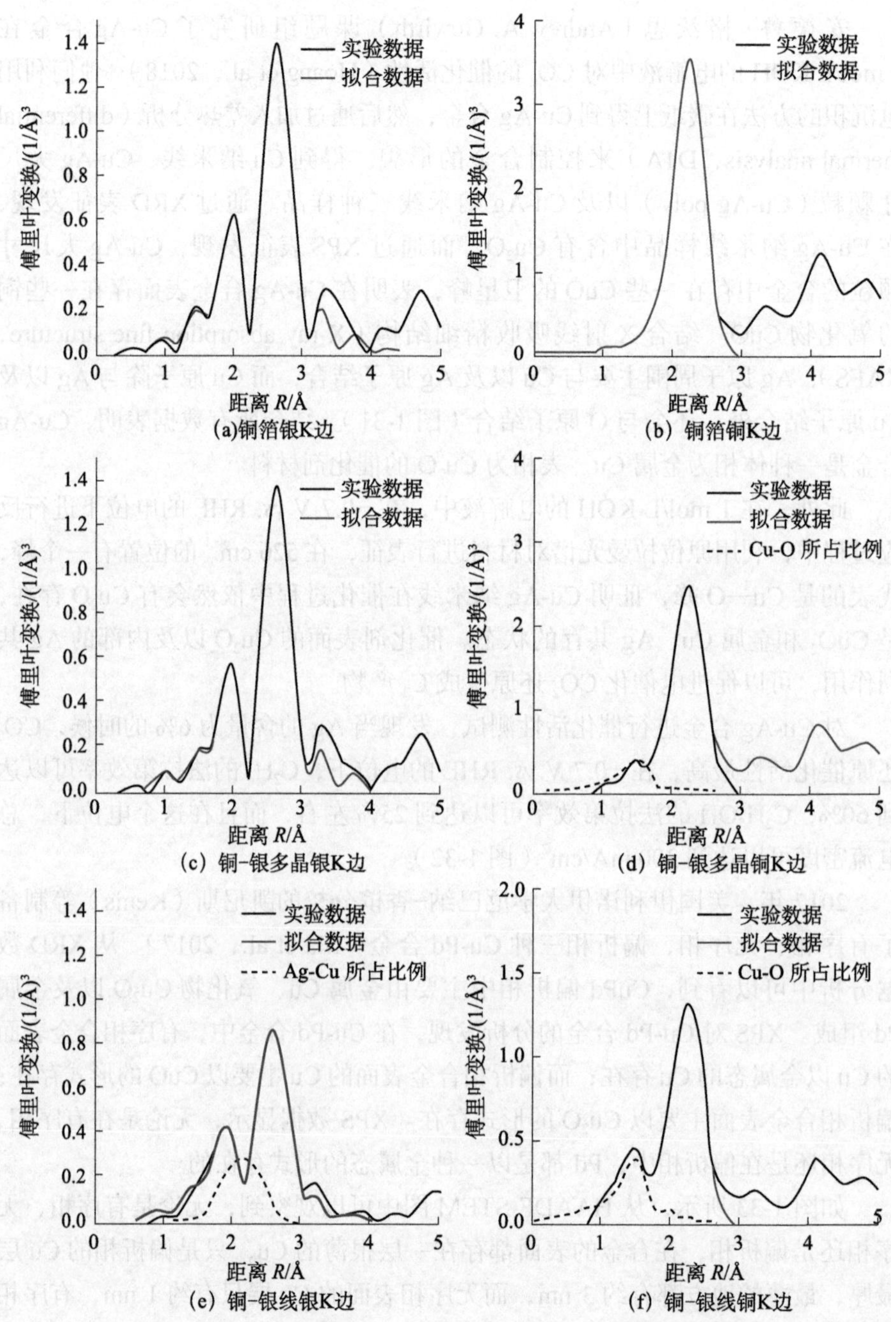

图 1-31 铜－银合金和铜－银线的银 K 边和铜 K 边 EXAFS 数据；铜－银线、铜－银多晶样品、铜箔、银箔的银 K 边和铜 K 边 EXAFS 数据（Hoang et al., 2018）

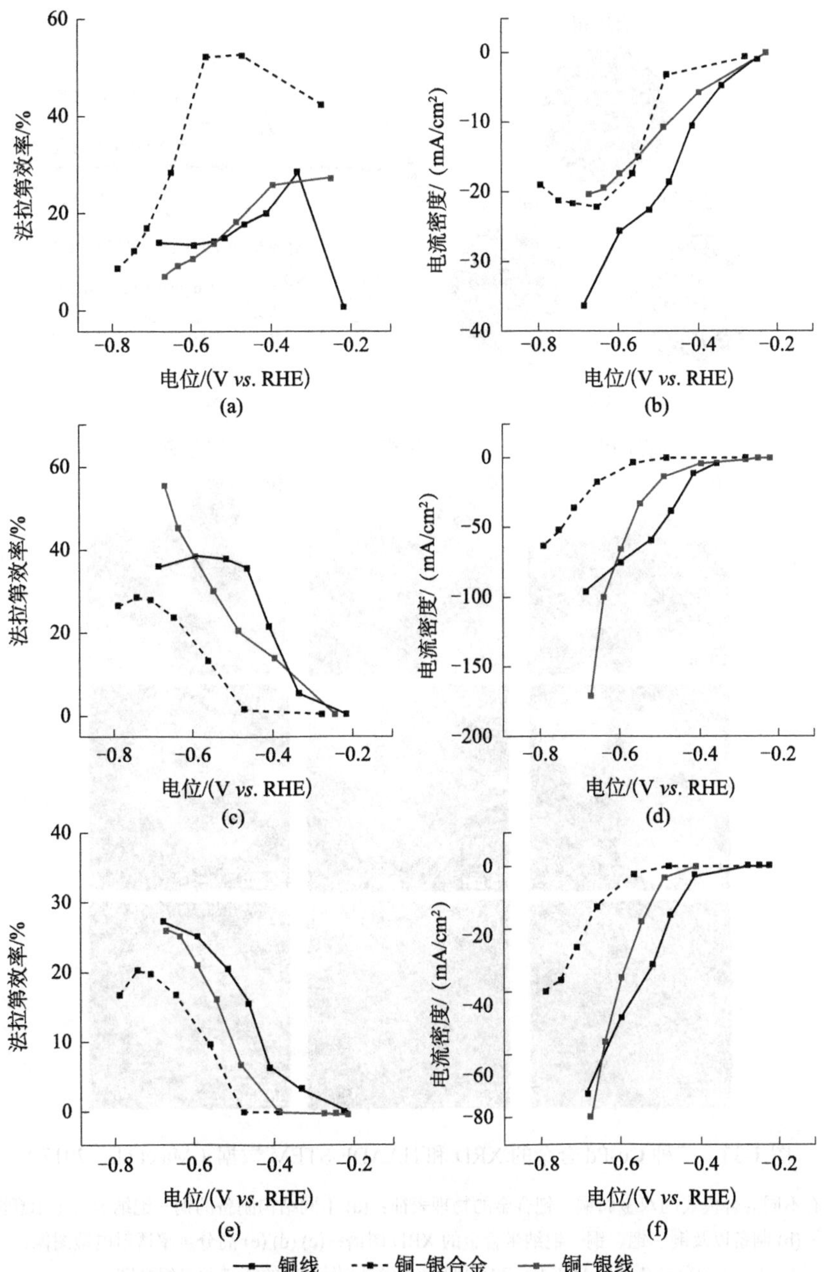

图 1-32　铜－银合金、铜线、铜－银线的催化活性对比（Hoang et al.，2018）

（a）CO 法拉第效率；（b）CO 电流密度；（c）C_2H_4 法拉第效率；（d）C_2H_4 电流密度；（e）C_2H_5OH 法拉第效率；（f）C_2H_5OH 电流密度

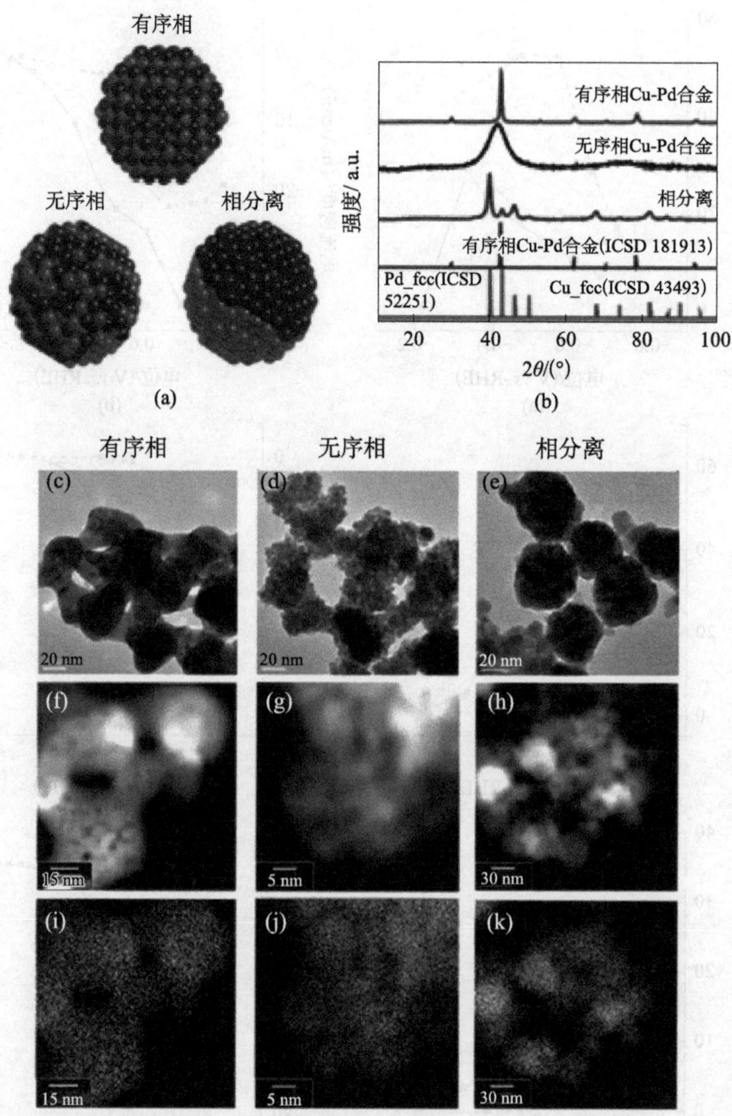

图 1-33　三种 Cu-Pd 合金的 XRD 和 HAADF-STEM 数据（Ma et al.，2017）

具有不同混合模式的双金属铜－钯合金的物理表征：(a) 不同结构制备的铜－钯纳米合金示意图；(b) 制备以及铜、钯、铜－钯纳米合金的 XRD 图谱；(c) (d) (e) 高分辨率透射电镜图像；(f) (g) (h) HAADF-STEM 图像；(i) (j) (k) 铜和钯的能谱元素组合图

CO_2 活性测试表明，相比于无序相和分离相，有序相的一碳产物的法拉第效率最高，可以达到 80% 以上；分离相的二碳产物的法拉第效率最高，可以达到 60% 以上（图 1-34）。

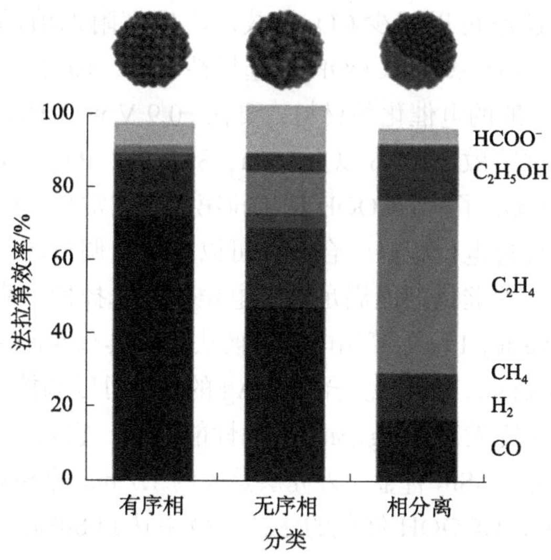

图 1-34 三种相的 Cu-Pd 合金催化活性对比（Ma et al.，2017）

杨培东课题组利用共还原金属前驱体的方法制备了不同 Au/Cu 比例的 Au-Cu 合金纳米颗粒（Kim et al.，2014）。电催化 CO_2 还原活性测试表明，在 −0.73 V *vs*. RHE 的电位下，对于 CO 的相对转换速率参数，Au_3Cu 是纯 Cu 的 93.1 倍，Au 是 Cu 的 83.7 倍，而 AuCu 是 Cu 的 40.4 倍（图 1-35）。

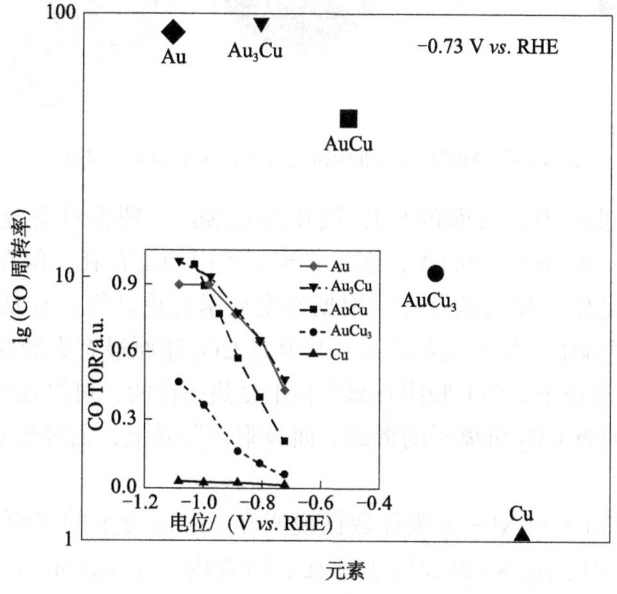

图 1-35 不同 Au-Cu 合金的催化活性图（Kim et al.，2014）

电催化 CO_2 还原可以减少 CO_2 排放，产生高附加值的化学品。然而，Cu 基催化剂将 CO_2 还原成 HCOOH 的选择性不高，而过电位很高，限制了其大规模应用。一般的电催化剂材料，要在 −0.9 V *vs*. RHE 左右电位下显示产 HCOOH 的法拉第效率 90% 以上，Cu、Sn、Pb、Pd、Co 等金属催化剂，其中 Sn 基催化剂显示了产 HCOOH 的法拉第效率非常高，然而，产 HCOOH 的电位窗口很窄且过电位很高。合金化可以有效地调节还原产物的选择性。目前，依然没有一种催化剂能满足低过电位、高选择性、高稳定性等标准。韦斯利·卢克（Wesley Luc）等用电流置换法制备具有 SnO_x 壳层的 Ag-Sn 双元合金材料（Luc et al.，2017）。合金中 Ag 的含量可以调控壳层 SnO_x 的厚度（图 1-36）。经过研究发现，$Ag_{76}Sn_{24}$ 催化性能最优，它是一个直径 20 nm 的球形结构，内核是 Ag_3Sn_2 合金，外壳层是大约 1.7 nm 的 SnO_x 层。在 −0.8 V *vs*. RHE 的电位下，HCOOH 催化的法拉第效率达到 80%，产 HCOOH 的法拉第电流密度是 16 mA/cm^2。

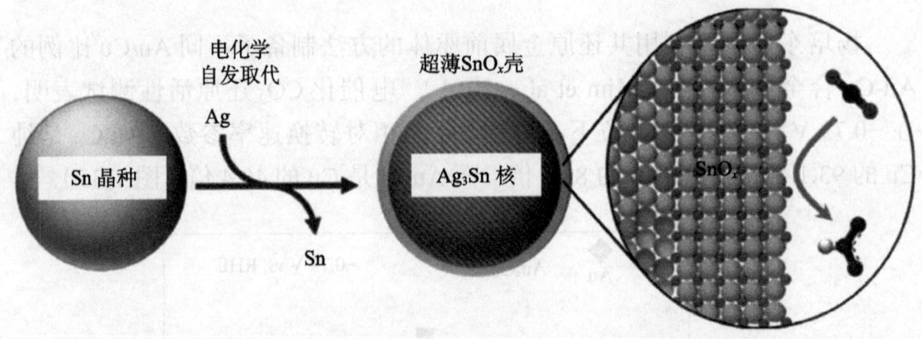

图 1-36　Ag-Sn 合金催化剂示意图（Luc et al.，2017）

在反应过程中，表面的 SnO_x 被还原成 Sn^{2+}，理论计算显示，在 SnO（101）界面，在 −0.6～−0.4 V，嵌入式羟基可以稳定存在，但在更高的负电位，氧空位是比较稳定的存在。根据密度泛函理论计算，在从 CO_2 转化为 HCOOH 的过程中，氧空位和嵌入式羟基是 CO_2 还原最有力的两种途径。在 0 V *vs*. RHE 电位下，CO_2 吸附在氧空位上是热中性的，可以观察到明显的电子转移，表明有 CO_2 负离子的形成。而吸附在羟基上，需经历 0.3 V 的自由能势垒。

TEM、EELS 与 XPS 的表征数据说明在 Ag-Sn 合金的表面有一层 SnO_x 存在，内核中的 Ag-Sn 以金属态存在。随着内核中 Ag-Sn 合金的 Sn 的浓度增加，表面 SnO 厚度和表面积增加。在 SnO 表面，相对于 $CO_2^{\cdot}$ 而言，$OCHO^{\cdot}$ 更加稳定，且被进一步催化还原成为甲酸。

孙予罕课题组发表了一篇关于 Pd-Sn 合金还原 CO_2 的文章（Bai et al., 2017）。通过润湿化学还原的方法制备了不同比例的 Pd-Sn 合金，通过 XRD 以及 XPS 分析发现，在合金中，除金属态的 Pd 和 Sn 以外，还存在两种氧化物形式——PdO 和 SnO_2。体相主要以金属 Pd 和金属 Sn 以及部分 SnO_2 的形式存在，而表相主要以 SnO_2 和 PdO 的形式存在。通过对不同的钯锡原子比的材料表征发现，随着 Sn/Sn^{4+} 的增加，Pd/Pd^{2+} 的比例会逐渐减小，这说明表相的一层 SnO_2 可以保护体相的 Pd 被氧化。

对不同比例的 Pd-Sn 合金进行催化活性测试发现，当 Pd/Sn 原子比例达到约 1∶1 时，CO_2 还原催化活性最高，过电位的绝对值最低，只有 −0.26 V，而法拉第效率可以达到 90% 以上（图 1-37）。

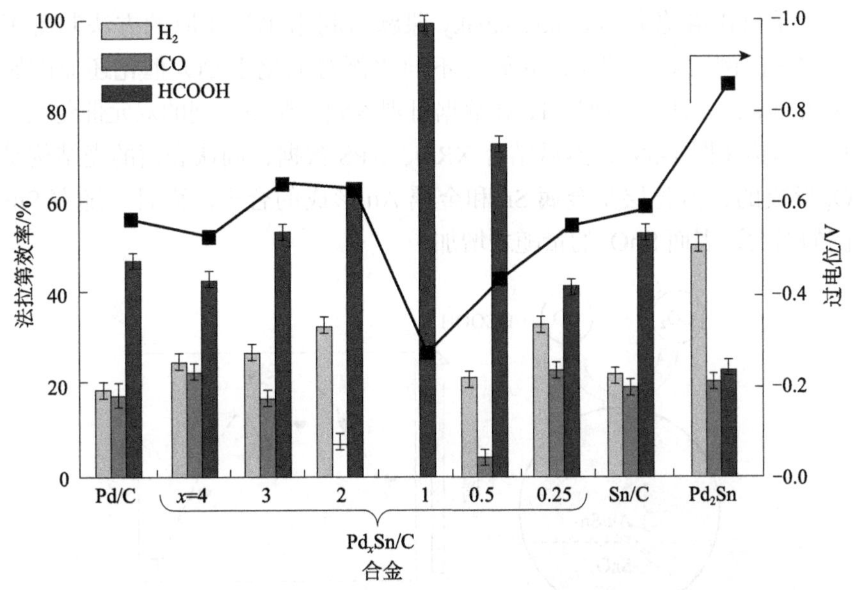

图 1-37　不同比例的 Pd-Sn 合金催化活性（Bai et al., 2017）

对合金的催化活性进行理论计算（图 1-38），考虑催化材料表面以一种 Pd-Sn-O 的形式存在，只有当催化材料的表面以一种 $PdSnO_2$ 的形式存在时，催化剂对 CO 的产生起到抑制作用，更有利于 HCOOH 的生成。

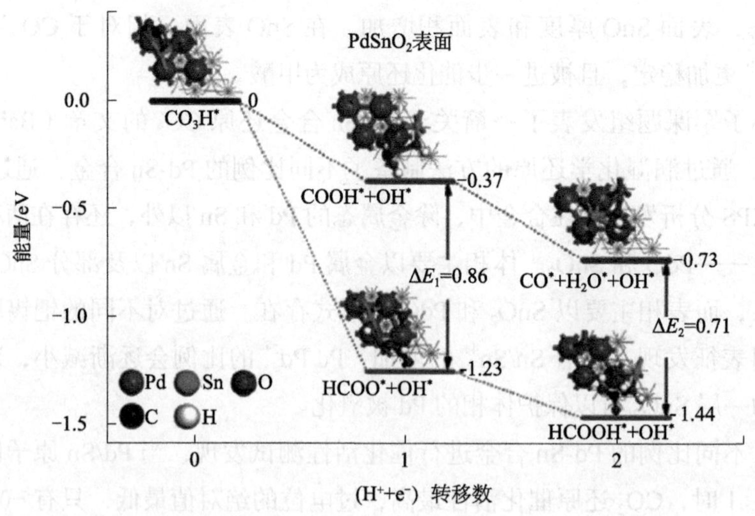

图 1-38 PdSnO$_2$ 催化剂 CO$_2$ 还原反应理论计算（Bai et al.，2017）

匈牙利赛格德大学 Csaba Janáky 课题组利用化学还原的方法制备了不同 Sn/Au 比例的双元催化，并研究不同比例对电化学 CO$_2$ 催化还原的影响（Ismail et al.，2019）。通过 TEM 数据可观察到，制备得到的双元催化剂是一种核壳结构（图 1-39），然后结合 XRD、XPS 数据，确认表面的壳结构是由 SnO$_x$ 形成的，内壳层是金属 Sn 和金属 Au 形成的合金。而且，随着 Sn/Au 比例的增大，表面 SnO$_x$ 的量随之增加。

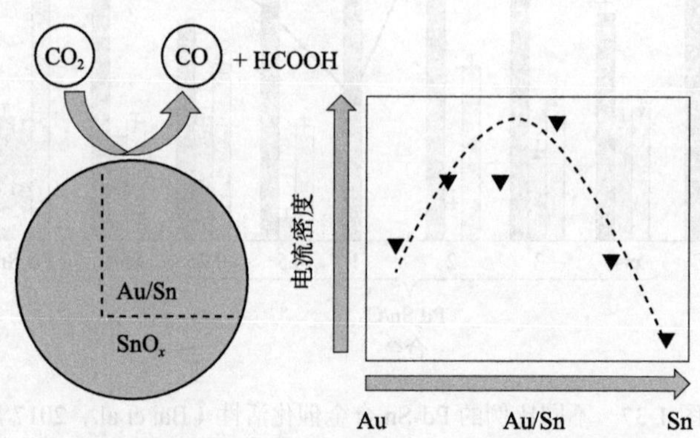

图 1-39 催化剂结构示意图（Ismail et al.，2019）

在催化过程中，他们利用原位拉曼光谱对 CO$_2$ 催化还原的中间产物进行观测，在原位拉曼光谱中，我们可以看到，在电化学还原过程中，催化剂表

面的 SnO_x 可以稳定存在。结合反应前后的 TEM 数据，双元催化剂表面的壳结构并没有消失，也进一步验证了原位拉曼光谱的结论。

对催化剂进行 CO_2 催化活性测试的结果表明，Au 本身是一个产氢效果比较好的催化剂，当与 Sn 结合形成双元催化剂后，随着 Sn 的含量增加，表面 SnO_x 的量随之增加，对 HCOOH 的选择性也逐渐提高，同时，对产氢的抑制效果也逐渐增强。在四种不同 Sn/Au 比例的催化剂中，它们对 CO 的法拉第效率全都是 10% 左右，也就说明 SnO_x 的厚度对 CO 的选择性没有明显的影响（图 1-40）。

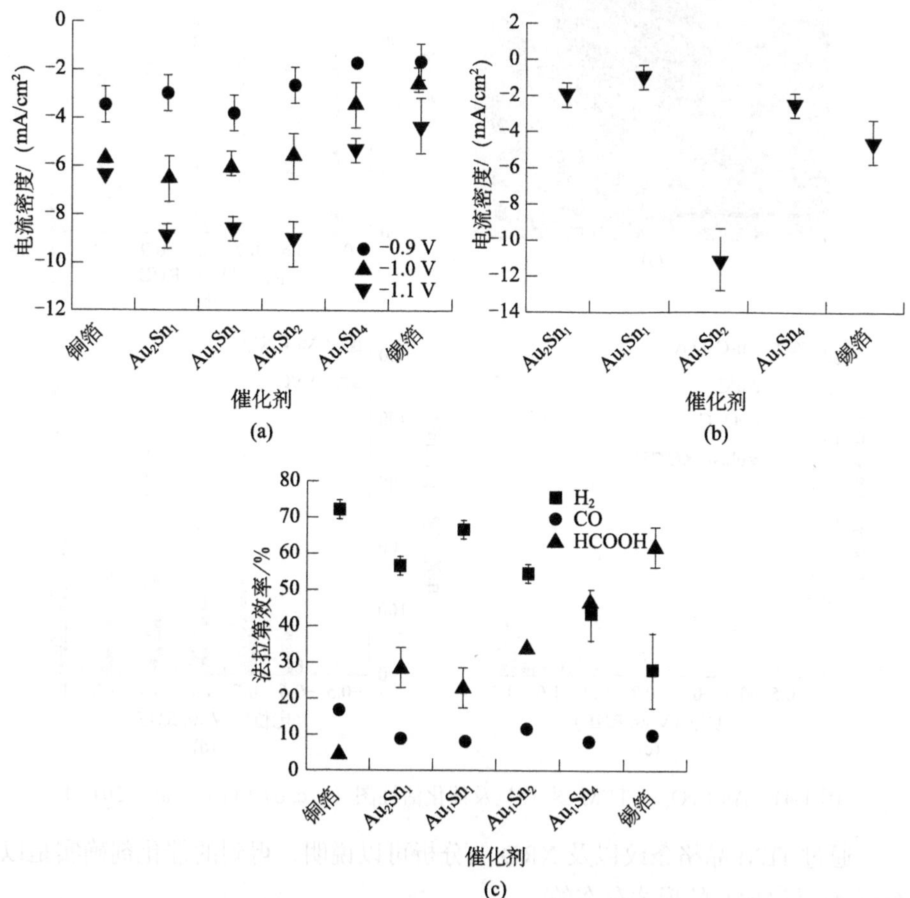

图 1-40 不同原子比的 Au-Sn 合金催化活性图；金－锡纳米粒子和母体金属的电化学二氧化碳还原活性（Ismail et al.，2019）

(a) 总电流密度；(b) 在 1.1 V *vs.* RHE 下测量的归一化电流密度；(c) 在 1.0 V *vs.* RHE 下的法拉第效率

3. 金属-氧化物复合电催化材料

我国科学家包信和等（Gao et al.，2017）的研究发现，在 −0.89 V *vs.* RHE 条件下，Au-CeO_x 电催化剂显示了 89% 的 CO 选择性，远高于单独的 Au 59% 和 CeO_x 9.8% 的选择性。另外，Au-CeO_x 的电流密度也高于单独 Au 的 1.6 倍，而 CeO_x 的电流密度几乎可以忽略（图 1-41）。

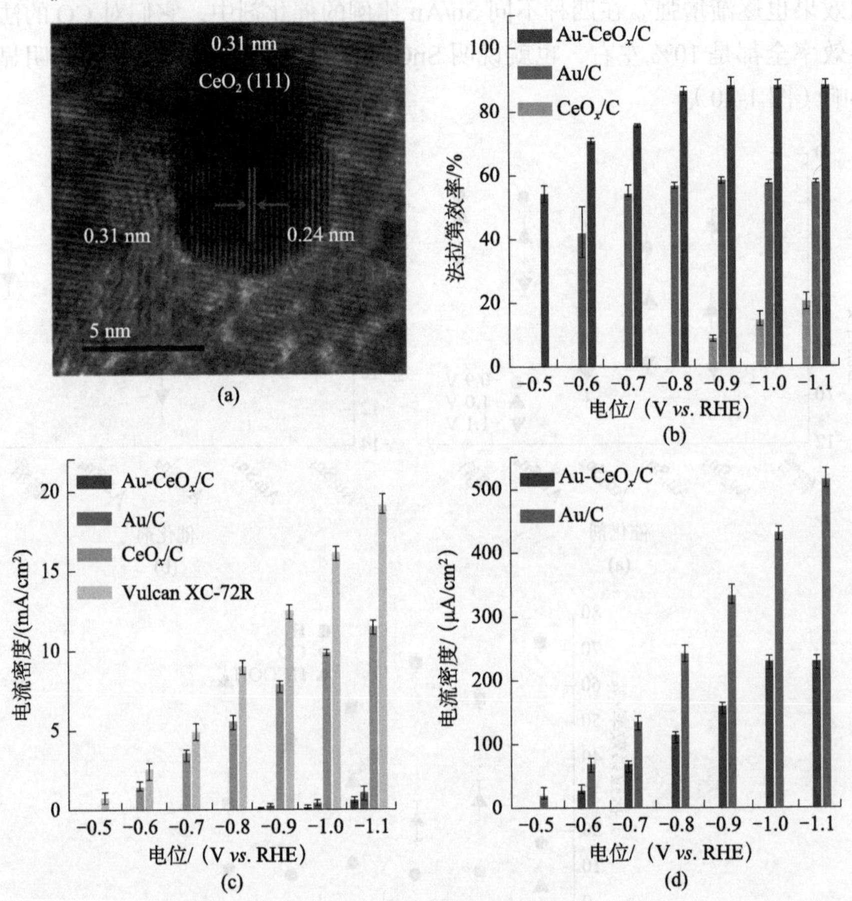

图 1-41　Au/CeO_x 的 TEM 图（a）及催化活性图（b, c, d）（Gao et al.，2017）

通过 TEM 晶格条纹以及 XRD 的分析可以说明，得到的催化剂确实是以金属 Au 与 CeO_2 的形式存在的。

在 Au（110）表面沉积岛状 CeO_x，通过原位 STM 表征 CO_2 的吸附行为。结果表明，CeO_x 上 CO_2^* 的吸附过程是先在 CeO_x 和 Au 的界面处吸附 CO_2，然后拓展到整个 CeO_x 表面。单独的 Au 表面则没有观察到 CO_2 的吸附。Au-CeO_x 不仅有利于 CO_2 的吸附，还有利于它的活化（图 1-42）。

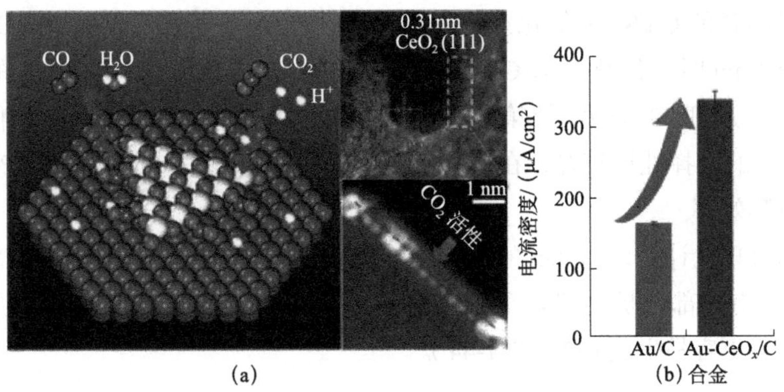

图 1-42 Au-CeO_x 的反应示意图（Gao et al., 2017）

孙守恒课题组利用热分解乙酰丙酮锡的方法制备出 SnO_2 包裹 Cu 的核壳结构，通过调节乙酰丙酮锡的量来控制表面 SnO_2 的厚度（Li et al., 2017）。TEM 数据显示，内壳层的 Cu 纳米颗粒直径约为 7 nm，制备出两种不同厚度的 SnO_2（图 1-43）。

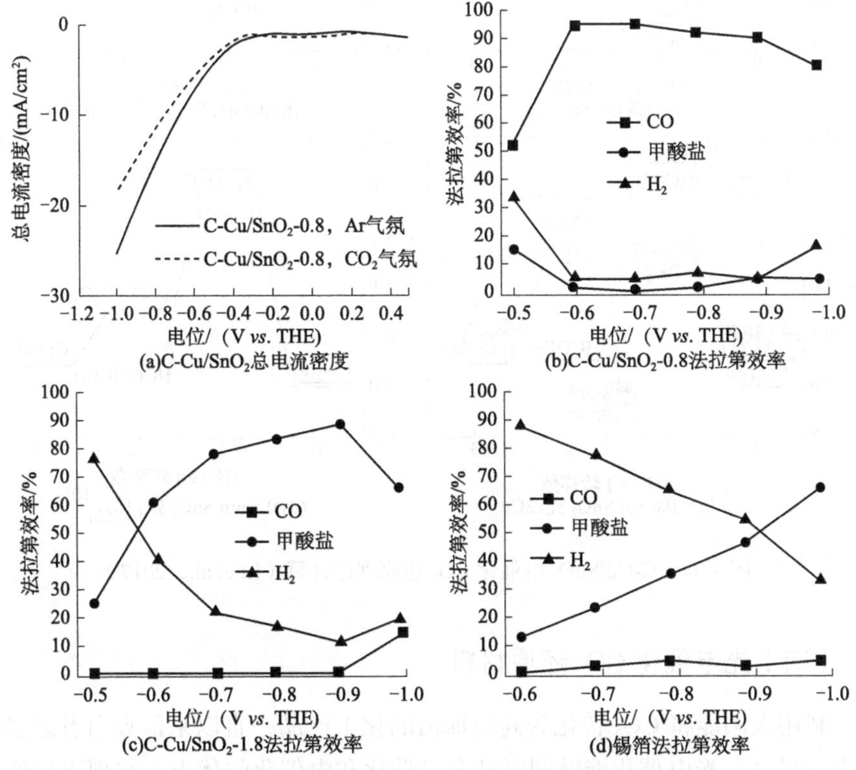

图 1-43 各种 SnO_2 厚度的 C-Cu/SnO_2 的催化活性和法拉第效率（Li et al., 2017）

对制备的 Cu/SnO_2 催化剂进行催化活性测试时发现，当表面 SnO_2 的厚度为 0.8 nm 时，催化剂对 CO 的选择性比较好，在 −0.6 V *vs*. RHE 的电位下，可以达到 93% 的法拉第效率；当 SnO_2 的厚度为 1.8 nm 时，催化剂对 HCOOH 的选择性比较好，在 −0.9 V *vs*. RHE 的电位下，可以达到 90% 左右的法拉第效率。

通过理论计算，0.8 nm 的 SnO_2 会对 CO 的选择性比较高是因为内部金属态的 Cu 有一部分扩散到表面上的 SnO_2 中，从而导致 SnO_2 的晶格被单向压缩，结果更容易生成 CO（图 1-44）。

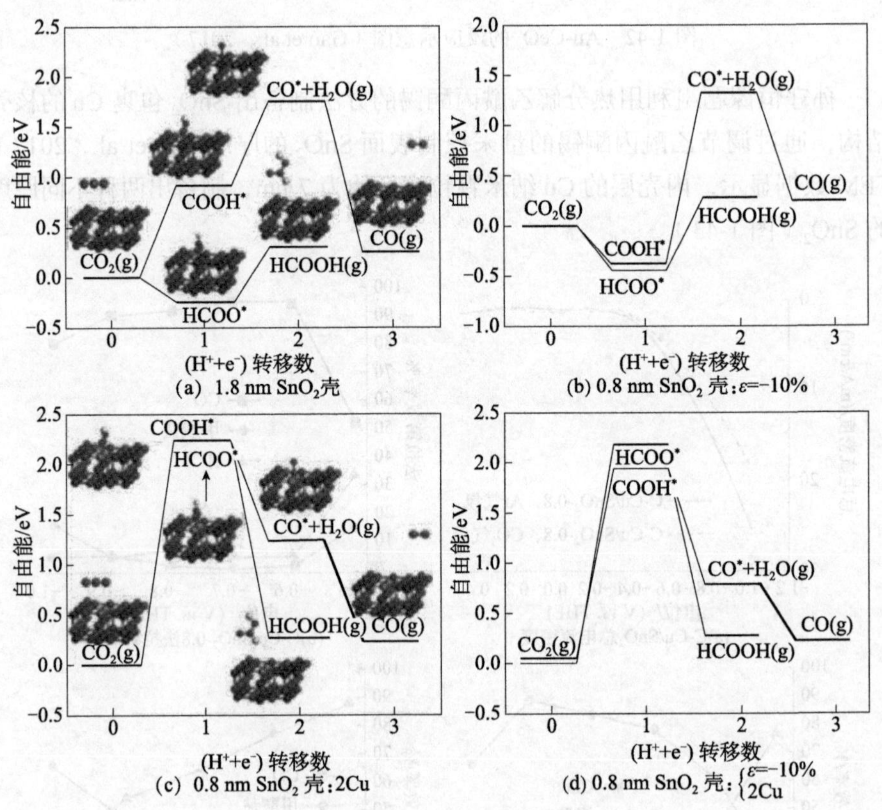

图 1-44 C-Cu/SnO_2 电催化 CO_2 还原理论计算（Li et al.，2017）

（四）光电催化 CO_2 还原材料

利用太阳能将 CO_2 转化为高附加值的化工产品一直以来都吸引着诸多研究者的目光。光电催化能够同时结合光催化和电催化的优点，实现更高效率

和更高选择性的 CO_2 还原。这里重点讨论半导体、半导体与金属复合、半导体与金属氧化物复合，以及金光伏叠层电池等在光电催化 CO_2 还原领域的最新成果，并分析光电催化 CO_2 还原面临的关键科学问题与挑战，以期对未来光电催化 CO_2 还原催化剂的设计提供建议。

非常多的 p 型光阴极包括 Si、ZnTe、CdTe、p-InP、Cu_2O、p-NiO 被用于光电催化 CO_2 还原（Kalamaras et al.，2018；Zhang et al.，2017）。光阴极侧复合金属助催化剂能够进一步地提升 CO_2 还原的活性和选择性。但如何在低的电位下自发地吸附 CO_2 并反应得到期望的产物仍然是一个挑战。纯金属催化剂只存在单功能的活性位点，通常与 CO_2 分子的连接较弱，不能提供多位点用于稳定中间产物，导致高的反应过电位和低的催化活性。

鉴于此，TiO_2、ZnO 与金属构成的金属氧化物－金属界面成为研究的热点。TiO_2 的引入不仅构成金属氧化物－金属界面，用来吸附活化 CO_2，同时 TiO_2 也作为钝化层对光阴极进行保护，提升材料的稳定性。原子层沉积（atomic layer deposition，ALD）TiO_2 在钝化 InP 表面的同时形成 p-n 结。1-乙基-3-甲基咪唑四氟硼酸盐（$[EMIM]BF_4$）离子液体溶解在乙腈中作为电解液，在 0.78 V 电位下，得到了 99% 的 CO 法拉第效率（Zeng et al.，2015）。在 0.10 mol/L $NaHCO_3$、pH =8 溶液、−0.60 V 电位条件下，Ti-TiO_2-CuO 电极光电催化 CO_2 还原有最高的性能，得到甲醇的选择性为 97%（Brito & Zanoni，2017）。TiO_2 钝化层保护 InP 纳米柱光阴极在 532 nm 波长光激发下进行光电催化 CO_2 还原产甲醇。TiO_2 层在提供一个稳定的光催化界面的同时，TiO_2 内的 O 缺陷也增强了光转化效率。平面波密度泛函理论计算确认了 TiO_2 表面的 O 缺陷作为催化的活性位点进行 CO_2 还原反应。Cu 纳米颗粒沉积到 TiO_2 上，作为助催化剂进一步提升了其选择性，得到了 8 倍的甲醇产量和 8.7% 的法拉第效率（Qiu et al.，2015）。

储升等制备了金属 / 氧化物进行持续的活化 CO_2 分子，同时稳定反应中间产物。Pt-TiO_2/GaN/n^+-p Si 基光阴极的太阳能转换效率为 0.87%，转化数为 24 800，稳定性达到 10 h。Pt/TiO_2 表面修饰的光阴极 CO 的选择性远高于单独 Pt 和 TiO_2 修饰的样品（图 1-45）。

理论计算表明，当没有 TiO_2 时，CO_2 在 Pt 表面的吸附较弱，CO_2 的键角约为 180°，没有明显变化，而有 TiO_2 团簇时，CO_2 吸附变强，键角变为 125°，这种扭曲导致了 CO_2 键的减弱，使其活化。金属 / 氧化物界面的协同作用加速了 CO_2 活化，稳定了中间物种 CO_2^*，因此提高了 CO_2 转化成 CO 的法拉第效率（Chu et al.，2018）。将 15 nm 的 p 型层、650 nm 的本征光吸收层、

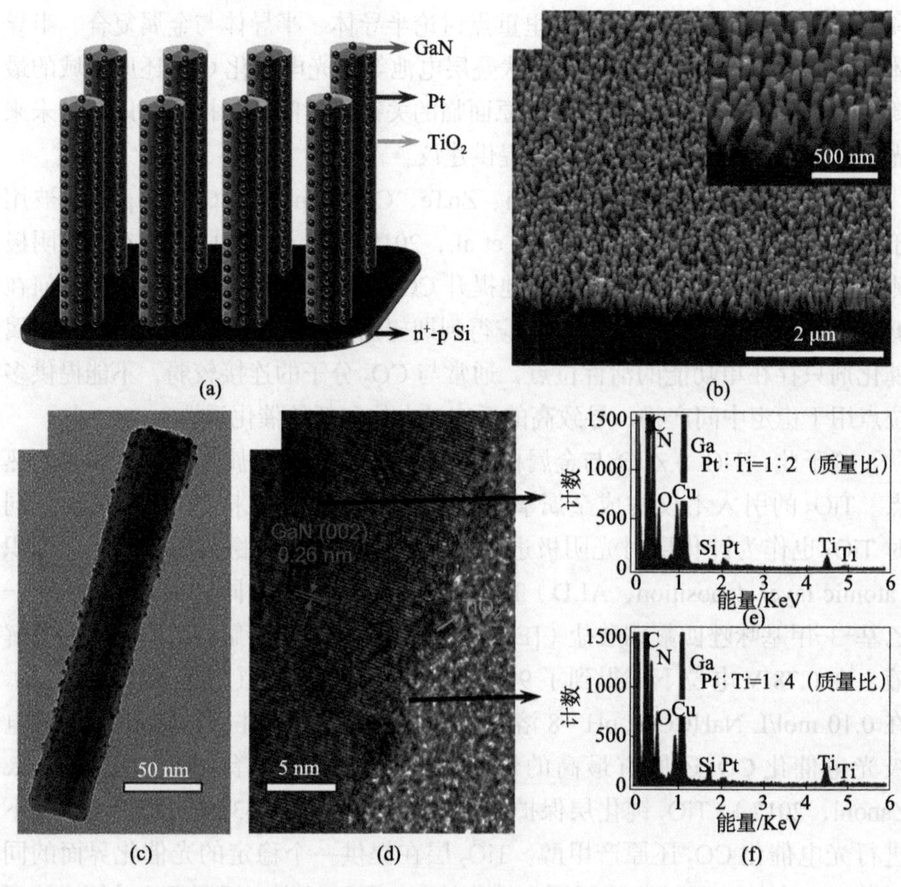

图 1-45　Pt-TiO_2/GaN/n^+-p Si 样品的表征（Qiu et al.，2015）

30 nm 的 n 型层（p-i-n）通过等离激元增强化学气相沉积（plasma enhanced chemical vapor deposition，PECVD）到掺氟的二氧化锡（FTO）导电玻璃上，再通过 ALD 技术在 Si 基光阴极上沉积 40 nm 厚的 TiO_2 保护层，最后通过电子束蒸发 Au 层到 TiO_2 上，制备得到了 a-Si/TiO_2/Au 光阴极。a-Si p-i-n 结提供了大的光电压，在 −0.4～−0.1 V *vs.* RHE 下，大晶粒的 Au 实现了高产量的合成气的合成（Li et al.，2019）。两步法合成 Cu-ZnO/GaN/n^+-p Si 基光阴极，首先采用离子辅助分子束外延技术在 p-n 结 Si 片上生长一维 GaN 纳米线。这种一维纳米线结构有利于光吸收和电子迁移。合成气的转化数为 1330，通过助催化剂 Cu 和 ZnO 的协同效应，得到的 CO/H_2 比值为 1∶2。CO 最高的法拉第效率为 70%，过电位为 180 mV（Chu et al.，2016）。

二、电催化 CO_2 还原材料研究目前面临的重大科学问题及前沿方向

虽然电催化 CO_2 还原反应已经有近百年的研究历史，而且取得了很大的进步，但是依旧存在许多问题和挑战。主要包括以下几个方面。

（1）高碳产物选择性依然很低。在目前的 CO_2 还原中，虽然 CO 和 HCOOH 的法拉第效率已经可以达到 90% 以上，但是其他的还原产物，如 CH_4、CH_3OH、C_2H_5OH、CH_3COOH、C_2H_4 等目前可以达到的最高法拉第效率只有 50% 左右。

（2）催化剂稳定性较差。稳定性是表征一种催化材料优劣的重要参数，也是实现工业化生产所必须克服的问题。目前的一些催化材料虽然已经可以达到上百小时的稳定性，但是距离实现工业化生产的目标还很远。尤其是在光电催化领域，由于催化剂的结构复杂，以及光照腐蚀等因素的影响，目前光电催化 CO_2 还原的最佳稳定性只有 200 h 左右。

（3）反应机理问题。（光）电催化 CO_2 还原反应生成各种复杂的化学燃料是一个十分复杂的反应过程。但是，到目前为止，所有的反应流程全都是根据理论计算模拟出来的，还没有在实验中明确观察到。另外，虽然已经确认催化剂对（光）电催化 CO_2 还原有催化效果，但是催化剂与 CO_2 分子之间的相互作用关系以及催化活性位点依旧没有在实验中得到准确的结果。

三、研究总结与展望

化石燃料的过度使用使空气中的 CO_2 浓度大幅度提高，导致全球气候变暖、环境遭到严重破坏。随着电催化与光电催化技术的发展，人们已经探索出一种利用（光）电催化 CO_2 还原反应生成具有高价值的化石燃料的方法来人工解决这一问题。随着近百年的发展，（光）电催化 CO_2 还原的技术已经逐渐成熟。科学家研究了不同的电解液对（光）电催化 CO_2 还原活性的影响（König et al.，2019），也探索出一系列的（光）电催化 CO_2 还原的催化剂材料，比如对 HCOOH 选择性较高的 Sn、Bi，对 CO 选择性较高的 Zn、Au、Ag 等，以及最为特殊的具有十几种产物的 Cu 以及其他用于（光）电催化的半导体材料，如 Si、TiO_2、ZnO 等。科学家发现，将不同的金属构成合金并调节合金中的组分比例对 CO_2 还原的催化活性都有很大影响。此外，有一些课题组发现，将两种不同的催化剂组合所形成的一种新型的复合催化剂会有不同的催化效果。在光（电）催化中，选择不同的半导体材料以及在半导体材料上担载

不同的助催化剂对光（电）催化的催化活性与产物的选择性都有很大影响。

（光）在电催化 CO_2 还原领域中人们已经取得了许多重大的成果，利用（光）电催化将 CO_2 还原转化成具有高价值的化学燃料必然能够实现工业化生产，达到环境保护以及资源循环利用的目的。

四、学科发展政策建议与措施

为了早日实现工业化生产，更深层次探索（光）电催化 CO_2 还原的反应机理，本书对（光）电催化领域的发展提出以下几点建议。

（一）探索新型的高性能催化材料

在（光）电催化 CO_2 还原的产物中，相较于 CO、HCOOH 这些产物，CH_4、CH_3OH、C_2H_5OH、C_2H_4 以及其他高碳产物更加具有工业化价值。因此，探索一系列对高碳产物具有高选择性的材料是目前（光）电催化 CO_2 还原领域中的研究热点。

（二）研究高稳定性催化材料

催化材料的稳定性是表征催化剂优劣的重要参数，目前的一些（光）电催化 CO_2 还原催化剂都是与水溶液直接接触的，在反应过程中对催化剂有很大的侵蚀作用，导致催化剂稳定性较差。虽然目前的一些催化剂可以达到上百小时的稳定性，但是还远远没有达到工业化生产的要求。探索一种高稳定性的催化材料或设计一种防侵蚀的方案是解决这一问题的主要方法。该问题的解决在工业化生产中可以减少原料损失，降低成本，有助于工业化生产的早日实现。

（三）大力发展新型表征技术

目前的许多研究依然是针对催化剂材料进行非原位表征，但是这些表征手段无法避免催化剂与空气中的 O_2、CO_2、水蒸气进行接触，进而会改变其特性。此外，非原位表征手段无法准确描述催化剂在催化过程中的实际状态，因此导致无法准确地研究（光）电催化 CO_2 还原反应的机理。应大力发展原位表征手段来准确探索（光）电催化 CO_2 还原反应机理，进一步指导催化剂材料的设计。

任斐隆（南京大学）、罗文俊（南京大学）

本节参考文献

Bai X F, Chen W, Zhao C C, et al. 2017. Exclusive formation of formic acid from CO_2 electroreduction by a tunable Pd-Sn alloy. Angewandte Chemie International Edition, 56: 12219-12223.

Baruch M F, Pander J E, White J L, et al. 2015. Mechanistic insights into the reduction of CO_2 on tin electrodes using *in situ* ATR-IR spectroscopy. ACS Catalysis, 5: 3148-3156.

Bevilacqua M, Filippi J, Folliero M, et al. 2016. Enhancement of the efficiency and selectivity for carbon dioxide electroreduction to fuels on tailored copper catalyst architectures. Energy Technology, 4: 1020-1028.

Brito J F D, Zanoni M V B. 2017. On the application of Ti/TiO_2/CuO n-p junction semiconductor: a case study of electrolyte, temperature and potential influence on CO_2 reduction. Chemical Engineering Journal, 318: 264-271.

Cavalca F, Ferragut R, Aghion S. 2017. Nature and distribution of stable subsurface oxygen in copper electrodes during electrochemical CO_2 reduction. Journal of Physical Chemistry C, 121: 25003-25009.

Centi G, Perathoner S. 2009. Opportunities and prospects in the chemical recycling of carbon dioxide to fuels. Catalysis Today, 148: 191-205.

Chen Y H, Kanan M W. 2012. Tin oxide dependence of the CO_2 reduction efficiency on tin electrodes and enhanced activity for tin/tin oxide thin-film catalysts. Journal of the American Chemical Society, 134: 1986-1989.

Chen Y H, Li C W, Kanan M W. 2012. Aqueous CO_2 reduction at very low overpotential on oxide-derived Au nanoparticles. Journal of the American Chemical Society, 134: 19969-19972.

Chu S, Fan S Z, Wang Y J, et al. 2016. Tunable syngas production from CO_2 and H_2O in an aqueous photoelectrochemical cell. Angewandte Chemie International Edition, 55: 14262-14266.

Chu S, Ou P F, Ghamari P, et al. 2018. Photoelectrochemical CO_2 reduction into syngas with the metal/oxide interface. Journal of the American Chemical Society, 140: 7869-7877.

Eilert A, Cavalca F, Roberts F S, et al. 2017. Subsurface oxygen in oxide-derived copper electrocatalysts for carbon dioxide reduction. Journal of Physical Chemistry Letters, 8: 285-290.

El-Khouly M E, El-Mohsnawy E, Fukuzumi S. 2017. Solar energy conversion: from natural to artificial photosynthesis. Journal of Photochemistry and Photobiology C: Photochemistry

Reviews, 31: 36-83.

Feng Y, Li Z, Liu H, et al. 2018. Laser-prepared CuZn alloy catalyst for selective electrochemical reduction of CO_2 to ethylene. Langmuir, 34: 13544-13549.

Gao D F, Arán-Ais R M, Jeon H S, et al. 2019. Rational catalyst and electrolyte design for CO_2 electroreduction towards multicarbon products. Nature Catalysis, 2: 198-210.

Gao D F, Zhang Y, Zhou Z W, et al. 2017. Enhancing CO_2 electroreduction with the metal-oxide interface. Journal of the American Chemical Society, 139: 5652-5655.

Halmann M. 1978. Photoelectrochemical reduction of aqueous carbon dioxide on p-type gallium phosphide in liquid junction solar cells. Nature, 275: 115-116.

Hoang T T H, Verma S, Ma S, et al. 2018. Nanoporous copper-silver alloys by additive-controlled electrodeposition for the selective electroreduction of CO_2 to ethylene and ethanol. Journal of the American Chemical Society, 140: 5791-5797.

Hori Y, Kikuchi K, Suzuki S. 1985. Production of CO and CH_4 in electrochemical reduction of CO_2 at metal electrodes in aqueous hydrogencarbonate solution. Chemistry Letters, 14: 1695-1698.

Hori Y, Wakebe H, Tsukamoto T, et al. 1994. Electrocatalytic process of CO selectivity in electrochemical reduction of CO_2 at metal electrodes in aqueous media. Electrochimica Acta, 39: 1833-1839.

Ismail A M, Samu G F, Balog Á, et al. 2019. Composition-dependent electrocatalytic behavior of Au-Sn bimetallic nanoparticles in carbon dioxide reduction. ACS Energy Letters, 4: 48-53.

Kalamaras E, Maroto-Valer M M, Shao M, et al. 2018. Solar carbon fuel via photoelectrochemistry. Catalysis Today, 317: 56-75.

Kim D, Resasco J, Yu Y, et al. 2014. Synergistic geometric and electronic effects for electrochemical reduction of carbon dioxide using gold-copper bimetallic nanoparticles. Nature Communications, 5: 4948.

Klingan K, Kottakkat T, Jovanov Z P, et al. 2018. Reactivity determinants in electrodeposited Cu foams for electrochemical CO_2 reduction. ChemSusChem, 11: 3449-3459.

Kondratenko E V, Mul G, Baltrusaitis J, et al. 2013. Status and perspectives of CO_2 conversion into fuels and chemicals by catalytic, photocatalytic and electrocatalytic processes. Energy & Environmental Science, 6: 3112-3135.

König M, Vaes J, Klemm E, et al. 2019. Solvents and supporting electrolytes in the electrocatalytic reduction of CO_2. iScience, 19: 135-160.

Kuhl K P, Hatsukade T, Cave E R, et al. 2014. Electrocatalytic conversion of carbon dioxide to methane and methanol on transition metal surfaces. Journal of the American Chemical Society, 136: 14107-14113.

Lee C W, Hong J S, Yang K D, et al. 2018. Selective electrochemical production of formate from carbon dioxide with bismuth-based catalysts in an aqueous electrolyte. ACS Catalysis, 8: 931-937.

Lee M Y, Park K T, Lee W, et al. 2019. Current achievements and the future direction of electrochemical CO_2 reduction: a short review. Critical Reviews in Environmental Science and Technology, 50: 1-47.

Lei F C, Liu W, Sun Y F, et al. 2016. Metallic tin quantum sheets confined in graphene toward high-efficiency carbon dioxide electroreduction. Nature Communications, 7: 12697.

Li C C, Wang T, Liu B, et al. 2019. Photoelectrochemical CO_2 reduction to adjustable syngas on grain-boundary-mediated a-Si/TiO_2/Au photocathodes with low onset potentials. Energy & Environmental Science, 12: 923-928.

Li C W, Kanan M W. 2012. CO_2 Reduction at low overpotential on Cu electrodes resulting from the reduction of thick Cu_2O films. Journal of the American Chemical Society, 134: 7231-7234.

Li Q, Fu J J, Zhu W L, et al. 2017. Tuning Sn-catalysis for electrochemical reduction of CO_2 to CO via the core/shell Cu/SnO_2 structure. Journal of the American Chemical Society, 139: 4290-4293.

Liu S B, Xiao J, Lu X F, et al. 2019. Efficient electrochemical reduction of CO_2 to HCOOH over sub-2 nm SnO_2 quantum wires with exposed grain boundaries. Angewandte Chemie International Edition, 58: 8499-8503.

Low Q H, Loo N W X, Calle-Vallejo F, et al. 2019. Enhanced electroreduction of carbon dioxide to methanol using zinc dendrites pulse-deposited on silver foam. Angewandte Chemie International Edition, 58: 2256-2260.

Luc W, Collins C, Wang S, et al. 2017. Ag-Sn bimetallic catalyst with a core-shell structure for CO_2 reduction. Journal of the American Chemical Society, 139: 1885-1893.

Luc W, Fu X B, Shi J J, et al. 2019. Two-dimensional copper nanosheets for electrochemical reduction of carbon monoxide to acetate. Nature Catalysis, 2: 423-430.

Lum Y, Ager J W. 2019. Evidence for product-specific active sites on oxide-derived Cu catalysts for electrochemical CO_2 reduction. Nature Catalysis, 2: 86-93.

Ma M, Trześniewski B J, Xie J, et al. 2016. Selective and efficient reduction of carbon dioxide to carbon monoxide on oxide-derived nanostructured silver electrocatalysts. Angewandte Chemie International Edition, 55: 9748-9752.

Ma S, Sadakiyo M, Heima M, et al. 2017. Electroreduction of carbon dioxide to hydrocarbons using bimetallic Cu-Pd catalysts with different mixing patterns. Journal of the American Chemical Society, 139: 47-50.

Mandal L, Yang K R, Motapothula M R, et al. 2018. Investigating the role of copper oxide in electrochemical CO_2 reduction in real time. ACS Applied Materials & Interfaces, 10: 8574-

8584.

Mikkelsen M, Jørgensen M, Krebs F C. 2010. The teraton challenge: a review of fixation and transformation of carbon dioxide. Energy & Environmental Science, 3: 43-81.

Mistry H, Varela A S, Bonifacio C S, et al. 2016. Highly selective plasma-activated copper catalysts for carbon dioxide reduction to ethylene. Nature Communications, 7: 12123.

Qiao J L, Liu Y Y, Hong F, et al. 2014. A review of catalysts for the electroreduction of carbon dioxide to produce low-carbon fuels. Chemical Society Reviews, 43: 631-675.

Qiu J, Zeng G T, Ha M A, et al. 2015. Artificial photosynthesis on TiO_2-passivated InP nanopillars. Nano Letters, 15: 6177-6181.

Teeter T E, van Rysselberghe P. 1954. Reduction of carbon dioxide on mercury cathodes. Journal of Chemical Physics, 22: 759-760.

Wu J H, Huang Y, Ye W, et al. 2017. CO_2 reduction: from the electrochemical to photochemical approach. Advanced Science, 4: 1700194.

Wu J J, Risalvato F G, Ma S G, et al. 2014. Electrochemical reduction of carbon dioxide Ⅲ. The role of oxide layer thickness on the performance of Sn electrode in a full electrochemical cell. Journal of Materials Chemistry A, 2: 1647-1651.

Zeng G T, Qiu J, Hou B Y, et al. 2015. Enhanced photocatalytic reduction of CO_2 to CO through TiO_2 passivation of InP in ionic liquids. Chemistry: A European Journal, 21: 13502-13507.

Zhang D D, Shi J Y, Zi W, et al. 2017. Recent advances in photoelectrochemical applications of silicon materials for solar-to-chemicals conversion. ChemSusChem, 10: 4324-4341.

Zhang N, Long R, Gao C, et al. 2018. Recent progress on advanced design for photoelectrochemical reduction of CO_2 to fuels. Science China Materials, 61: 771-805.

Zhang R, Lv W X, Lei L X. 2015. Role of the oxide layer on Sn electrode in electrochemical reduction of CO_2 to formate. Applied Surface Science, 356: 24-29.

Zhang X, Hou X F, Zhang Q, et al. 2018. Polyethylene glycol induced reconstructing Bi nanoparticle size for stabilized CO_2 electroreduction to formate. Journal of Catalysis, 365: 63-70.

Zhao S L, Li S, Guo T, et al. 2019. Advances in Sn-based catalysts for electrochemical CO_2 reduction. Nano-Micro Letters, 11: 62.

Zheng X L, Ji Y F, Tang J, et al. 2019. Theory-guided Sn/Cu alloying for efficient CO_2 electroreduction at low overpotentials. Nature Catalysis, 2: 55-61.

Zhu D D, Liu J L, Qiao S Z. 2016. Recent advances in inorganic heterogeneous electrocatalysts for reduction of carbon dioxide. Advanced Materials, 28: 3423-3452.

第四节　光电催化产氢材料

一、光电催化产氢材料研究概述

（一）光电催化产氢材料研究的意义

上千年来，生物质一直都是人类最主要的可利用能量来源之一。然而到了最近的两百年，随着人类社会工业化水平的不断提高，化石能源的消耗占据了人类社会能源总消耗量的主导地位。当前全世界的年能源消耗量约为 14 TW，据估算在 2050 年将达到 28 TW，而在 21 世纪末将增加至 43 TW（Lewis & Nocera，2006）。如果人类社会继续当前的发展模式，不但会导致不可再生化石能源的急剧消耗与短缺，而且温室气体与污染物的大量排放也会引发严峻的生态和环境问题。因此，开发可再生的清洁能源以降低乃至消除对化石能源的依赖，是实现人类经济社会可持续发展的必由之路。可再生的清洁能源主要包括风能、水能、生物质能、地热能、氢能和太阳能等，但以上能源（除地热能外）都直接或间接来自太阳，且其他能源的总和不及太阳能的 1%。而且太阳能分布广泛，无须运输且用之不竭。目前，人类利用太阳能最有效的手段是通过带隙合适的半导体吸收太阳能，将太阳能转换成电能。目前，多节太阳能电池的太阳能转换效率已经达到 47.1%，远远高于自然界光合作用（小于 2%）。但是电能的大规模存储成本高，且无法保证随时随地提供稳定、低成本、高质量的电能。氢气的能量密度高，可规模化生产，其使用不会造成温室气体的排放，是一种极具前景的新型绿色能源。通过模拟自然界的光合作用，藤岛昭（Fujishima Akira）和本多健一（Kenichi Honda）提出利用太阳能和半导体直接分解水，从而大量、低成本地生产氢气（Fujishima & Honda，1972）。由于光电催化水分解池结构简单、成本低廉，在太阳能制氢领域具有巨大的产业化潜力。

（二）光电催化产氢的基本原理

1. 光电催化水分解池的基本结构

无论光电催化水分解池选用何种材料与配置方式，其核心结构与普通电解池类似，都由阳极、阴极、外电路及电解液组成（图 1-46）。与普通电解池类似，光电催化水分解池的阳极表面发生氧化反应，即水分子被氧化为氧气；阴极电极表面发生还原反应，即水分子被还原为氢气。与普通电解池的区别在于，光电催化水分解池的阳极或阴极或二者同时能捕获太阳光，以代替部分或全部水分解反应所需的电能，其中具有光响应的阳极或阴极称为光阳极或光阴极。

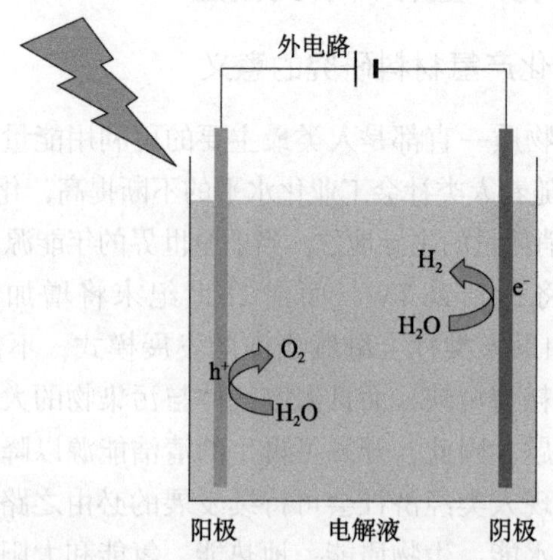

图 1-46　光电催化水分解池的基本结构

2. 光电催化水分解池的工作原理

光电催化水分解池的工作原理如图 1-47 所示，这里以 n 型半导体为例进行说明（Grätzel，2001；Walter et al.，2010）。光电催化水分解池完成一个反应循环共需以下 3 步：①半导体吸收能量大于或等于其禁带宽度的光子并产生电子－空穴对；②电子－空穴对扩散至空间电荷层并在其内建电场作用下发生分离，电子迁移至半导体体相，空穴迁移至半导体表面；③迁移至半导体体相的电子被导电基底收集，并传输至对电极驱动水的还原反应，而迁移至半导体表面的空穴则驱动水的氧化反应 。对于 p 型半导体而言，其表面发生的是水的还原反应，而光生空穴则迁移至电极引发水的氧化反应。在光电催化水分解池中发生的物理与化学过程主要有

$$\text{半导体} + h\nu \longrightarrow h^+ + e^- \tag{1-7}$$

$$H_2O + 2h^+ \longrightarrow (1/2)O_2 + 2H^+ \tag{1-8}$$

$$2H^+ + 2e^- \longrightarrow H_2 \tag{1-9}$$

光电催化水分解池的净反应过程为

$$H_2O + h\nu \longrightarrow (1/2)O_2 + H_2 \tag{1-10}$$

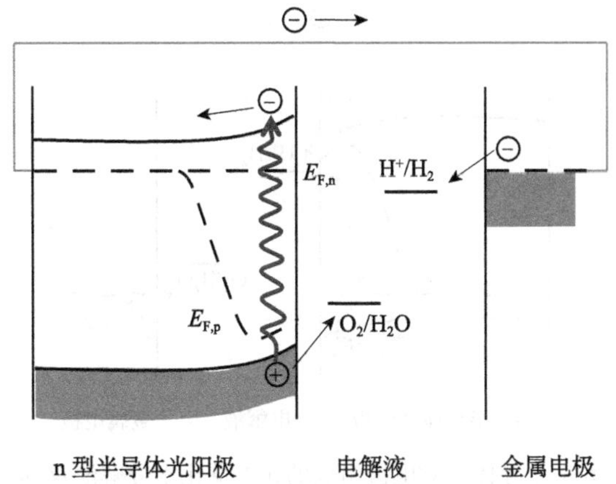

图 1-47　光电催化水分解池的工作原理（以 n 型半导体为例）

其中，$E_{F,n}$：电子准费米能级；$E_{F,p}$：空穴准费米能级

从热力学角度来说，完成水的完全分解反应，半导体的带隙至少应为 1.23 eV（对应于 O_2/H_2O 的标准电极电位与 H^+/H_2 的标准电极电位之差）。考虑到产氢与产氧反应的过电位及电路阻抗损失，半导体的带隙至少应为 1.8 eV（Walter et al.，2010）。此外，半导体光生电子的电位必须负于（高于）H^+/H_2 的标准电极电位才能驱动水还原反应。相应地，为了实现水氧化反应，半导体的光生空穴电位必须正于（低于）O_2/H_2O 的标准电极电位。因此，光电催化水分解池不仅要求半导体有合适的带隙，还要求其能级位置与水分解的电极电位相匹配，即半导体的价带与导带位置要完全跨越水的氧化与还原电位。

3. 光电催化水分解池的主要类型

根据半导体对太阳光的吸收利用方式及半导体的类型，太阳能光电催化水分解池可分为以下几种类型（Walter et al.，2010）。

（1）n 型或 p 型半导体与对电极组成的光电催化水分解池（photoelectrochemical cells，PECs），如图 1-47、图 1-48 所示。在这种类型的光电催化水分解池中，半导体的价带与导带位置要完全跨越水的氧化与还原电位，并能克服产氧与产氢反应的过电位。由于 n 型与 p 型半导体与电解液接触后所形成的半导体 / 电解液结不同，在 n 型半导体表面发生的是水氧化反应，而在 p 型半导体表面发生的是水还原反应，即迁移至半导体与电解液界面并驱动氧化还原反应的总是半导体的少数载流子。

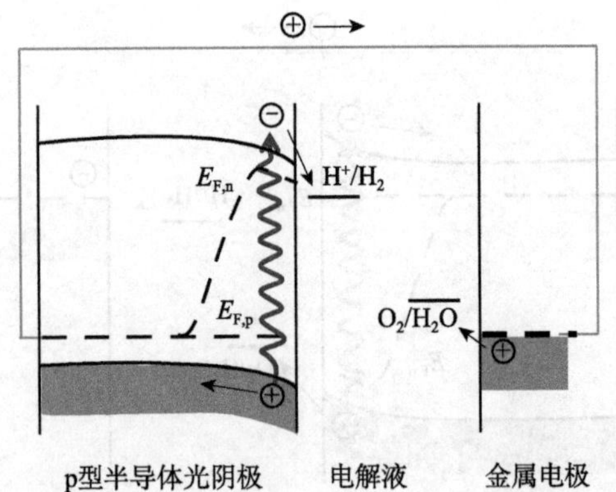

图 1-48　p 型半导体与对电极组成的光电催化水分解池及其工作原理

$E_{F,n}$：电子准费米能级；$E_{F,p}$：空穴准费米能级

（2）n 型半导体与 p 型半导体串联组成的 p/n 双光子光电催化水分解池，如图 1-49 所示。在此类型的光电催化水分解池中，并不要求 n 型与 p 型半导体的价带与导带位置同时跨越水的氧化与还原电位。工作过程中，价带位置足够正（低）的 n 型半导体驱动水氧化反应，导带位置足够负（高）的 p 型半导体驱动水还原反应，而前者产生的还原电位不足的电子与后者产生的氧化电位不足的空穴在电路中复合。

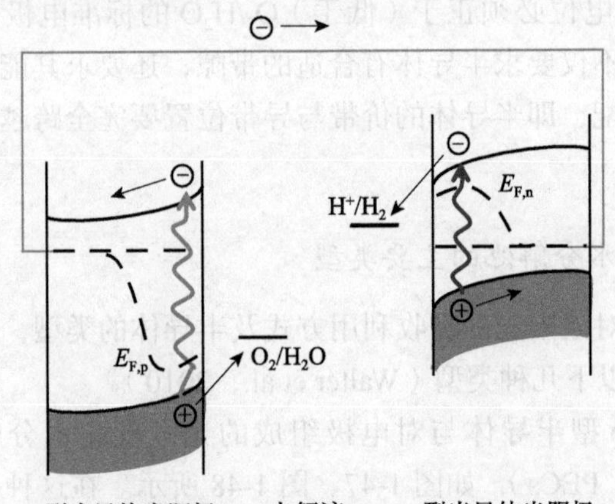

图 1-49　n 型与 p 型半导体串联组成的 p/n 双光子光电催化水分解池及其工作原理

$E_{F,n}$：电子准费米能级；$E_{F,p}$：空穴准费米能级

（3）n 型或 p 型半导体与太阳能电池（photovoltaic，PV）串联组成的 PV/n 或 PV/p 双光子光电催化水分解池，如图 1-50、图 1-51 所示（Khaselev &

Turner，1998；Brillet et al.，2012）。以 PV/n 双光子光电催化水分解池为例，n 型半导体在光激发下驱动水氧化反应，而其产生的还原电位不足的电子与太阳能电池的空穴发生复合，后者产生的光生电子则迁移至金属对电极，引发水的还原反应。需要指出的是，PV/n 或 PV/p 双光子光电催化水分解池并不要求 n 型半导体的价带位置足够正（低）或 p 型半导体的导带位置足够负（高），其太阳能电池模块能够提供额外的电位以满足反应要求。

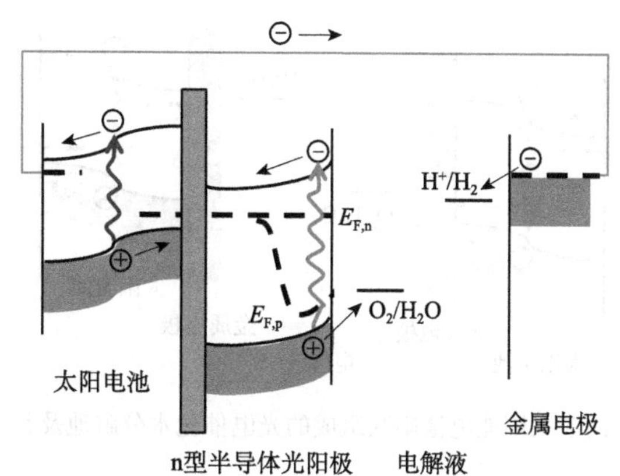

图 1-50　n 型半导体与太阳能电池串联组成的 PV/n 双光子光电催化水分解池及其工作原理

$E_{F,n}$：电子准费米能级；$E_{F,p}$：空穴准费米能级

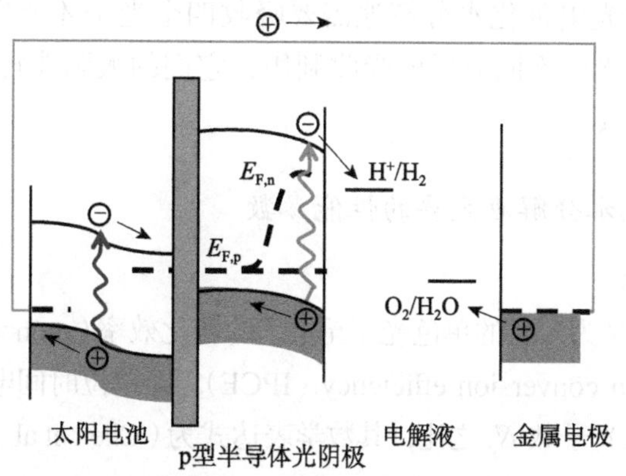

图 1-51　p 型半导体与太阳能电池串联组成的 PV/p 双光子光电催化水分解池及其工作原理

$E_{F,n}$：电子准费米能级；$E_{F,p}$：空穴准费米能级

（4）两节太阳能电池串联组成的光电催化水分解池，如图 1-52 所示。在这种光电催化水分解池中，涂覆在太阳能电池表面的金属电极作为产氢与产

氧电催化剂，在太阳能电池产生的电能驱动下完成水分解反应。严格来说，这种构型的水分解池不能归类为光电催化水分解池，因为后者的核心是半导体 / 电解液结。两节太阳能电池串联组成的光电催化水分解池与太阳能电池加电解池的组合无本质区别（Sivula & van de Krol，2016）。

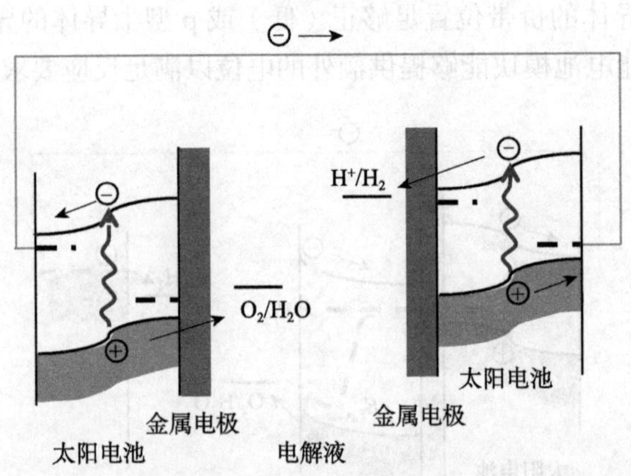

图 1-52　两节太阳能电池串联组成的光电催化水分解池及其工作原理

需要指出的是，n 型或 p 型半导体与对电极组成的光电催化水分解池需要吸收两个光子生成一分子氢气，但因半导体的带隙过大（以满足水分解反应的热力学与动力学要求）造成此光电催化水分解池的太阳能转换效率低。其他三种构型的光电催化水分解池需要吸收四个光子才能生成一分子氢气，因可实现对太阳光谱不同波段的吸收利用，它们的太阳能转换效率高，具有更广泛的应用前景。

4. 光电催化水分解池主要的性能参数

1）量子效率

量子效率定义为入射的单色光子至电子的转化效率（monochromatic incident photon-to-electron conversion efficiency，IPCE)，即单位时间电路中产生的电子数 N_e 与入射单色光子数 N_p 之比，其数学表达式为（Chen et al.，2010）

$$\mathrm{IPCE} = \frac{N_e}{N_p} = \frac{1240 \times J_p(\lambda)}{\lambda \times P} \tag{1-11}$$

其中，λ 为入射的单色光子的波长（单位为 nm），P 为光强（单位为 μW/cm^2），$J_p(\lambda)$ 为光电流密度（单位为 μA/cm^2）。

2）太阳能转换效率

在光电催化水分解池中，太阳能转换效率 η 定义为太阳能转换为化学能的效率（Walter et al.，2010）。也有研究者将太阳能转换效率 η 称为 ABPE（applied bias photon-to-current efficiency）。对于 n 型半导体光阳极，其太阳能转换效率 η 的数学表达式为

$$\eta = \frac{q(V_{\text{ox}} - V_{\text{app}})}{I_0}\int_0^{1240/E_g} \text{IPCE}(\lambda)\, N(\lambda)\text{d}\lambda = \frac{J_{\text{p-ox}}(1.23 - V_{\text{app}})}{I_0} \tag{1-12}$$

其中，q 为电子的电量（单位为 C）；V_{ox} 为氧化产物的标准电极电位（单位为 V），由于光电催化水分解池的氧化产物为氧气，V_{ox} 取值为 1.23 V；V_{app} 为外加的偏压（相对于可逆氢电极，单位为 V）；E_g 为半导体的禁带宽度（单位为 eV）；$N(\lambda)$ 为波长为 λ 的单色光子的数目；I_0 为入射太阳光的功率（通常采用 AM 1.5 G = 100 mW /cm^2）；$J_{\text{p-ox}}$ 为光阳极氧化水所产生的光电流密度（单位为 mA/cm^2）。

对于 p 型半导体光阴极，其太阳能转换效率 η 的数学表达式为

$$\eta = \frac{q(V_{\text{app}} - V_{\text{re}})}{I_0}\int_0^{1240/E_g} \text{IPCE}(\lambda)\, N(\lambda)\text{d}\lambda = \frac{J_{\text{p-re}} V_{\text{app}}}{I_0} \tag{1-13}$$

其中，V_{re} 为还原产物的标准电极电位（单位为 V），由于光电催化水分解池的还原产物为氢气，V_{re} 取值为 0；E_g 为半导体的禁带宽度（单位为 eV）；$J_{\text{p-re}}$ 为光阴极还原水所产生的光电流密度绝对值（单位为 mA/cm^2）。

以上太阳能转换效率的计算方法适用于光阳极和光阴极而非整个光电催化水分解池。光电催化水分解池整体的太阳能转换效率的计算方法与式（1-12）相同，但电流密度-电压曲线的测量需要在两电极体系中（只包含阴极与阳极两个电极）进行。光电催化水分解池整体太阳能转换效率 η 的计算公式可以理解为

$$\eta = \frac{q(V_{\text{ox}} - V_{\text{re}} - V_{\text{app}})}{I_0}\int_0^{1240/E_g} \text{IPCE}(\lambda)\, N(\lambda)\text{d}\lambda = \frac{J_{\text{op}}(1.23 - V_{\text{app}})}{I_0} \tag{1-14}$$

其中，J_{op} 为水分解池所产生的光电流密度（单位为 mA/cm^2）。

3）太阳能转换氢能效率

若光电催化水分解池的运行不需要任何外加偏压，所投入的资源仅有太阳光与水，此时其太阳能转换效率可称为太阳能转换氢能效率（solar-to-hydrogen efficiency，STH），其数学表达式为（Walter et al.，2010；Chen et al.，2010）

$$\text{STH} = \frac{1.23 \times J_{\text{op}}}{I_0} \tag{1-15}$$

二、光电催化产氢材料研究的历史及现状

（一）光电催化产氢材料研究的历史

贝克勒尔（Becquerel）在 1839 年发现光生伏打效应，即将涂覆卤化银颗粒的两个相同的金属电极浸入稀酸溶液中，并用光辐照其中一个电极时会产生电流（Becquerel，1839）。1955 年，布拉顿（Brattain）等研究了锗电极浸入电解液后的光电化学性质（Brattain & Garrett，1955），该工作标志着光电化学的诞生。随后在 1966 年，格里舍尔（Gerischer）详细研究了半导体/电解液接触界面的电化学与光电化学特性（Gerischer，1966）。格里舍尔与布拉顿的工作被认为是现代光电化学的基石。藤岛昭和本多健一在 1972 年利用紫外光激发二氧化钛电极将水成功分解为氢气与氧气，该工作在光电化学领域内具有里程碑意义（Fujishima & Honda，1972）。2001 年，邹志刚等首次开发了具有可见光活性的光催化分解水材料体系，将光电催化材料的太阳光谱从紫外光区拓展至紫外-可见光区。可见光响应光催化剂的成功开发，为后续高效光电催化材料的选择提供了新的机遇，标志着光电催化新时代的到来（Zou et al.，2001）。

（二）光电催化产氢材料国内外发展现状

1. 光阳极材料的研究现状

OER 是一个四电子-质子耦合过程，该过程动力学缓慢，是整个水分解反应的速率控制步骤（Walter et al.，2010）。因此，开发高活性和高稳定性的 OER 光阳极是提升 PECs 水分解太阳能转换效率的关键，也是近年来 PECs 制氢研究的重点与热点（Li et al.，2013）。一方面，利用廉价易得的半导体构建基于直接半导体/电解液结的 PECs 水分解电极具有大规模应用的成本优势，围绕此目标，研究人员已研制出金属氧化物、金属（氧）氮化物等半导体材料用于光阳极的构建；另一方面，考虑到光伏领域已开发出数种高效捕获太阳光、带隙合适的高质量半导体，如 Si、Ⅲ-Ⅴ族化合物和Ⅱ-Ⅵ族化合物，以及相应的 p-n 同质结、金属-绝缘体-半导体（MIS）结、异质结等固体结构建技术用于高效分离光生电子-空穴对，直接将上述光伏领域的半导体材料用于构建水分解电极的设想由来已久。基于上述讨论，光阳极材料的研究现状将围绕 Si、Ⅲ-Ⅴ族化合物、Ⅱ-Ⅵ族化合物、金属（氧）氮化物、金属氧化物这五类半导体材料进行阐述。

1）Si 基光阳极材料

单晶 Si 的带隙宽度仅为 1.12 eV，能捕获太阳光谱中波长小于 1100 nm 的光子，且 Si 在地球表面储量丰富。单晶 Si 半导体的价带位置较浅，约 0.7 V *vs.* RHE，本身产生的光生空穴并不具备足够的氧化能力来驱动 OER 过程，当其作为 OER 光阳极时需要外加偏压的辅助。分解水反应所需的总光电压为 1.6～1.8 V，而使用 Si 作为 OER 光阳极（以及产氢的光阴极）在理想情况下能提供大约 0.55 V 的光电压。然而，单晶 Si（以及多晶 Si、无定形 Si）半导体在电解液中不稳定，易生成 SiO_2 绝缘层，外加正偏压也会加剧其表面的氧化腐蚀反应。因此，使用 Si 基光阳极高效驱动 OER，首先需要解决的问题就是抑制其表面氧化腐蚀反应。目前的策略是在 Si 基光阳极表面引入一层保护物质，以阻隔其与电解液的直接接触。

（1）p-n 同质结单晶 Si 光阳极。目前，基于单晶 Si 的 p-n 同质结在 Si 基太阳能电池的构造中占据主导地位。当以该单晶 Si 基 p-n 同质结来构建光阳极时，其内部的 p-n 同质结能在标准太阳光下产生几乎恒定的光电压值，而不受表面保护层以及电解液成分的影响。基于此特点，p-n 结单晶 Si 光阳极可以作为一个理想的研究平台来考察多种保护层的可靠性，也能为其他器件构型 Si 光阳极以及其他易光腐蚀光阳极保护层的选择与优化提供有价值的参考。针对 p-n 结单晶 Si 光阳极保护层的选择，除了要考虑其在电解液中的使役稳定性，良好的透光性以减少竞争性光吸收、体相电荷传输电阻小，以及与 Si 形成有利的界面能级状态以促进光生空穴从光阳极转移至电解液等也是须予以重视的原则（Bae et al.，2001）。这些面向 p-n 结单晶 Si 光阳极的保护层选择原则，实际上也适用于其他光阳极乃至光阴极。当前可用作 p-n 结单晶 Si 光阳极保护层的材料，根据其具体物化特性可以分为金属、透明导电氧化物、半导体、具产氧催化活性的金属氧化物、聚合物和碳等。显而易见，这些保护层的耐腐蚀性、透光性、电荷传输机制、与 Si 形成的界面能级状态等不尽相同。根据导电性的区别，上述保护层又可简单分为导体、半导体以及绝缘体三类。后续关于 p-n 同质结单晶 Si 光阳极保护层的讨论也将分为导体保护层、半导体保护层、绝缘体保护层这三部分。

金属是最常见的一种导体，因此利用金属薄膜作为保护层来稳定 p-n 同质结单晶 Si 光阳极是很自然的想法。在实施过程中，一方面，要尽量降低金属薄膜厚度，进而减少其带来的不利光学损失（如竞争性光吸收或者光反射效应）；另一方面，金属薄膜需要达到一定的厚度才能均匀覆盖 Si 光阳极，进而实现保护层的功效。理想情况下，制备纳米级厚度的无针孔、裂纹等缺陷的超薄金属薄膜能够在达到保护目标的同时最大限度地允许入射光的穿透。也有研究者为了确保对 Si 光阳极的有效保护而使用了厚的金属保护膜，此时光入射面与金属保护膜分别位于 Si 光阳极的两侧，由此避免了厚的金属保护膜带来的不利光学损失（Guo et al., 2018）。此外，如前所述的保护层的选择原则，金属保护层还需兼顾优异的电荷提取和传导能力。金属通常具有良好的导电性，允许电荷在其内部快速传输，然而其能否有效提取 p-n 结单晶 Si 光阳极产生的光生空穴则在很大程度上取决于 Si/ 金属膜二者之间的界面能级状态。一方面，Si 半导体与金属直接接触后，可能会在二者之间的界面区域形成缺陷态，继而造成光生电子-空穴对在此区域复合严重。针对 Si/ 金属膜界面缺陷态的问题，可以采取的策略是在二者之间（即沉积金属薄膜保护层之前）引入一层钝化层。另一方面，Si 半导体与金属膜之间的能级排列也是需要着重考虑的关键，其对光生空穴从 Si 转移至电解液的效率影响极大。理想情况下，在 Si 光阳极、金属保护层、电解液之间形成级联能级排列，能最大限度地促进光生空穴从 Si 至电解液的定向传输与转移。具体分析 p-n 结单晶 Si 光阳极，其功函较高的 p 型 Si 层在外，更靠近电解液侧。此时，选择高功函的金属保护层继而与 p 型 Si 层形成空穴选择性接触或欧姆接触，能够实现光生空穴的高效分离与收集［图 1-53（a）］（Feng et al., 2020）。

上述金属保护层的选择原则与优化策略也适用于透明导电氧化物、碳及导电聚合物基的保护层。然而，相比于种类繁多的金属及合金，这几类导体保护层的功函可调性较差，与功函较高的 p 型 Si 层接触后可能会引发后者形成不利的能带弯曲［图 1-53（b）］（Feng et al., 2020）。此时需要对 p 型 Si 层进行表面修饰，以使其与低功函保护层接触后形成允许空穴向外传输的界面能级分布［图 1-53（c）］（Feng et al., 2020）。诺塞拉（Nocera）等研究了锡掺杂氧化铟锡（ITO）透明导电氧化物薄膜作为保护层稳定 p-n 同质结单晶 Si 光阳极。由于 ITO 的功函（4.4～4.7 eV）比 p 型 Si（5.0～5.2 eV）低，二者接触后会在 p 型 Si 侧形成向下的能带弯曲，且 p 型 Si 较低的受主掺杂浓度导致该能带弯曲扩展至其体相相当的深度（可达 1 μm），因此，以 ITO 作为保护层直接与 p-n 同质结单晶 Si 接触，将诱导后者表面形成相当厚度的肖特基势垒，阻碍光生空穴向 ITO 的传输，继而降低光电极的整体性能。针对这个问题，诺塞拉等在 p-n 同质结单晶 Si 的表面引入了一层重掺杂 p^+ 型 Si，当其与 ITO 接触时虽然也形成向下的能带弯曲，但由于 p^+ 型 Si 的受主掺杂浓度很高，该能带弯曲的厚度仅为数纳米。在这种情况下，光生空穴可经隧穿效应高效传输至 ITO，p^+ 型 Si 与 ITO 之间的界面表现出欧姆接触的特征

[图 1-53（d）]（Pijpers et al.，2011）。

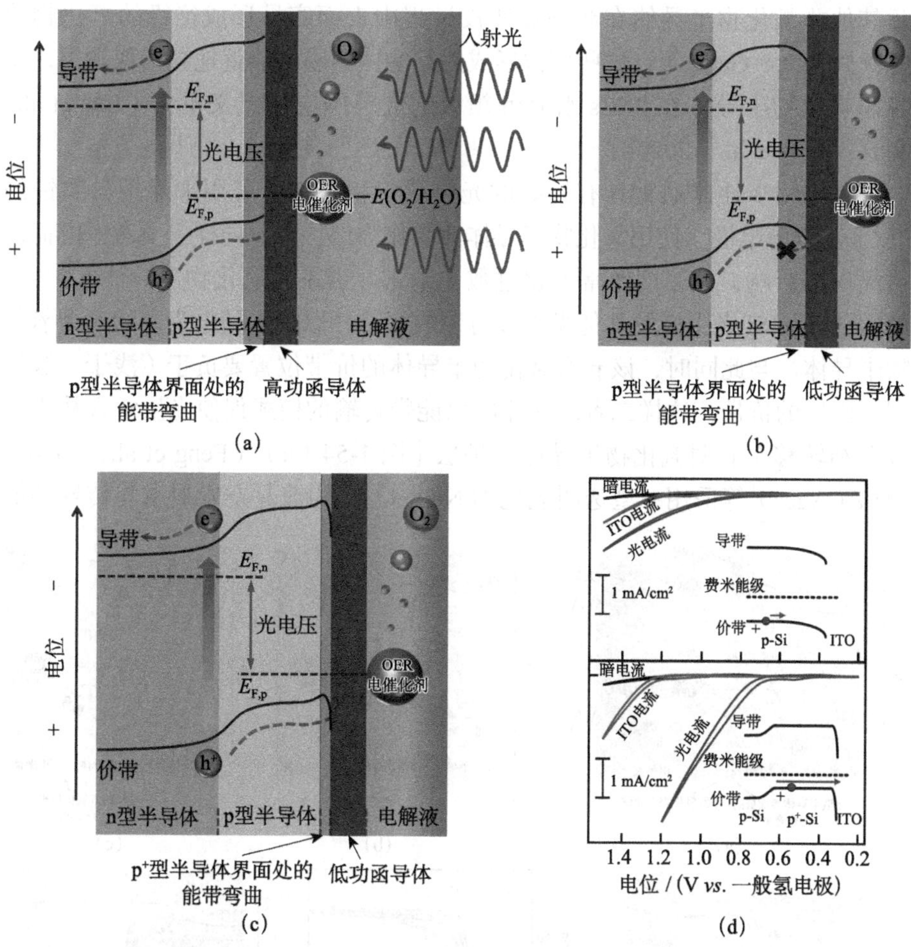

图 1-53　(a) p-n 同质结光阳极与高功函导体接触形成的能级排列状况；(b) p-n 同质结光阳极与低功函导体接触形成的能级排列状况；(c) p-n 同质结光阳极表面引入重掺杂 p^+ 层后与低功函导体接触形成的能级排列状况；(d) np-Si/ITO/Co-Pi（上图）与 npp^+-Si/ITO/Co-Pi（下图）光阳极的空穴传输机制及电流密度 - 电压特性曲线

其中，$E_{F,n}$：电子准费米能级；$E_{F,p}$：空穴准费米能级

基于对电荷传输效率的考虑，较少有研究会专注于使用绝缘体薄膜作为 PECs 水分解电极的保护层。然而，当绝缘体薄膜厚度减小至某一临界值后，光生载流子（包括空穴和电子）可经由隧穿效应高效通过该绝缘体薄膜。因此，选择能耐受电解液化学腐蚀及运行过程中光电化学腐蚀的绝缘体，并采用适宜的成膜方法将这些绝缘体以极薄的厚度沉积在待保护的 PECs 水分解

电极表面，有助于推动绝缘体保护层在 PECs 水分解领域的应用。此外，Si 及其他非氧化物半导体在制备及使役过程中表面容易形成绝缘的氧化物薄层，如 SiO_x、GaO_x 等。这些原位形成的绝缘体薄膜若能就地充当保护层，将对非氧化物半导体在 PECs 水分解领域的大范围应用意义重大（Bae et al., 2001；Feng et al., 2020）。

氧化物半导体通常具有较好的光电化学稳定性，可用来保护易腐蚀的 PECs 水分解电极。使用氧化物半导体薄膜作为保护层，同样需要考虑电荷传输损失的影响，其厚度通常不超过数十纳米。当采用氧化物半导体保护 p-n 同质结单晶 Si 光阳极及其他光阳极时，最好选择以空穴导电为主的 p 型氧化物半导体。与此同时，该 p 型氧化物半导体的价带位置要负于（浅于）被保护光阳极的价带，这样二者之间才能以能带传输的模式促使光生空穴从光阳极顺利转移至 p 型氧化物半导体保护层［图 1-54（a）］（Feng et al., 2020）。阿格（Ager）等采用 p 型透明氧化物 $NiCo_2O_4$ 作为兼具空穴提取与传导功能

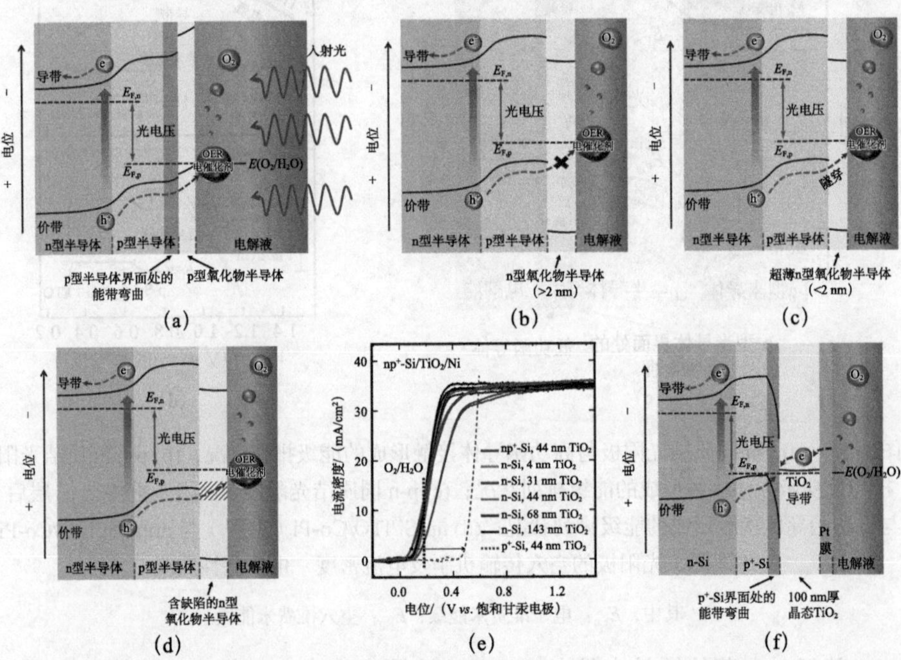

图 1-54　p-n 同质结光阳极与 (a) p 型氧化物保护层、(b) 厚的 n 型氧化物保护层、(c) 超薄 n 型氧化物保护层，以及 (d) 含缺陷的厚 n 型氧化物保护层形成的能级排列状况及空穴传输机制；(e) n-Si/X nm ALD-TiO_2/Ni (X = 4, 31, 44, 68, 143) 及 np$^+$-Si/44 nm ALD-TiO_2/Ni 光阳极在 1 mol/L KOH 电解液中的电流密度－电压曲线，光源为 ELH 型卤钨灯，1.25 个太阳光照强度；(f) np$^+$-Si/ 晶态 TiO_2/Pt 光阳极的能级排列状况及氧化水过程中涉及的电子传输机制

的保护层，外加 NiFe 基 OER 电催化剂后使得 np^+-Si 光阳极在大于 30 mA/cm^2 的光电流密度下稳定性超过 72 h（Chen et al.，2015）。

然而，耐光电化学腐蚀的 p 型金属氧化物半导体数量有限，如何确保它们在待保护的光阳极表面形成无针孔、裂纹的薄膜也颇具挑战。相比之下，相当数量的 n 型金属氧化物半导体在水氧化条件下具备优异的稳定性，相应的成膜方法也多种多样，并且这些材料的带隙较大，不会引入明显的竞争性光吸收。因此，开发 n 型金属氧化物半导体作为 p-n 同质结单晶 Si 光阳极及其他光阳极的保护层具有重要的实际应用价值，但颇具挑战。原因是这些 n 型金属氧化物半导体的价带位置通常正于（深于）需要进行保护的 Si 等非氧化物光阳极，这样一来，后者产生的光生空穴将不能通过能带传输模式有效转移至前者［图 1-54（b）］（Feng et al.，2020）。若想将 n 型金属氧化物半导体用于保护 Si 等非氧化物光阳极，第一种策略与绝缘体基保护层类似，将 n 型金属氧化物半导体制备成超薄的薄膜，以便 Si 等非氧化物光阳极产生的光生空穴能够借助隧穿效应高效通过［图 1-54（c）］（Feng et al.，2020）。一方面，由于载流子的隧穿效率随隧穿距离呈现指数衰减，要求 n 型金属氧化物半导体保护层应尽可能薄；另一方面，过薄的保护层厚度可能会诱导针孔、裂纹等缺陷的形成，不利于保护层有效阻隔光阳极与电解液的直接接触。因此，开发高效的薄膜沉积技术，如原子层沉积，以实现纳米级厚度可控、均匀致密的 n 型金属氧化物半导体保护膜，对稳定易发生光腐蚀的水分解光阳极意义重大。实现 n 型金属氧化物半导体保护 Si 等非氧化物光阳极功能的第二种策略是在其内部引入相当含量的缺陷态［图 1-54（d）］（Feng et al.，2020）。此时，n 型金属氧化物半导体保护层的厚度可在一定范围内增加，以抑制针孔、裂纹等缺陷的形成，继而实现对光阳极的有效保护，同时被保护的光阳极产生的光生空穴也能借助保护层的缺陷态实现高效传导。例如，已知 TiO_2 在较宽的 pH、电位范围以及多种溶液中均表现了优异的耐化学腐蚀及光电化学腐蚀的特性，如何将其作为保护层来稳定 Si 等非氧化物光阳极引起了众多研究者的兴趣。路易斯（Lewis）课题组以 ALD 技术在 np^+-Si 光阳极表面引入 4～143 nm 厚的 TiO_2 作为保护层，外加 Ni 基岛状 OER 电催化剂后，可在大于 30 mA/cm^2 的光电流密度下稳定性超过 100 h［图 1-54（e）］（Hu et al.，2014）。如前所述，从能带角度分析 TiO_2 的价带位置正于（深于）Si，二者之间的能带位置排列决定了 TiO_2 适合作为空穴阻挡层而非空穴选择层；而当 TiO_2 厚度达到百纳米时，载流子的隧穿机制将失效。路易斯等指出，赋予厚 TiO_2 保护层体相空穴传输特性的关键是引入免烧结、富含缺陷的 ALD-TiO_2，此时 Si 光阳极产生的光生空穴可遵循缺陷介导传输机制高效通过该 TiO_2 厚保护层。采用富含缺陷的 ALD-TiO_2 厚保护层策略，不仅能实现对平面型 p-n 同质结单晶 Si 光阳极的有效保护，也使得 np^+-Si 径向结阵列光阳极持续驱动水氧化反应超过 2200 h（Shaner et al.，2015）。

上述两种赋予 n 型金属氧化物半导体保护光阳极功能的策略，其依据的模型是电极表面水氧化法拉第电流来源于光生空穴向电解液的定向迁移。有别于这两种策略所遵循的空穴传输模型，乔根多夫（Chorkendorff）等利用厚度约 100 nm 的晶态 TiO_2 薄膜作为 np^+-Si 光阳极的保护层，其表面水氧化法拉第电流来源于水分子的电子经由 OER 电催化剂、晶态 TiO_2 保护层向 p^+-Si 的定向迁移［图 1-54（f）］（Mei et al.，2015）。为了在晶态 TiO_2 光阳极保护层中实现导带（或靠近导带的缺陷态）电子传输机制，需要满足以下三个条件：① p^+-Si 的受主掺杂浓度足够高，以便其与 TiO_2 保护层接触后形成的耗尽层厚度足够薄，继而允许来自 TiO_2 导带的电子有效隧穿并与 np^+-Si 同质结产生的空穴复合；② OER 电催化剂与晶态 TiO_2 保护层之间的接触是欧姆接触，继而 OER 电催化剂从水分子夺取的电子能顺利转移至 TiO_2 的导带；③避免使用具“夹断效应”的 OER 电催化剂，以确保水分子的电子经由 OER 电催化剂进入晶态 TiO_2 保护层。对于第三个条件，究其原因是，晶态 TiO_2 保护层与电解液之间存在肖特基势垒，采用具“夹断效应”的 OER 电催化剂将使该不利的肖特基势垒在电极表面占据主导，继而阻碍水分子的电子注入晶态 TiO_2 保护层的导带。理想情况下，晶态 TiO_2 保护层应被 OER 电催化剂完全覆盖，以此避免其与电解液接触后形成不利的肖特基势垒。

易发生光腐蚀的 PECs 水分解电极表面除必要的保护层外，通常也需负载电催化剂来加速表面的氧化还原反应。若能采用兼具保护层功效与电催化活性的双功能薄膜，将有助于简化相应光电极的结构与制备流程。夏普（Sharp）课题组使用 ALD 技术在 np^+-Si 光阳极表面沉积了 CoO_x 保形涂层，该光阳极在大于 30 mA/cm^2 的光电流密度下稳定性超过 72 h。该 CoO_x 保形涂层中包含两种 Co 的化合物，第一种是位于底层并与 Si 光阳极接触的致密且连续的 Co_3O_4 层，其不具备离子可渗透性且在光阳极运行过程中不发生相变，借此可以确保腐蚀性电解液与 Si 光阳极的有效隔离；第二种是位于表层的结构无序且处于化学非稳态的 $Co(OH)_2$，用以提供丰富的活性位点，继而加快表面产氧反应动力学速率［图 1-55（a）］（Yang et al.，2016）。前述章节讨论的金属薄膜保护层，在制备或使役过程中，假如其表面被氧化，继而原位生成了 OER 电催化剂，则此包含了内部金属保护层及外部（氢）氧化物 OER 电催化剂的复合薄膜也属于兼具保护层功效与电催化活性的双功能薄膜。例如，石高全课题组在 np^+-Si 光阳极表面电沉积制备了 Ni-Fe 合金保护层，后者在运行过程中表面原位生成了 Ni(Fe)OOH 基 OER 电催化剂。该双功能保护层使得 np^+-Si 光阳极在 1 mol/L KOH 电解液及近 30 mA/cm^2 的光电流密度下可稳定运行约 13 h [图 1-55（b）]（Yu et al.，2017）。路易斯课题组证明，其在 np^+-Si 光阳极表面沉积的 NiO_x 层不仅具备传导空穴功能、保护层功能、OER 电催化剂功能，甚至能充当减反膜来增加 Si 光阳极的光收集效率。基于此 p 型导电、透明、电催化活性、抗反射的 NiO_x 多功能保护层，np^+-Si 光阳极在 1 mol/L KOH 电解液及大于 30 mA/cm^2 的光电流密度下可连续运行

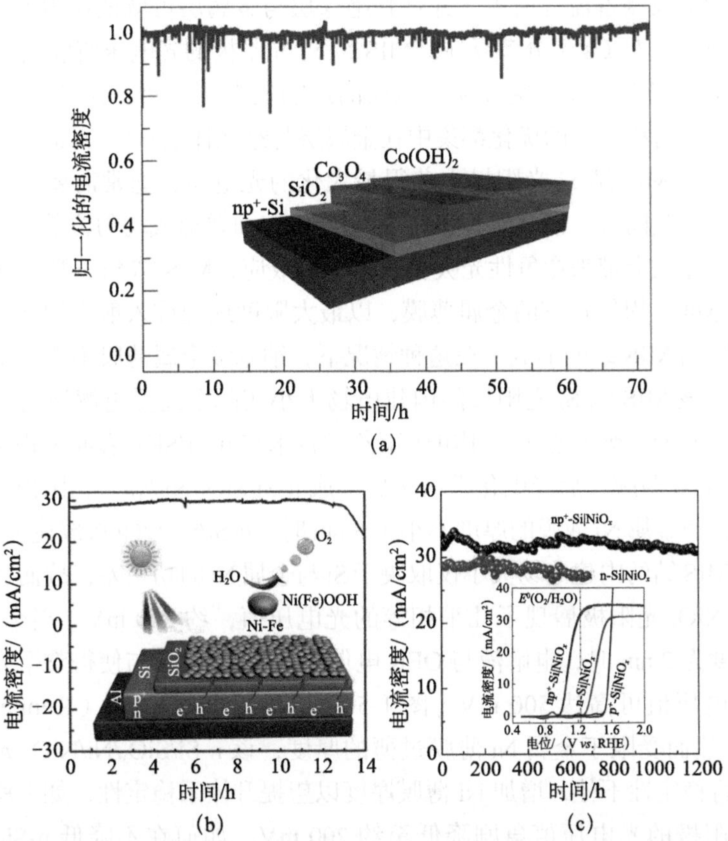

图 1-55　(a) np^+-Si/Co_3O_4/$Co(OH)_2$ 光阳极在 1 mol/L NaOH 电解液中的电流密度 - 时间曲线，外加偏压为 1.8 V *vs.* RHE，插图为电极构造示意图；(b) np^+-Si/SiO_x/NiFe 光阳极在 1 mol/L KOH 电解液中的电流密度 - 时间曲线，外加偏压为 1.85 V *vs.* RHE，插图为电极构造示意图；(c) np^+-Si/NiO_x 与 n-Si/NiO_x 光阳极在 1 mol/L KOH 电解液中的电流密度 - 时间曲线，外加偏压为 1.73 V *vs.* RHE，插图为二者及 p^+-Si/NiO_x 电极的电流密度 - 电压曲线

超过 1200 h［图 1-55（c）］。需要指出的是，虽然上述 NiO_x 保护层具备多种优异特征，可能有望用于保护所有易发生光腐蚀的光阳极，但实际上其使用范围是有限制的。这是因为目前的薄膜沉积技术并不能确保 NiO_x 保护层无任何针孔、裂纹等缺陷，继而无法实现被保护光阳极与电解液的有效隔离。因此，NiO_x 基保护层仅适用于在反应条件下形成不溶性表面氧化物的光阳极，不溶的表面氧化物可充当钝化膜，阻碍光阳极的进一步腐蚀及持续氧化。对于发生溶解性腐蚀的光阳极，则需制备无针孔、裂纹等缺陷的保形涂层，以确保在光阳极与电解液之间建立有效的物理屏障（Sun et al., 2015）。

（2）MIS 结 Si 光阳极。相比于 p-n 同质结单晶 Si 光阳极，MIS 结 Si 光阳极的构造及制备流程简单，并且其绝缘层与金属层可就地作为保护层实现高效、稳定的 PECs 水分解反应。MIS 结 Si 光阳极的光电压值取决于 Si 与金属薄膜的功函差，并受绝缘体层性质的影响较大。因此，针对 MIS 结 Si 光阳极的效率与稳定性的优化都集中在金属层与绝缘体层这两方面。

为了在 MIS 结 Si 光阳极中获得最大化的光电压，通常需要选择高功函的金属，与此同时，该金属薄膜在 Si 半导体表面要形成均匀覆盖。考虑到金属薄膜的存在会带来竞争性光吸收或光反射效应，MIS 结 Si 光阳极中可选用纳米级厚度、均匀致密的金属薄膜，以最大限度地允许入射光的穿透，与此同时不影响 MIS 结的形成。在这种情况下，纳米级金属薄膜不能完全屏蔽 Si 半导体，该 MIS 结 Si 光阳极的内建电场大小实际上还受电解液与电催化剂的影响。例如，戴宏杰等采用电子束蒸发技术在 n-Si/SiO_x 表面沉积了不同厚度的金属 Ni 薄膜，后经电化学活化后形成 n-Si/SiO_x/Ni/NiO_x 光阳极。测试结果显示，当金属 Ni 薄膜的厚度不小于 5 nm 时，n-Si/SiO_x/Ni 形成掩埋 MIS 结；该掩埋 MIS 结的内建电场大小仅取决于 Si 与金属 Ni 的功函差，故而这些 n-Si/SiO_x/Ni/NiO_x 光阳极展现了几乎相等的光电压值，约 200 mV。当金属 Ni 薄膜的厚度为 2 nm 时，电解液与 OER 电催化剂 NiO_x 的参与使得整个光阳极的输出光电压值可高达 500 mV［图 1-56（a）、图 1-56（b）］（Kenney et al., 2013）。然而受限于金属 Ni 薄膜过薄的厚度，该 n-Si/SiO_x/Ni/NiO_x 光阳极的长期运行稳定性不佳，增加 Ni 薄膜厚度以望提升体系稳定性，如上所述，会导致光阳极的光电压值急剧降低至约 200 mV。如何在不降低 n-Si/SiO_x/Ni/NiO_x 光阳极效率（光电压）的前提下提升其使役稳定性是一个难题。史密斯（Smith）等采用双金属策略构建了 MIS 结 Si 光阳极，其中内层 2 nm 厚的高功函 Pt 金属膜用以和 n-Si/SiO_x/Al_2O_3 形成高质量的 MIS 结，外层 4 nm 厚的 Ni 金属膜充当了保护层以及原位生成高效 OER 电催化剂 NiO_x。该 n-Si/SiO_x/

Al_2O_3/Pt/Ni/NiO_x 光阳极在 1 mol/L KOH 电解液及 28 mA/cm^2 的光电流密度下可连续运行 200 h［图 1-56（c）］（Digdaya et al.，2017）。

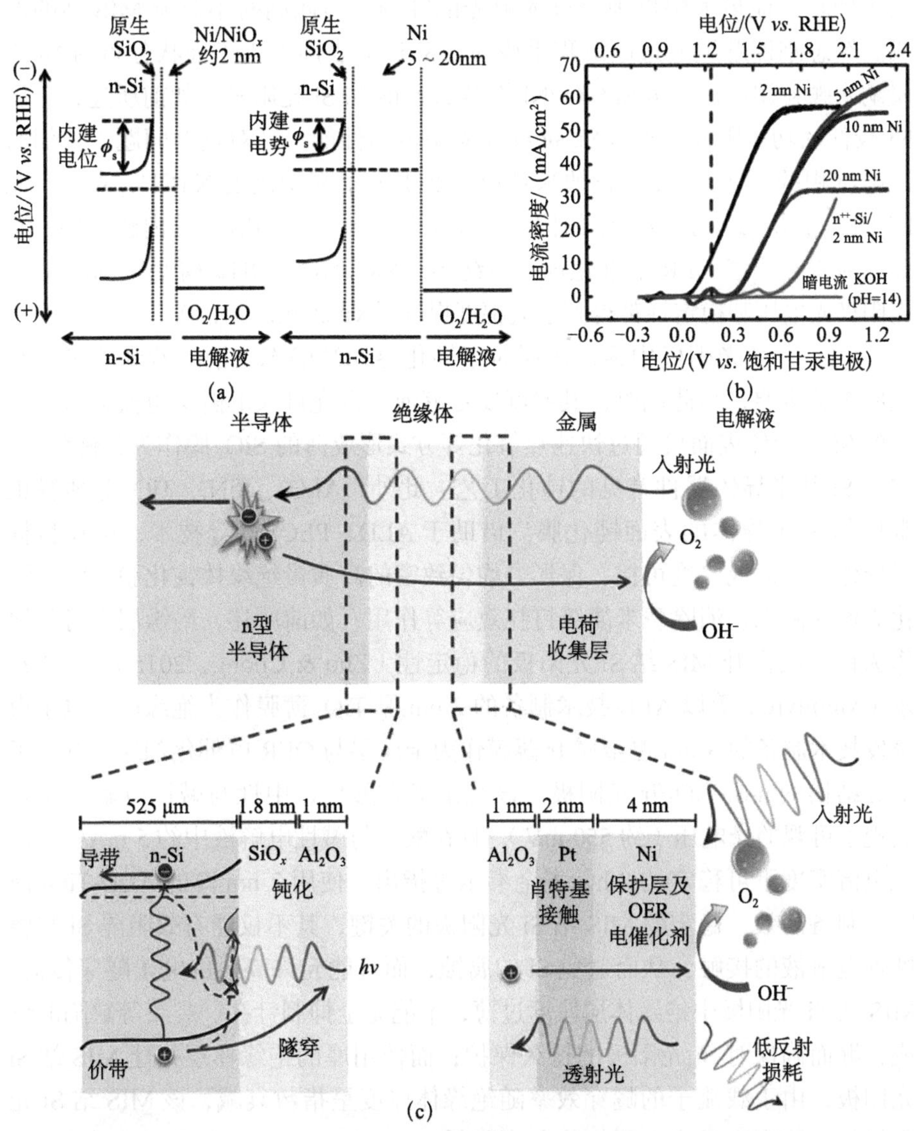

图 1-56　2 nm Ni 及 5～20 nm Ni 与 n-Si 形成的 n-Si/Ni/NiO_x MIS 结光阳极的 (a) 能级结构示意图与 (b) 电流密度－电压曲线；(c) n-Si/SiO_x/Al_2O_3/Pt/Ni MIS 结光阳极的构造示意图及各层的功能

优化金属薄膜的厚度与功函（成分），可有效平衡光吸收过程与高质量MIS结的构建。除此之外，在n-Si半导体表面引入岛状分布的高功函金属纳米颗粒，能最大限度地允许入射光的穿透，与此同时不影响MIS结的形成。邹志刚课题组通过电沉积手段在n-Si/SiO_x表面引入了岛状分布的Ni金属纳米颗粒用以形成n-Si/SiO_x/Ni构型的MIS结Si光阳极。如前所述，金属Ni较低的功函限制了此n-Si/SiO_x/Ni光阳极的性能。针对这个问题，他们通过电沉积的方法在Ni金属纳米颗粒表面引入了高功函的$Ni(OH)_2$壳层。此$Ni(OH)_2$壳层参与了MIS结的构建，并显著增大了该MIS结的内建电场，与此同时也充当了OER电催化剂，最终n-Si/SiO_x/Ni/$Ni(OH)_2$构型的MIS结Si光阳极光电压和稳定性都得到了大幅度提升（Xu et al.，2017）。

MIS结Si光阳极中绝缘体层对于钝化界面缺陷态、消除费米能级钉扎效应至关重要，与此同时，其厚度要足够薄，以允许大的隧穿电流通过。洁净的Si半导体表面可通过快速热氧化等方式形成薄的SiO_x膜作为其钝化膜，这是Si基半导体器件常见的钝化工艺。此外，Al_2O_3、SiN_x、TiO_x等薄膜也常用作Si半导体的表面钝化膜。借助于ALD、PECVD等技术，可在目标半导体表面形成厚度可控、保形、均匀致密的高质量绝缘体钝化膜。除了钝化界面缺陷态、消除费米能级钉扎效应等作用，如前所述，绝缘层也可就地作为保护层提升MIS结Si光阳极的稳定性（Zhu & Chong，2015）。麦金泰尔（McIntyre）等以ALD技术制备的2 nm厚TiO_2薄膜作为绝缘层，电子束蒸发技术制备的3 nm厚金属Ir薄膜作为金属层与OER电催化剂层，制备了MIS结构的n-Si/TiO_2/Ir光阳极。该光阳极在酸性、中性与碱性电解液中均展现了可观的光电压（约550 mV），且在酸性与碱性电解液中约5 mA/cm^2的光电流密度下可稳定约8 h。麦金泰尔等指出，使用2 nm厚的ALD-TiO_2薄膜是制备高效、稳定的MIS结Si光阳极的关键，其不仅能有效阻隔Si与腐蚀性电解液的接触，从而避免Si的腐蚀，而且能允许高效的电子隧穿传输。MIS结Si光阳极中绝缘体层厚度过薄，不能完全抑制针孔、裂纹等缺陷的形成，继而确保对Si光阳极的持久保护；而使用厚的绝缘体层构建MIS结Si光阳极，由于载流子的隧穿效率随绝缘体厚度呈指数衰减，该MIS结Si光阳极的电流密度及光电压也将急剧降低（Chen et al.，2011）。通过对比np^+-Si/TiO_2/Ir光阳极，麦金泰尔等研究了MIS结构n-Si/TiO_2/Ir光阳极光电压随TiO_2绝缘体层厚度（1～12 nm）的变化规律。测试结果显示，随着TiO_2厚度的增加，n-Si/TiO_2/Ir光阳极的光电压急剧降低；通过对比同等TiO_2厚度的p^+-Si/TiO_2/Ir电极，n-Si/TiO_2/Ir光阳极的光电压随TiO_2厚度的降低速率是

TiO_2 引起的欧姆损失的 3 倍以上。麦金泰尔等由此认为，MIS 结 Si 光阳极中绝缘体的存在会引入空穴提取势垒，继而对光电压产生不利影响；只有当 Si 与绝缘体界面处空穴密度较高时（如通过构建 np^+-Si/TiO_2/Ir 光阳极），该光阳极的光电压与绝缘体层的厚度（1～10 nm）无关（图 1-57）（Scheuermann et al.，2016）。

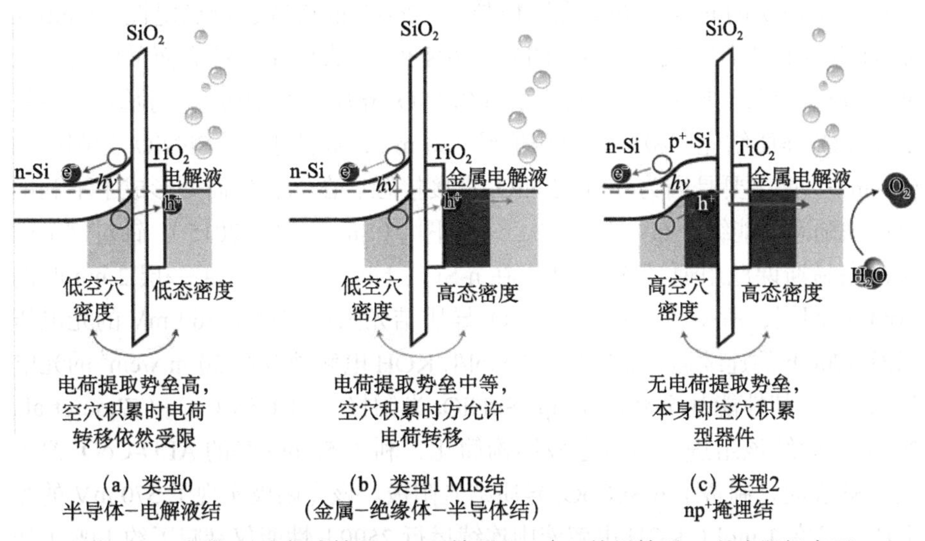

图 1-57 直接 Si/ 电解液结型、MIS 结型和 np^+-Si 掩埋结型 Si 基光阳极表面引入绝缘体保护层后电荷提取势垒对比

（3）异质结 Si 光阳极。除了构建 p-n 同质结、MIS 结等，两种半导体接触后在界面区域形成的内建电场同样可用以高效分离光生电子–空穴对，此即异质结的概念。类似于 MIS 结 Si 光阳极，异质结 Si 光阳极的电极结构与制备流程也较简单。以 n-Si 作为光吸收层构建高效的异质结 Si 光阳极时，需要考虑以下四个要素：①选择高功函的半导体作为覆盖层，这样其与 n-Si 半导体接触后可促使后者表面形成向上的能带弯曲，继而驱动光生空穴由 n-Si 向半导体覆盖层传输；② n-Si 与半导体覆盖层界面处的能级排列应促进或允许光生空穴由 n-Si 向半导体覆盖层高效转移；③半导体覆盖层应具备优良的光生空穴传导能力，以降低电荷传输损失；④采用有效手段钝化 n-Si 与半导体覆盖层之间的界面缺陷态，消除费米能级钉扎效应。此外，半导体覆盖层也可就地作为保护层提升异质结 Si 光阳极的稳定性。基于上述原因，优化半导体覆盖层的选择与沉积方法是提升异质结 Si 光阳极效率和稳定性的关键。

以 n-Si 作为光吸收层构建异质结 Si 光阳极时，p 型金属氧化物是首选的半导体覆盖层，这是因为其可同时满足高功函、能带结构匹配、传导空穴及耐光电化学腐蚀的要求。然而，n-Si 与 p 型金属氧化物覆盖层之间可能会存在界面缺陷态，造成光生电子-空穴对在此复合严重，需要以合适的手段消除或钝化。例如，鉴于 p 型金属氧化物半导体 NiO_x 优异的特性，如功函适宜、能带匹配、透明、空穴导电、电催化活性、抗反射、碱性稳定等，以及已实验证实的 np^+-Si/NiO_x 光阳极超过 1200 h 的连续运行稳定性（1 mol/L KOH 电解液及大于 30 mA/cm^2 的光电流密度），直接在 n-Si 表面沉积该 NiO_x 薄膜理论上可以获得高效、稳定的 n-Si/NiO_x 异质结光阳极。实验结果显示，n-Si/NiO_x 异质结光阳极的光电压值约 180 mV，远低于 np^+-Si/NiO_x 光阳极的 510 mV。这主要是因为 n-Si 与 NiO_x 直接接触后在二者界面处形成了不利的界面缺陷态或能级分布［图 1-55（c）插图］（Sun et al.，2015）。针对这个问题，路易斯课题组通过 ALD 技术在 n-Si 与 NiO_x 之间引入了一层 2 nm 厚的 CoO_x 中间层，所得的 n-Si/CoO_x/NiO_x 异质结光阳极实现了 560 mV 的光电压以及 1700 h 的连续运行稳定性（1 mol/L KOH 电解液及约 30 mA/cm^2 的光电流密度），其性能甚至略优于 np^+-Si/NiO_x 光阳极［图 1-58（a）］（Zhou et al.，2015）。后续该组进一步将电极结构简化，利用 50 nm 厚的 ALD-CoO_x 覆盖在 n-Si 表面，制备了 n-Si/CoO_x 异质结光阳极。该光阳极实现了 570 mV 的光电压，且在 1 mol/L KOH 电解液中连续运行 2500 h 性能仅衰减了约 14%（初始光电流密度约为 30 mA/cm^2）。n-Si/CoO_x 异质结光阳极在含有不同氧化还原电对的电解液中均产生了接近的光电压值，证明其掩埋异质结的属性。上述现象表明，50 nm 厚的 ALD-CoO_x 覆盖层足够致密，针孔、裂纹等缺陷极少，这将有效阻隔 n-Si 与腐蚀性电解液的接触，继而避免 Si 的腐蚀，同时也能抑制多孔 CoOOH 在其晶界处的生成（Zhou et al.，2016）。

假如 n 型金属氧化物功函适宜，富含缺陷态以提取与传导空穴，则可用作半导体覆盖层与 n-Si 构建异质结 Si 光阳极。路易斯课题组在研究厚 4～143 nm、免烧结、富含缺陷的 ALD-TiO_2 作为保护层稳定 np^+-Si 光阳极时，也考察了 n-Si/ALD-TiO_2/Ni 异质结光阳极的性能。该异质结光阳极的饱和光电流密度与 np^+-Si/ALD-TiO_2/Ni 光阳极基本一致，其产生的光电压值约为 400 mV，低于后者的 520 mV ± 30 mV。由于 n-Si/ALD-TiO_2/Ni 异质结光阳极的开启电位与所用的 ALD-TiO_2 厚度（4～143 nm）无关，这表明该光阳极掩埋异质结的属性，以及 ALD-TiO_2 低的体相空穴传输电阻［图 1-54（e）］（Hu et al.，2014）。王旭东课题组发现，免烧结的 ALD-TiO_2 保护层含有亚

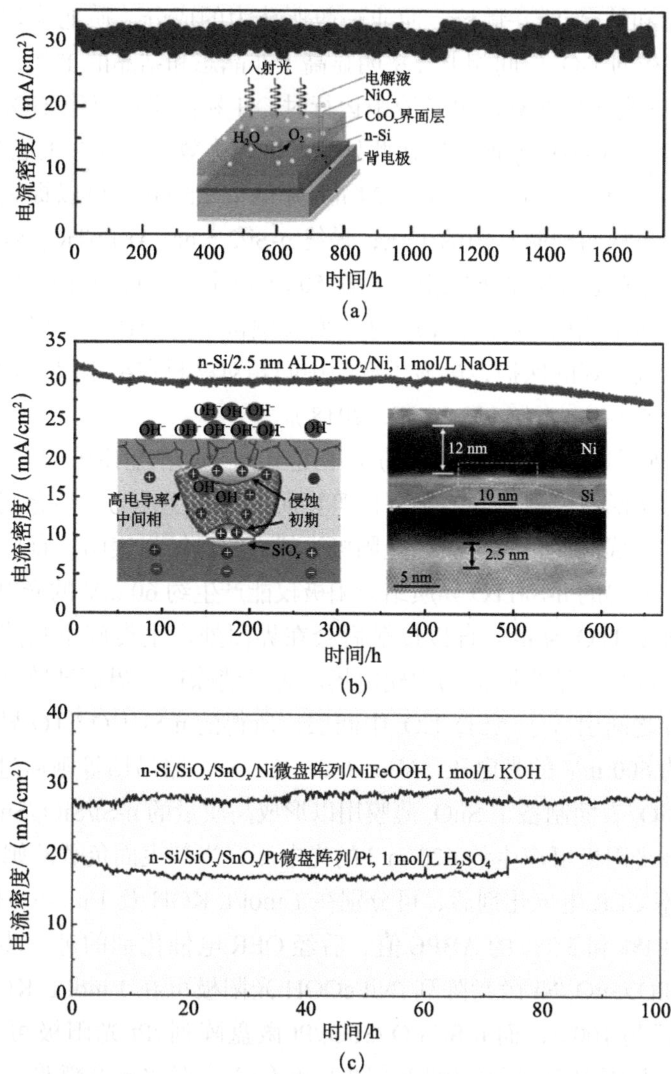

图 1-58　(a) n-Si/SiO_x/CoO_x/NiO_x 光阳极在 1 mol/L KOH 电解液中的电流密度－时间曲线，外加偏压为 1.63 V *vs*. RHE，插图为电极构造示意图；(b) n-Si/2.5 nm ALD-TiO_2/Ni 光阳极在 1 mol/L NaOH 电解液中的电流密度－时间曲线，外加偏压为 1.8 V *vs*. RHE，插图为高电导率的中间相诱导的 ALD-TiO_2 保护层失效机制及电极的截面 TEM 表征；(c) n-Si/SiO_x/SnO_x/Ni 微盘阵列 /NiFeOOH 光阳极在 1 mol/L KOH 电解液中的电流密度－时间曲线，外加偏压为 1.61 V *vs*. RHE，以及 n-Si/SiO_x/SnO_x/Pt 微盘阵列 /Pt 光阳极在 1 mol/L H_2SO_4 电解液中的电流密度－时间曲线，外加偏压为 1.94 V *vs*. RHE

稳的中间相和结晶 TiO_2 颗粒，而非一般所认为的成分、质地均匀的无定形态。这些亚稳的 TiO_2 中间相电导率明显高于非晶态和结晶的 TiO_2，使得 n-Si/ALD-TiO_2/Ni 光阳极在碱性电解液中运行时 OH^- 易富集在亚稳 TiO_2 中间相附近，继而引发 ALD-TiO_2 保护层在此处的腐蚀与失效。对此，王旭东等提出，降低 ALD-TiO_2 保护层的厚度（从 24 nm 降低至 2.5 nm）可以提高亚稳中间相的成核势垒从而抑制它们的生成；最终 n-Si/2.5 nm ALD-TiO_2/Ni 光阳极在 30 mA/cm^2 的光电流密度下能稳定工作 500 h 以上，远优于 n-Si/24 nm ALD-TiO_2/Ni 光阳极的稳定性（70 h）。这一发现颠覆了传统认知，即更厚的保护层给予光电极更好的保护，彰显了亚稳中间相对材料在纳米尺度物化特性的重要影响［图 1-58（b）］（Yu et al.，2018）。

ITO 透明导电氧化物常用作保护层稳定 p-n 同质结光阳极（以及光阴极），而其较高的功函值、透明度、导电性等特质使其有望作为 n 型金属氧化物覆盖层与 n-Si 形成高质量的异质结光阳极。然而实验结果表明，直接在 n-Si 表面沉积 ITO 制得的 n-Si/ITO 异质结光阳极仅能产生约 60 mV 的光电压值。李灿课题组证实 ITO 与 n-Si 直接接触后会在界面处产生类施主缺陷，极大削弱了 n-Si/ITO 异质结光阳极的内建电场。为了消除这些界面缺陷态，他们在 ITO 与 n-Si 之间引入了 ALD-TiO_x 中间层，所得的 n-Si/TiO_x/ITO 异质结光阳极可产生约 600 mV 的光电压（Yao et al.，2016）。路易斯等则通过喷雾沉积法在 n-Si/SiO_x 表面制备了 SnO_x 薄膜用以形成高质量的 n-Si/SiO_x/SnO_x 异质结光阳极。该光阳极可产生约 620 mV 的光电压，当其表面负载欧姆接触的 Ni 及 Pt/IrO_x 基 OER 电催化剂后，可分别在 1 mol/L KOH 及 1 mol/L H_2SO_4 电解液中实现 4.1% 和 3.7% 的 ABPE 值。后经 OER 电催化剂的进一步优化，所得的 n-Si/SiO_x/SnO_x/Ni 微盘阵列 /NiFeOOH 光阳极可在 1 mol/L KOH 电解液中稳定运行超 100 h，而 n-Si/SiO_x/SnO_x/Pt 微盘阵列 /Pt 光阳极可在 1 mol/L H_2SO_4 电解液中稳定运行超 100 h［图 1-58（c）］。对比 n 型简并半导体 ITO，喷雾沉积法制备的 SnO_x 其费米能级更低，功函更大，导致 n-Si/SiO_x/SnO_x 异质结的势垒高度（1.07 eV）大于前述的 n-Si/TiO_x/ITO 异质结（0.95 eV）（Moreno-Hernandez et al.，2018）。

2）Ⅲ-Ⅴ族化合物光阳极材料

Ⅲ-Ⅴ族半导体材料是由元素周期表中Ⅲ和Ⅴ主族元素形成的化合物，主要包含 GaAs、InP、GaN 等二元化合物及 GaInP、GaInN、AlGaAs 等三元化合物。以 GaAs 为例，将其与 Si 进行对比，GaAs 的禁带宽度为 1.42 eV，是直接跃迁型半导体，而单晶 Si 是禁带宽度为 1.12 eV 的间接跃迁型半导体；GaAs 的电子迁移率是单晶 Si 的 6 倍以上，且耐热、抗辐射性能优于 Si。目前，单结 GaAs 太阳能电池的光电转换效率突破了 29.1%（1 个太阳光强度下），而 Si 基太阳能电池中效率最高的异质结和单晶 Si 太阳能电池光电转换效率分别达到了 26.7% 与 26.1%。与 Si 基光阳极类似，Ⅲ-Ⅴ族半导体用作光阳极时也容易发生光腐蚀反应。因此，Ⅲ-Ⅴ族半导体光阳极表面也需要引入保护层，以阻隔其与电解液的直接接触进而实现高效、稳定的 OER。以 Si 基光阳极作为研究平台考察的多种保护层，也有望用于稳定Ⅲ-Ⅴ族半导体光阳极。

路易斯课题组证实，使用富含缺陷的 ALD-TiO_2 保护层，不仅能稳定 Si 基光阳极（包括平面 p-n 同质结 np^+-Si/ALD-TiO_2/Ni、径向 p-n 同质结 np^+-Si/ALD-TiO_2/$NiCrO_x$ 阵列，以及异质结 n-Si/ALD-TiO_2/Ni 光阳极），也能实现对平面 np^+-GaAs/ALD-TiO_2/Ni 光阳极的有效保护。该 np^+-GaAs/ALD-TiO_2/Ni 光阳极在 1 mol/L KOH 电解液及大于 12 mA/cm^2 的光电流密度下可连续运行超过 25 h。采用同样的保护层手段，n-GaP/ALD-TiO_2/Ni 异质结光阳极在 1 mol/L KOH 电解液及 2.3 mA/cm^2 的光电流密度下可连续运行超过 5 h［图 1-59（a）、图 1-59（b）］（Hu et al.，2014）。阿格等证实，兼具空穴提取与传导功能的 p 型透明氧化物 $NiCo_2O_4$ 保护层不仅能用于保护 np^+-Si 光阳极，也能在一定程度上提升 n-InP/$NiCo_2O_4$/NiFe 异质结光阳极在碱性电解液中的运行稳定性（在大于 13 mA/cm^2 的光电流密度下连续运行约 4 h）（Chen et al.，2015）。

相比于 InP、GaAs、GaInP 等Ⅲ-Ⅴ族化合物，InGaN 的抗化学及光电化学腐蚀能力强，更适合用于构建光阳极。易柏德（Ebaid）等指出，已报道的 $In_xG_{a1-x}N$ 基光阳极普遍具有较小的 STH 效率，主要原因在于之前所用的衬底材料（如蓝宝石、Si 片）电导率较低，影响了电荷传输，以及缺乏合适手段有效钝化高密度的表面缺陷态。针对上述问题，该课题组通过分子束外延法在 Ti/Mo 基底上生长了高质量的 $In_{0.33}Ga_{0.67}N$ 纳米棒阵列，并使用乙二硫醇对其进行表面处理，以实现表面缺陷态钝化以及提供 Ir 基 OER 电催化剂的附着位点，最终所得的 Mo/Ti/$In_{0.33}Ga_{0.67}N$/Ir 光阳极 ABPE 值达到了 1.9%（Ebaid et al.，2017）。

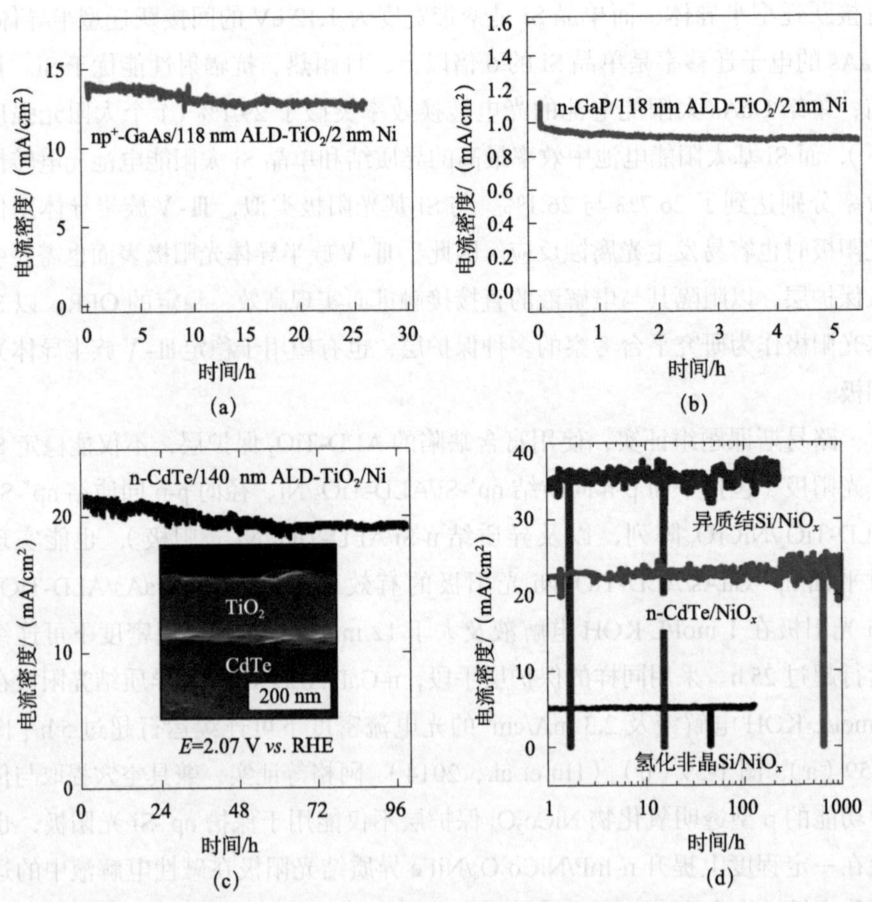

图 1-59 (a) np^+-GaAs/118 nm ALD-TiO_2/2 nm Ni 及 (b) n-GaP/118 nm ALD-TiO_2/2 nm Ni 光阳极在 1 mol/L KOH 电解液中的电流密度－时间曲线，外加偏压为 0.19 V *vs*. 饱和甘汞电极；(c) n-CdTe/140 nm ALD-TiO_2/Ni 光阳极在 1 mol/L KOH 电解液中的电流密度－时间曲线，外加偏压为 2.07 V *vs*. RHE，插图为电极的截面 TEM 表征；(d) 异质结 Si/NiO_x、n-CdTe/NiO_x 及氢化非晶 Si/NiO_x 光阳极在 1 mol/L KOH 电解液中的电流密度－时间曲线，外加偏压为 1.73 V *vs*. RHE

3）Ⅱ-Ⅵ族半导体光阳极材料

Ⅱ-Ⅵ族半导体材料是由元素周期表中Ⅱ副族（Zn、Cd、Hg）和Ⅵ主族元素（O、S、Se、Te）形成的化合物。Ⅱ-Ⅵ族化合物通常是直接跃迁型半导体，相比于Ⅲ-Ⅴ族半导体，其熔点高，离子键成分大。除 CdTe 半导体外，其他Ⅱ-Ⅵ族化合物均为单极性半导体（且大多数为 n 型半导体），较难通过掺杂的手段改变导电类型。与Ⅲ-Ⅴ族半导体类似，Ⅱ-Ⅵ族半导体用作光阳极时表面也需要引入保护层，以抑制光生空穴引发的氧化腐蚀反应。

CdS 是Ⅱ-Ⅵ族半导体材料的典型代表，其禁带宽度为 2.4 eV，导带和价带位置完全跨越了水的氧化还原电位，理论上能在无外加偏压条件下实现水的完全分解反应。然而，CdS 作为光阳极驱动 OER 时，其光生空穴容易氧化自身晶格中的 S 离子，继而造成 CdS 光阳极表面发生不可逆的腐蚀降解。虽然采用保护层策略可在一定程度上提升 CdS 光阳极的稳定性，然而当前已报道的 CdS 光阳极普遍稳定性不佳。原因之一可能是 CdS 半导体的带隙较大，导致其理论 STH 效率低（约 9.1%），不足以引起研究人员的重视。原因之二可能是当前的 CdS 薄膜质量不佳，难以在其表面形成致密、保形的高质量保护膜以有效阻隔 CdS 与腐蚀性电解液的接触。原因之三可能是 CdS 光阳极发生光腐蚀反应后生成可溶性产物（如亚硫酸根、硫酸根），不能对后续腐蚀反应形成有效遏制。

相比于 CdS，CdTe 的带隙较小，约为 1.43 eV，是理想的太阳能电池材料，受到众多研究人员的关注。目前能通过多种方法获得高质量的多晶、单晶 CdTe 薄膜。此外，在 OER 过程中，CdTe 表面可生成一层不溶性（或难溶性）$CdTeO_4$ 钝化层，阻碍了 CdTe 的持续氧化腐蚀进程。上述优点使得 CdTe 有望用于构建高效、稳定的分解水光阳极。路易斯课题组在 n-CdTe 单晶表面引入 140 nm 厚、免烧结、富含缺陷的 ALD-TiO_2 保护层与 Ni 金属膜 OER 电催化剂，据此构建的 n-CdTe/ALD-TiO_2/Ni 光阳极在 1 mol/L KOH 电解液中产生的光电压值为 435 mV ± 15 mV，且能在大于 18 mA/cm^2 的光电流密度下连续运行超过 100 h［图 1-59（c）］（Lichterman et al.，2014）。后续该组证实，p 型导电、透明、电催化活性、抗反射的 NiO_x 多功能保护层，能确保 n-CdTe/NiO_x 异质结光阳极在 1 mol/L KOH 电解液及约 22 mA/cm^2 的光电流密度下持续运行超过 1000 h［图 1-59（d）］（Sun et al.，2015）。该组也尝试以 50 nm 厚的 ALD-CoO_x 保护层覆盖在 n-CdTe 单晶表面制备了 n-CdTe/CoO_x 异质结光阳极。然而该光阳极的开启电位不佳（约 1.45 V *vs*. RHE），饱和光电流密度仅为 14 mA/cm^2。造成该 n-CdTe/CoO_x 异质结光阳极性能不佳的原因可能是 n-CdTe 与 CoO_x 晶格匹配度低，二者界面处形成了不利的界面缺陷态或能级分布；ALD 生长 CoO_x 保护层时使用的臭氧也可能对 n-CdTe 单晶表面产生有害的氧化损伤（Zhou et al.，2016）。

4）金属（氧）氮化物光阳极材料

我们以经典金属氧氮化物 TaON 和金属氮化物 Ta_3N_5 来讨论金属（氧）氮化物半导体材料的特性。TaON 与 Ta_3N_5 通常是由 Ta_2O_5 作为前驱体高温氨解获得的。根据半导体能带理论，Ta_2O_5、TaON 与 Ta_3N_5 的导带底主要由 Ta 的空 d 轨道构成，因此它们的导带底位置较接近；它们的价带顶则分别由 O 2p 轨道、O 2p 与 N 2p 的杂化轨道，以及 N 2p 轨道构成。由于 N 的电负性（3）小于 O（3.5），N 2p 轨道的能级要高于（负于）O 2p 轨道，因此，从 Ta_2O_5 至 TaON 再至 Ta_3N_5，价带顶依次负移，最终造成三种材料的禁带宽度依次减小（Ta_2O_5，3.9 eV，TaON，2.4 eV，Ta_3N_5，2.1 eV），吸收带边逐步红移（图 1-60）（Chun et al.，2003）。三种材料中带隙最小的 Ta_3N_5 可以吸收波长至 600 nm 左右的可见光，基于 AM 1.5 G 100 mW/cm^2 的标准太阳光谱计算可知，Ta_3N_5 的理论 STH 效率约为 16.2%。除了具备吸收可见光的能力，TaON 与 Ta_3N_5 的能级位置也完全适合于水的分解电位。因此，TaON 与 Ta_3N_5，尤其是后者，在太阳能光电催化分解水制氢领域潜力巨大。除 TaON 与 Ta_3N_5 之外，还存在一些含 Ti、Nb、Ta 等金属的钙钛矿型氧氮化物，如 $LaTiO_2N$、$SrNbO_2N$、$LaTaON_2$ 等，它们的禁带宽度及能级位置与 Ta_3N_5 比较接近，因此也是一类极具潜力的光电化学水分解电极材料（Feng et al.，2020）。

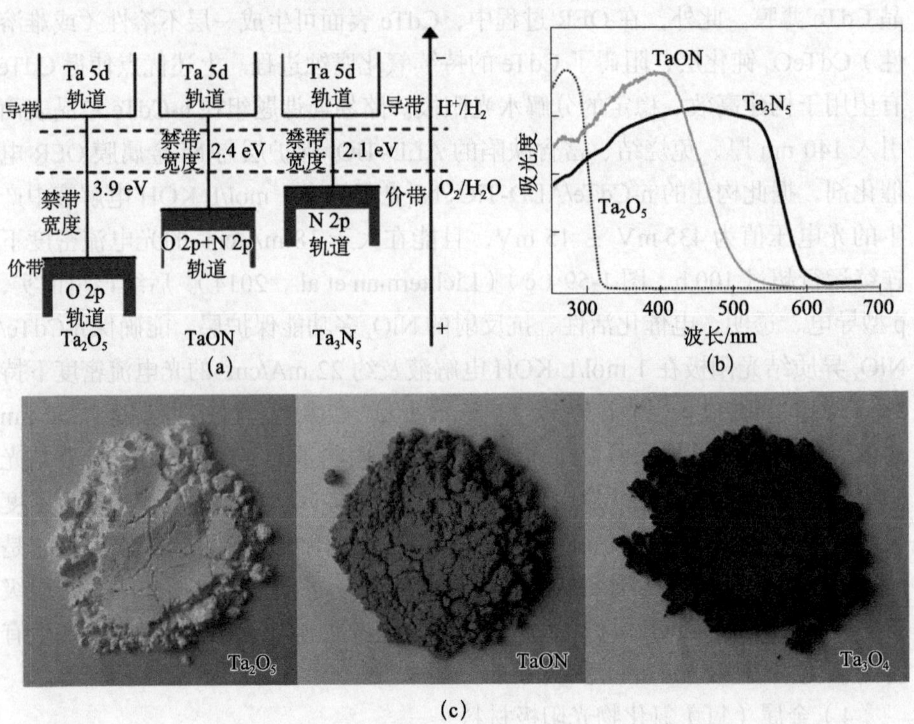

图 1-60 (a) Ta_2O_5、TaON 与 Ta_3N_5 的能带结构示意图；Ta_2O_5、TaON 与 Ta_3N_5 粉末的 (b) 紫外 - 可见吸收光谱及 (c) 照片

金属（氧）氮化物半导体的缺陷以阴离子空位、氧杂质等施主缺陷为主，因此通常情况下，它们表现为 n 型半导体，适合作为驱动 OER 的光阳极。然而，金属（氧）氮化物半导体在高温氮化制备过程中容易形成大量的缺陷，继而造成其构建光阳极时电荷分离效率低。与其他非氧化物光阳极类似，金属（氧）氮化物光阳极也容易发生光腐蚀反应。因此，为了在金属（氧）氮化物光阳极表面实现高效、稳定的 OER，相关的研究都集中在提升其电荷分离效率与光电化学稳定性两方面。

（1）电荷分离效率。由于金属（氧）氮化物半导体通常是由氧化物前驱体高温氮化制备的（850～1000 ℃），其理想衬底需具备以下几个特性，如晶格失配度低、高电导率、功函适宜、耐高温氨解等，然而符合上述条件的衬底颇少。早期 Ta_3N_5 及其他金属（氧）氮化物光阳极的制备通常选择金属 Ta 箔或 Ti 箔作为衬底，例如以金属 Ta 箔作为导电衬底及 Ta 源将其表面氧化而后高温氨气氛围氮化即可得到 Ta_3N_5 薄膜。上述金属 Ta 箔制备 Ta_3N_5 薄膜的过程历经 Ta 至 Ta_2O_5 再至 Ta_3N_5 的晶体转变，由于 Ta、Ta_2O_5 和 Ta_3N_5 的晶体结构差别，最终所形成的 Ta_3N_5 薄膜会因残余应力而产生裂纹，甚至从 Ta 衬底上剥落。堂免一成（Kazunari Domen）课题组于 2004 年最早研究了金属 Ta 箔制备的 Ta_3N_5 光阳极，但并未讨论影响该光阳极电荷分离效率的因素（Ishikawa et al.，2004）。2012 年，哈拉米洛（Jaramillo）课题组通过控制 Ta 箔的氧化时间，在其表面制备了不同厚度的 Ta_3N_5 薄膜光阳极，并据此详细分析了它们的电子与空穴传输特性。对于薄的 Ta_3N_5 光阳极，由于电子至 Ta 衬底及空穴至固液界面的传输距离都较短，此时电子与空穴在传输过程中体相复合较少，然而该 Ta_3N_5 光阳极的性能受限于其较差的光捕获能力（薄的薄膜厚度）。Ta_3N_5 光阳极薄膜厚度的增加会增强其光捕获能力。与此同时，由于残余应力产生的裂纹也将增大电极的比表面积并减小空穴至固液界面的传输距离，最终该光阳极的光电流密度明显提升。进一步增加薄膜厚度则会导致 Ta_3N_5 光阳极薄膜疏松、颗粒间连接性变差，这将不利于电子有效传输至 Ta 衬底，限制了 Ta_3N_5 光阳极性能的继续提升［图 1-61（a）］（Pinaud et al.，2012）。

受限于高温氮化的苛刻合成条件，如何在金属箔衬底上原位制备高质量的金属（氧）氮化物薄膜光阳极颇具挑战。2019 年，邹志刚课题组开发了一种无机蒸气反应沉积法，在 Ta 箔衬底上原位合成了高质量的 $SrTaO_2N$ 薄膜，该薄膜由致密的、贯穿整个薄膜厚度的八面体晶粒组成［图 1-61（b）］。在 $SrTaO_2N$ 薄膜与金属 Ta 导电衬底之间存在一层高导电性的金属相 TaN 过渡层，有利于光生电子从 $SrTaO_2N$ 薄膜高效传输至金属 Ta 衬底。该 $SrTaO_2N$ 光阳极在模拟太阳光源的照射下，光电流密度在 1.23 V *vs.* RHE 处达到了 3 mA/cm^2，而且展现了良好的光电化学稳定性。此外，该无机蒸气反应沉积法也能用于制备其他高质量的钙钛矿型氧氮化物薄膜，包括 $CaTaO_2N$、$BaTaO_2N$、$SrNbO_2N$、$BaNbO_2N$ 等（Fang et al.，2019）。

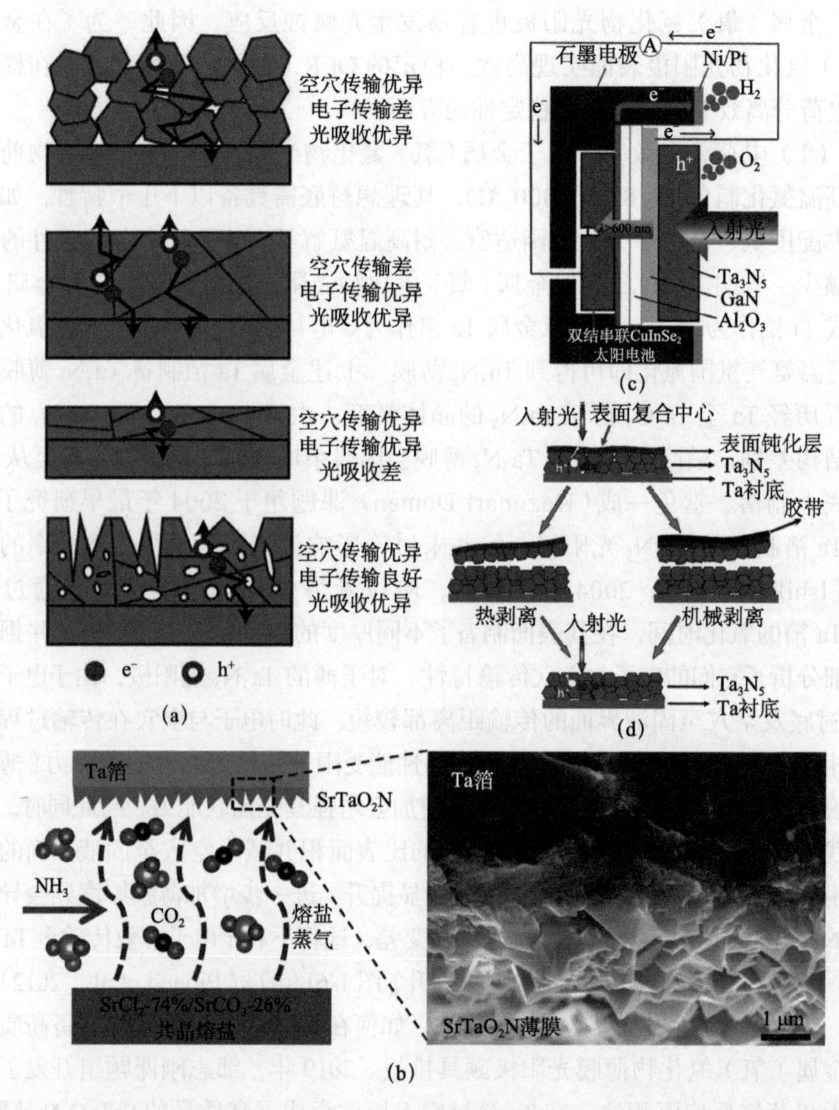

图 1-61 (a) Ta_3N_5 薄膜形貌对其光捕获能力及电子与空穴传输特性的影响；(b) 无机蒸汽反应沉积法在 Ta 箔衬底上原位制备 $SrTaO_2N$ 薄膜的机理示意图及 $SrTaO_2N$ 薄膜的截面 SEM 表征图；(c) $CuInSe_2$ 太阳能电池 /Ta_3N_5 光阳极叠层全水分解器件的构造示意图；(d) 剥离策略去除 Ta_3N_5 光阳极表面钝化层的过程示意图

开发合适的衬底用以制备透明的金属（氧）氮化物光阳极对后续构建叠层结构的光电化学水分解电池意义重大。2016 年，哈曼（Hamann）课题组采用 ALD 技术制备的 Ta 掺杂 TiO_2 薄膜作为衬底获得了世界上首个透明 Ta_3N_5 光阳极。该透明 Ta_3N_5 光阳极在模拟太阳光源的照射下，光电流密度在 1.23 V *vs.* RHE 处达到了 0.77 mA/cm^2（Hajibabaei et al.，2016）。2019 年，哈曼课题组采用 $TaCl_5$ 与 NH_3 作为原料以 ALD 技术在 FTO 衬底上低温（550 ℃）制备了透明、晶态、膜厚可控的 Ta_3N_5 光阳极。在含空穴捕获剂亚铁氰化钾的电解液中 103 nm 厚的 Ta_3N_5 光阳极展现了最优的性能，其开启电位约为 0.3 V *vs.* RHE，光电流密度在 1.23 V *vs.* RHE 处达到了 2.4 mA/cm^2（Hajibabaei et al.，2019）。2019 年，堂免一成课题组以透明导电 n-GaN 作为衬底制备了高结晶度的透明 Ta_3N_5 光阳极，与双结 $CuInSe_2$ 太阳能电池串联后 STH 效率可达 7%［图 1-61（c）］（Higashi et al.，2019）。

除了衬底的选择与优化，构筑微纳结构也是提升金属（氧）氮化物光阳极电荷分离效率的一个重要策略。纳米线（或棒、管、片）阵列电极结构在光吸收方面能增强电极的光捕获能力，在载流子输运方面能同时缩短少数载流子的传输距离并建立多子传输至导电衬底的连续通道，是光电催化领域研究的热点。2009 年，米斯拉（Misra）课题组以阳极氧化而后氮化的策略在 Ta 箔上制得 TaON 纳米管阵列光阳极，该光阳极在模拟太阳光源的照射下，光电流密度在 1.52 V *vs.* RHE 处达到了 2.6 mA/cm^2（Banerjee et al.，2009）。2010 年，格兰姆斯（Grimes）课题组以类似的策略在 Ta 箔上制得 Ta_3N_5 纳米管阵列光阳极，获得了当时 Ta_3N_5 光阳极性能的世界最佳值（1 mol/L KOH 为电解液，Pt 为对电极的两电极体系，0.5 V 偏压下 450 nm 处 IPCE 达 5.3%）（Feng et al.，2010）。然而为了避免 Ta_3N_5 纳米管阵列的坍塌，其氮化过程温度较低，故最终所得的 Ta_3N_5 纳米管阵列光阳极缺陷密度较高，性能远低于 Ta_3N_5 的理论值。2013 年，堂免一成课题组以 Al_2O_3 纳米管阵列为掩模版阳极氧化 Ta 箔获得 Ta_2O_5 纳米棒阵列，后经氮化得到 Ta_3N_5 纳米棒阵列光阳极。相比于前述高缺陷密度的 Ta_3N_5 纳米管阵列光阳极，该 Ta_3N_5 单晶纳米棒阵列光阳极性能有了大幅度的提升（表面担载 IrO_2 后，1.23 V *vs.* RHE 偏压下，400～520 nm 波长范围内 IPCE 高于 35%）（Li et al.，2013）。2013 年，成会明课题组通过氢氟酸水热腐蚀的手段在 Ta 箔上制备出 Ta_2O_5 纳米棒阵列，后经氮化也获得了 Ta_3N_5 单晶纳米棒阵列光阳极）（Zhen et al.，2013）。

需要指出的是，通常的研究认为光阳极，包括金属（氧）氮化物光阳极，空穴扩散长度短，是导致其体相电子-空穴复合严重、电荷分离效率低的主要因素。因此，为了提升光阳极的水分解性能，诸多研究聚焦于减小光阳极的特征尺寸，以期缩短空穴至固液界面的传输距离，光阳极特征尺寸的减小也有利于增加空间电荷区的体积占比。然而，光阳极的不当纳米化会导致光生电子传输至导电基底的过程中受到的散射作用增大，反而使得电子传输过程成为光阳极的性能瓶颈。因此，构建高性能光阳极，一方面，要减小其特征尺寸以缩短空穴至固液界面的传输距离并增加空间电荷区的体积占比；另一方面，要建立电子传输至导电基底的连续通道。为了证实上述观点，邹志刚课题组构建了结晶性好、晶界密度低、三维交联、多孔结构的 $LaTiO_2N$ 微米颗粒组装膜光阳极。在 pH = 7 的磷酸钾缓冲溶液中，该光阳极在 2 V *vs*. RHE 处光电流密度达到了 4.2 mA/cm^2，同等条件下 $LaTiO_2N$ 纳米颗粒组装膜光阳极的光电流密度仅为 0.47 mA/cm^2。$LaTiO_2N$ 微米颗粒组装膜光阳极表面引入 Co_3O_4 基 OER 电催化剂后，其饱和光电流密度值可达 6.5 mA/cm^2，在 1.23 V *vs*. RHE 处光电流密度为 4.45 mA/cm^2，是当时世界上性能最佳的 $LaTiO_2N$ 光阳极（Feng et al.，2014）。

缺陷是影响金属（氧）氮化物光阳极电荷分离效率的第三个因素。金属（氧）氮化物中主要有氮空位（V_N）、氧占氮位（O_N）及还原态的金属离子（如 Nb^{4+}、Ti^{3+}、Ta^{4+}）等缺陷。其中，O_N 缺陷的形成能为负值，这导致 Ta_3N_5 中往往有相当含量的O杂质，金属氧氮化物则通常为非计量比化合物。此外，O_N 缺陷通常在金属（氧）氮化物中引入浅施主能级，因此大多数金属（氧）氮化物是以电子导电为主的 n 型半导体。邹志刚课题组指出，上述缺陷中 V_N 所带的正电荷会使其成为电子的深陷阱，进而不利于金属（氧）氮化物光阳极的电子传输过程并引发载流子（电子和空穴）的复合。他们发现，当 Ta_2O_5 前驱体中引入少量的 Ge，高温氨解制备 Ta_3N_5 时，Ge 可能作为助熔剂减少 V_N 浓度，进而加快 Ta_3N_5 光阳极的电子传输动力学过程（Feng et al.，2014）。邹志刚课题组还关注了金属氧氮化物的非化学计量比特性（其氧氮比例可在一定范围内变化）对其电荷分离效率的影响。他们通过控制前驱体用量和氨气流速，精确地调节了 TaON 的氧氮比例，并以此为基础阐述了氧氮比例变化对 TaON 光电催化分解水性能的影响规律。实验结果表明，控制氧氮比例能够调控 TaON 电极的空间电荷层厚度和电导率，合适的氧氮比例能够使得 TaON 光阳极展现最优的 OER 性能。这一研究表明，理解并精确调控金属（氧）氮化物的化学计量比对进一步优化其太阳能转换效率尤为重要（Feng et al.，2019）。此外，金属（氧）氮化物表面形成的无定形或亚化学计量比物质也可能会阻碍光生空穴向电解液或 OER 电催化剂的输运。针对这个问题，邹志刚课题组提出机械剥离及化学刻蚀的策略以去除金属（氧）氮化物表面的这些缺陷物质，大幅提升了金属（氧）氮化物光阳极的性能［图1-61（d）］（Li et al.，2013）。金属（氧）氮化物经氢气处理后阴离子缺陷（氧空位与氮空位）浓度有所增加，这将提升其电导率，并促进光生电子与空穴的体相传输。阿部（Abe）课题组与邹志刚课题组分别以 $BaTaO_2N$ 与 $SrTaO_2N$ 光阳极为例证明氢气处理能够降低金属（氧）氮化物光阳极的体相电荷传输电阻，实现其电荷分离效率的大幅度提升（Higashi et al.，2013；Zhong et al.，2016）。

预先制备金属（氧）氮化物粉末而后以合适方法负载在导电衬底上也能获得相应的光阳极。众多研究领域，如染料敏化太阳能电池使用的刮涂法并不适于制备金属（氧）氮化物薄膜。这是因为配制刮涂法的浆料时通常会引入一些高沸点的有机溶剂或高分子材料，后续需以高温氧化的方法去除。然而，高温氧化过程会导致金属（氧）氮化物的氧化分解。阿部课题组采用电泳沉积法在 FTO 衬底上制备了 TaON 光阳极，薄膜制备过程采用的丙酮分散剂可经常温挥发去除。此时得到的 TaON 光阳极性能较差，主要原因是 TaON 颗粒之间及 TaON 与 FTO 之间连接较差，缺乏光生电子传输至 FTO 衬底的连续通道。由于 TaON 属于硬质难熔化合物，很难通过简单的热处理促进 TaON 颗粒间的交联。针对这个问题，阿部课题组提出 $TaCl_5$ 交联而后氨气氛围低温热处理的策略，在 TaON 颗粒之间及 TaON 与 FTO 之间引入 Ta_2O_5（N 掺杂的 Ta_2O_5）作为桥联，以此在电泳沉积法制备的 TaON 光阳极中构建了光生电子传输至 FTO 衬底的连续通道（Abe et al.，2010）。目前以这种方法可以制得多种高活性的金属（氧）氮化物光阳极，如 TaON、Ta_3N_5、$SrTaO_2N$、$LaTaON_2$、$LaTiO_2N$ 等。

（2）光电化学稳定性。金属（氧）氮化物光阳极表面的水氧化反应与其自身的氧化腐蚀反应互相竞争，一旦水氧化反应进行不畅，其自身的氧化腐蚀反应就会占据消耗空穴的主导地位。因此，提高金属（氧）氮化物光阳极的光电化学稳定性与加快其表面产氧反应动力学速率（即提高其表面电荷注入效率）息息相关。堂免一成课题组于 2004 年首次报道了 Ta_3N_5 薄膜在光电化学分解水反应中的应用。然而 Ta_3N_5 薄膜在分解水进程中性能不断衰减，其表面的氮元素比例也急剧下降，表明光生空穴诱致 Ta_3N_5 表面晶格中的 N^{3-} 被氧化。采用亚铁氰根离子（$[Fe(CN)_6]^{4-}$）作为空穴捕获剂，Ta_3N_5 薄膜光阳极实现了 27 h 的稳定性（Ishikawa et al.，2004）。基于上述认识，2011 年该课题组在 Ta_3N_5 光阳极表面担载 OER 电催化剂 IrO_2 以加快光生空穴氧化水反应速率，由此 Ta_3N_5 光阳极表面的氧化腐蚀得以改善，分解水稳定性有所提升（Yokoyama et al.，2011）。巴德（Bard）课题组于 2012 年比较了不同 OER 电催化剂 IrO_2、Co_3O_4、Co-Pi（磷酸钴）、Pt，以及吸附的 Co^{2+} 对 Ta_3N_5 光阳极电流密度-电压曲线及光电化学稳定性的影响。研究结果表明，合适的 OER 电催化剂能大幅提升 Ta_3N_5 光阳极的性能与稳定性，主要原因可归结于 OER 电催化剂能加快 Ta_3N_5 光阳极表面的水氧化速率，Ta_3N_5 自身的氧化腐蚀反应得以抑制（Cong et al.，2012）。堂免一成课题组在 2013 年采用光辅助电沉积手段在 Ta_3N_5 光阳极表面负载了 Co-Pi OER 电催化剂，该光阳极在 1.2 mA/cm^2 的光电流密度下可稳定约 20 min（Li et al.，2013）。后续该课题组采用 Co-Pi OER 电催化剂修饰的 Ba 掺杂 Ta_3N_5 光阳极，在碱性电解液及 4.5 mA/cm^2 的光电流密度下可连续运行 20 min（Li et al.，2013）。2015 年，施穆基（Schmuki）课题组在 Ta_3N_5 纳米棒阵列光阳极表面负载三元 OER 电催化剂 Co-Pi、$Co(OH)_x$ 及 NiFe-LDH（镍和铁的层状双氢氧化物），可在 1 mol/L KOH 电解液及 5 mA/cm^2 的光电流密度下稳定工作 2 h（Wang et al.，2015）。2016 年，王敦伟课题组讨论了阻隔 Ta_3N_5 与电解液（及其他活性氧物质）接触对 Ta_3N_5 光阳极稳定性的重要性，并指出单纯 OER 电催化剂的负载对提升 Ta_3N_5 光阳极稳定性效果有限。据此，他们采用 ALD 技术在 Ta_3N_5 纳米管阵列光阳极表面负载了一层致密 MgO 保护层，以此隔绝 Ta_3N_5 与电解液的直接接触，外加 $Co(OH)_x$ 基 OER 电催化剂后，该 $Ta_3N_5/MgO/Co(OH)_x$ 光阳极可在大于 6 mA/cm^2 的光电流密度下持续工作 0.5 h（He et al.，2016）。2017 年，堂免一成课题组在 Ta_3N_5 光阳极表面负载一层 GaN 作为保护膜，外加光辅助电沉积的 Co-Pi 作为 OER 电催化剂，该 Ta_3N_5/GaN/Co-Pi 光阳极在 8 mA/cm^2 的光电流密度下可稳定工作 8 h（Zhong et al.，2017）。

在金属氧氮化物光阳极方面，阿部课题组于 2012 年在 TaON 光阳极表面以浸渍法引入了 CoO_x 基 OER 电催化剂，在 pH = 8 的磷酸钠缓冲溶液及 2 mA/cm^2 的光电流密度下，该 TaON/CoO_x 光阳极可稳定至少 1 h；相比之下，TaON/IrO_x 光阳极在持续光照 1 h 后光电流密度衰减至初始值的 1/3（Higashi et al.，2012）。2013 年，该组在氢气还原的 $BaTaO_2N$ 表面依次引入 CoO_x 与 RhO_x 作为 OER 电催化剂，最终该 $BaTaO_2N/CoO_x/RhO_x$ 光阳极在 pH = 8 的磷酸钠缓冲溶液及 2 mA/cm^2 的光电流密度下可稳定至少 1 h（Higashi et al.，2013）。2021 年，阿部课题组在 TaON 光阳极表面分别担载了 CoO_x 与 RhO_x 基 OER 电催化剂。实验结果显示，当电解液 pH 大于 8 时，CoO_x 与 RhO_x 均能明显提升 TaON 光阳极的稳定性；而当电解液 pH 小于 8 时，仅 RhO_x 对提升 TaON 光阳极的稳定性有帮助（Higashi et al.，2021）。

国内对金属（氧）氮化物光阳极光电化学稳定性的研究始于 2012 年左右。2012 年，邹志刚课题组采用廉价的 $Co(OH)_x$ 及 Co_3O_4 基 OER 电催化剂实现了 Ta_3N_5 光阳极光电化学稳定性的大幅度提升，获得了当时世界上最稳定的 Ta_3N_5 光阳极（1 mA/cm^2 的光电流密度下持续光照 2 h 后性能衰减至初始值的 75%）（Liao et al.，2012）。2013 年，成会明课题组采用 $Co(OH)_x$ 基 OER 电催化剂使得 Ta_3N_5 纳米棒阵列光阳极在 2.8 mA/cm^2 的光电流密度下稳定工作 10 min（Zhen et al.，2013）。同年，朱鸿民课题组在 Ta_3N_5 纳米棒阵列光阳极表面引入 $Co_3O_4/Co(OH)_2$ 双层 OER 电催化剂，最终该 Ta_3N_5 光阳极在 2.78 mA/cm^2 的光电流密度下稳定工作 2 h（Hou et al.，2013）。李灿课题组在 2014 年采用水铁石作为空穴储存层、Co_3O_4 作为 OER 电催化剂，使得 Ta_3N_5 光阳极在 5.2 mA/cm^2 的光电流密度下稳定工作 6 h（Liu et al.，2014）。基于上述工作，2016 年，李灿课题组在 Ta_3N_5 光阳极表面依次引入 TiO_x 电子阻挡层、$Ni(OH)_x$/水铁石复合空穴储存层、Co/Ir 双组分分子基 OER 催化剂，最终该光阳极可在 6.5 mA/cm^2 的光电流密度下稳定工作 20 min（Liu et al.，2016）。

近年来，以石墨相氮化碳（g-C_3N_4）为代表的聚合物半导体光催化材料引起了众多研究者的兴趣，其优势如下：由碳、氮等非金属元素组成，廉价易得；结构多样且形貌可调，易于成膜与器件化；能带结构与表面性质可调，光、热及化学稳定性好等。这些优势使 g-C_3N_4 等聚合物半导体有望成为新型、高效的 OER 光阳极材料（Fang et al.，2022）。然而，当前 g-C_3N_4 等聚合物基光阳极性能普遍较低，可能的限制因素包括如下几方面：材料的本征导电性差，载流子迁移率低，寿命短；材料的缺陷态密度高，光生载流子复合严重；聚合物之间及聚合物与导电衬底之间接触不佳，光生电子在电极体相及背电极处传输电阻较大。

5）金属氧化物光阳极材料

单晶 TiO_2 光阳极里程碑研究之后（Fujishima & Honda，1972），巴德课题组指出，要实现光电催化分解水制氢的大规模应用，亟须开发低成本、易制备、高稳定、吸光好的新型光电极材料。由于单晶样品制备过程复杂，巴德课题组采用化学气相沉积法制备了 α-Fe_2O_3 及 TiO_2 多晶薄膜作为光阳极，获得了与单晶样品接近的光电转换性能（Hardee & Bard，1977）。自此，开发金属氧化物半导体基光电催化材料成为光电催化领域的研究热点与前沿。

顾名思义，金属氧化物半导体是由金属与氧组成的半导体材料。金属氧化物半导体的化学及光稳定性好、储量丰富、价格低廉、易于制备，是PECs分解水制氢研究领域的重点与热点材料体系。其中代表性的材料有宽带隙的 TiO_2 (3.0～3.2 eV)、ZnO (3.2 eV)、$SrTiO_3$ (3.2 eV) 及具有可见光响应的 α-Fe_2O_3 (2.0～2.2 eV)、$BiVO_4$ (2.4 eV)、WO_3 (2.7 eV) 等。由于宽带隙的氧化物半导体对太阳光谱的捕获有限、理论STH效率较低，当前的研究重点是具可见光响应的金属氧化物半导体光阳极。本部分将就 α-Fe_2O_3 与 $BiVO_4$ 这两种经典材料展开讨论。

（1）α-Fe_2O_3 光阳极。α-Fe_2O_3 的禁带宽度约为2.1 eV，无毒、廉价、来源广泛，在水分解过程中稳定性好，是目前n型半导体光阳极研究的热点材料。在AM 1.5 G 100 mW/cm^2 的标准太阳光谱下，α-Fe_2O_3 的理论光电流密度约为12.6 mA/cm^2，理论STH效率约为15.5%。然而 α-Fe_2O_3 的光吸收系数较小，需要约500 nm厚的薄膜才能实现对入射光子的充分吸收，这与其短的空穴扩散长度（2~4 nm）形成了矛盾。针对这个问题，常用的策略是构建纳米线（或棒、管、片）阵列或三维交联、纳米多孔结构 α-Fe_2O_3 电极，以此增加空间电荷区在整个 α-Fe_2O_3 电极的体积占比，缩短空穴至电解液的传输距离，与此同时，确保电极对入射光子的充分吸收（Sivula et al., 2011)。格拉兹尔（Grätzel）课题组以常压化学气相沉积法在FTO衬底上制备出花椰菜结构的 α-Fe_2O_3 光阳极，其性能是当时的世界纪录（饱和光电流密度约4.3 mA/cm^2）[图1-62（a）、图1-62（b）]（Warren et al., 2013）。

除了空穴扩散长度短这一瓶颈因素，α-Fe_2O_3 光阳极差的电子传输性质也严重制约了其性能。针对这个问题，策略之一是在 α-Fe_2O_3 光阳极中引入一定浓度的施主缺陷。目前已有Ru、Ce、Pt、Si、Ge、Sn、Ti、Zr、Hf、V、Nb、Ta、Mo、W等元素被用作施主杂质掺杂于 α-Fe_2O_3，以提高其电子浓度及薄膜电导率。在 α-Fe_2O_3 中引入氧空位也能实现上述目的（Li et al., 2013）。前述花椰菜结构的 α-Fe_2O_3 光阳极性能优异的原因有三：①构造了纳米化的 α-Fe_2O_3 颗粒，用以解决少数载流子空穴扩散长度短的问题；②建立了多数载流子电子传输至导电基底的连续通道；③引入了施主杂质Si，以提升 α-Fe_2O_3 薄膜的载流子浓度及电导率。在高导电支架或基体上制备 α-Fe_2O_3 电极以便光生电子迅速传输至导电基底，是促进 α-Fe_2O_3 光阳极电子传输的第二种策略）(Sivula et al., 2009)。也有研究采用 Al^{3+}、In^{3+}、Cr^{3+} 等三价阳离子对 α-Fe_2O_3 进行掺杂，这些杂质的引入不会引起 α-Fe_2O_3 载流子浓度的变化。然而它们的离子半径与 Fe^{3+} 不同，会导致 α-Fe_2O_3 晶体结构发生畸变并产生局域极化场，对 α-Fe_2O_3 光阳极的电荷分离过程可能有一定的促进作用

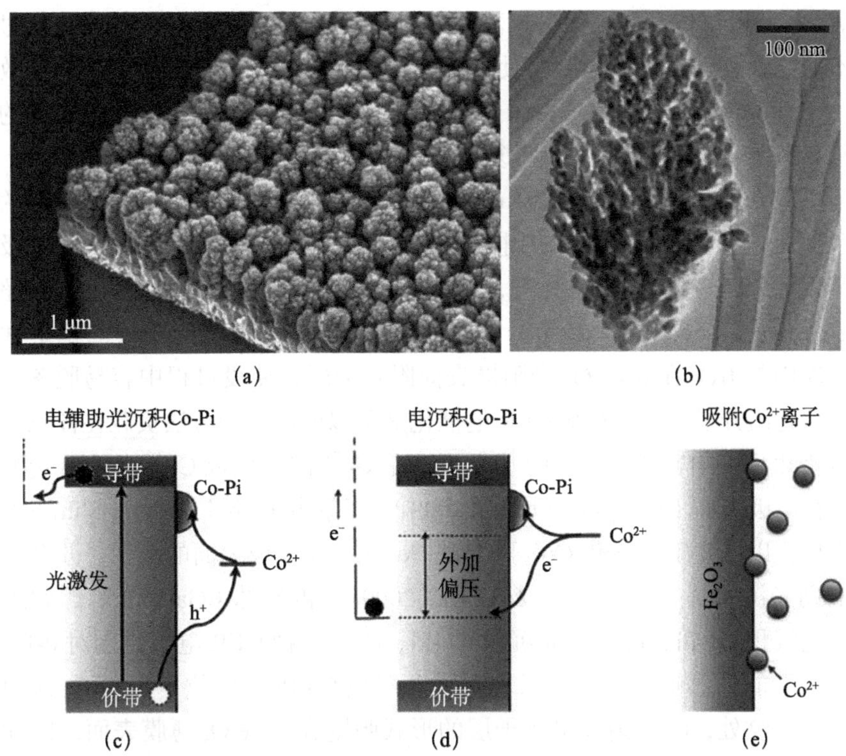

图 1-62　常压化学气相沉积法制备的花椰菜结构 α-Fe_2O_3 电极的 (a) SEM 和 (b) TEM 表征图；(c) 电辅助光沉积 Co-Pi、(d) 电沉积 Co-Pi 及 (e) 吸附 Co^{2+} 三种 OER 电催化剂在 α-Fe_2O_3 光阳极表面的形成过程示意图

（Li et al.，2013）。需要指出的是，过高的离子掺杂浓度会导致 α-Fe_2O_3 晶格加剧畸变或生成偏析相，从而降低光生电子与空穴的迁移率与寿命。

构筑异质结是提升光阳极与光阴极电荷分离效率的重要策略。毕迎普课题组构建了“三明治”结构的 Pt/α-Fe_2O_3/Fe_2TiO_5 异质结光阳极，由于三种材料间有利的能级排列，光生电子和空穴在该异质结光阳极中可实现高效定向分离与迁移（Wang et al.，2017）。此外，α-Fe_2O_3 与导电衬底直接接触后可能会形成不利的界面缺陷态或能级分布，继而阻碍电子转移至导电衬底并引发载流子复合。格拉兹尔课题组证实，FTO 衬底表面预先沉积的单层 SiO_x 可作为晶格应变缓冲层实现高质量超薄 α-Fe_2O_3 薄膜的成核与生长。该 FTO/SiO_x/α-Fe_2O_3 光阳极在 1.5 V *vs.* RHE 处的光电流密度达到了 0.63 mA/cm^2，同等条件下 FTO/α-Fe_2O_3 光阳极几乎无光电响应（Formal et al.，2010）。该课题组也研究了 Ga_2O_3 和 Nb_2O_5 作为衬层对超薄 α-Fe_2O_3 光阳极性能的影响。结果显示，

Ga_2O_3 衬层可以作为模板促进超薄 α-Fe_2O_3 薄膜的成核与生长，但其厚度过薄时本身结晶性较差不能有效行使模板作用，厚度过厚又会阻碍电子的有效传输（Hisatomi et al.，2012）；而 Nb_2O_5 作为衬层可有效阻止转移至 FTO 的电子与 α-Fe_2O_3 光阳极的空穴发生复合（Hisatomi et al.，2012）。

负载 OER 电催化剂是提升 α-Fe_2O_3 光阳极表面产氧反应动力学速率的有效手段。格拉兹尔课题组采用电泳沉积法在花椰菜结构的 α-Fe_2O_3 光阳极表面负载了 IrO_2 基 OER 电催化剂，使得 α-Fe_2O_3 光阳极的光电流密度首次超过 3 mA/cm^2（Tilley et al.，2010）。然而该 α-Fe_2O_3/IrO_2 复合光阳极的稳定性不佳，原因是 IrO_2 在 α-Fe_2O_3 光阳极表面附着力弱，使役过程中容易脱落。诺塞拉课题组开发的 Co-Pi 基 OER 电催化剂可以电沉积的方法负载在任意形状的电极表面。电沉积 Co-Pi 的基本机理是 Co^{2+} 被氧化成 Co^{3+} 而后与磷酸根沉淀在电极表面（Kanan & Nocera，008）。崔景申（Choi Kyoung-Shin）课题组提出，以光生空穴氧化 Co^{2+} 可实现 Co-Pi 在光阳极表面的沉积，并命名为电辅助光沉积法（Steinmiller & Choi，2009）。盖米林（Gamelin）课题组比较了电沉积 Co-Pi、电辅助光沉积 Co-Pi、Co^{2+} 三种 OER 电催化剂对 α-Fe_2O_3 光阳极性能的影响。研究结果表明，电沉积 Co-Pi 主要位于 α-Fe_2O_3 薄膜的孔洞、裂纹处，Co^{2+} 则以单离子层的形式吸附在 α-Fe_2O_3 薄膜表面，因此二者对 α-Fe_2O_3 光阳极性能的提升有限。相比之下，电辅助光沉积 Co-Pi 是由空穴引发的，可最大限度地减少光生空穴至 Co-Pi 基 OER 电催化剂的迁移距离，因此其对 α-Fe_2O_3 光阳极性能及稳定性的提升最大［图 1-62（c）］（Zhong et al.，2011）。

半导体材料的表面由于存在不饱和键（悬挂键），受此影响，其禁带中会出现表面能级，此即表面态。如前所述，纳米化是解决 α-Fe_2O_3 光阳极空穴扩散长度短这一缺点的重要策略，然而纳米化也会引入过多的表面态。虽然众多研究人员关注了 α-Fe_2O_3 光阳极的表面态，但表面态的本质、来源以及影响存在争议。例如，王连洲课题组发现，一定量的氧空位可以作为施主缺陷显著提升 α-Fe_2O_3 光阳极的性能；而当氧空位浓度过高时，一方面，会在 α-Fe_2O_3 光阳极表面引入大量的缺陷，继而提升其产氧反应动力学速率；另一方面，过多的氧空位充当了表面缺陷态，引发了严重的载流子复合（Wang et al.，2019）。格拉兹尔课题组采用 ALD 技术在花椰菜结构的 α-Fe_2O_3 光阳极表面引入了一层 Al_2O_3 覆盖层，钝化了 α-Fe_2O_3 光阳极的表面态，继而使其开启电位负移了 100 mV（Formal et al.，2011）。后续该课题组发现，化

学浴沉积制备的 Ga_2O_3 覆盖层可有效释放超薄 α-Fe_2O_3 与 FTO 衬底之间的晶格应力，并降低超薄 α-Fe_2O_3 光阳极的表面态密度；最终该 α-Fe_2O_3/Ga_2O_3 光阳极的开启电位相比 α-Fe_2O_3 负移了 200 mV（Hisatomi et al.，2011）。邹志刚课题组证实，将 α-Fe_2O_3 光阳极置于 NaCl 溶液中进行光电化学处理，也可以钝化其表面态（Zhang et al.，2012）。

综上可知，影响 α-Fe_2O_3 光阳极太阳能转换效率的因素主要有：①薄膜电导率低；②空穴扩散长度短；③表面产氧反应动力学速率缓慢；④表面与界面缺陷密度大。目前针对 α-Fe_2O_3 光阳极的研究主要围绕上述几个方面展开。除了 α-Fe_2O_3 这一热点光阳极材料，邹志刚课题组首次证实亚稳相的 β-Fe_2O_3 也可用于构建光阳极，并且其理论 STH 效率高达 20.9%（α-Fe_2O_3 为 15.5%）（Zhang et al.，2020）。后续该课题组采用喷雾热裂解法在 FTO 衬底上制备了亚稳相 β-Fe_2O_3 薄膜，由于 β-Fe_2O_3 与 FTO 衬底之间存在界面应力，该 β-Fe_2O_3 薄膜展现了优异的热及激光辐照稳定性，且在模拟太阳光辐照下可稳定工作 110 h（Li et al.，2021）。

（2）$BiVO_4$ 光阳极。$BiVO_4$ 的常见晶体结构共三种，即四方白钨矿、单斜白钨矿及锆石。其中，单斜白钨矿结构的 $BiVO_4$ 带隙最小（2.4 eV），具备可见光吸收能力，其理论 STH 效率为 9.2%。工藤（Kudo）课题组于 1998 年首次报道了 $BiVO_4$ 粉末用于光催化分解水反应（Kudo et al.，1998）。2003 年，佐山（Sayama）课题组首次报道了 $BiVO_4$ 薄膜在光电化学分解水反应中的应用（Sayama et al.，2003）。目前影响 $BiVO_4$ 光阳极太阳能转换效率的因素主要有：①薄膜电导率低；②表面产氧反应动力学速率缓慢；③光电化学稳定性差。

2011 年，邹志刚课题组首次以 Mo 和 W 作为施主杂质掺杂于 $BiVO_4$ 光阳极的 V 位点。研究结果表明，Mo 和 W 的掺杂（原子比为 3）使得 $BiVO_4$ 光阳极的载流子浓度分别提升了约 9 倍和 2.7 倍。相应地，Mo 和 W 的引入使得 $BiVO_4$ 光阳极在 1.23 V *vs.* RHE 处的光电流密度从约 0.2 mA/cm^2 分别提升至约 2.2 mA/cm^2 和 1.45 mA/cm^2，其中 Mo 掺杂 $BiVO_4$ 光阳极的性能是当时的世界纪录（Luo et al.，2011）。克罗尔（Krol）课题组采用喷雾热裂解法在 $BiVO_4$ 光阳极中引入了梯度分布的 W 杂质，其掺杂浓度由体相到表面逐渐降低；该 W 梯度掺杂的 $BiVO_4$ 光阳极其内建电场分布于整个电极，因而可实现光生载流子的定向、高效传输与分离［图 1-63（a）］（Abdi et al.，2013）。通过贫氧氛围热处理或电化学还原等手段在 $BiVO_4$ 光阳极中引入氧空位，也能有效提升其载流子浓度与薄膜电导率。

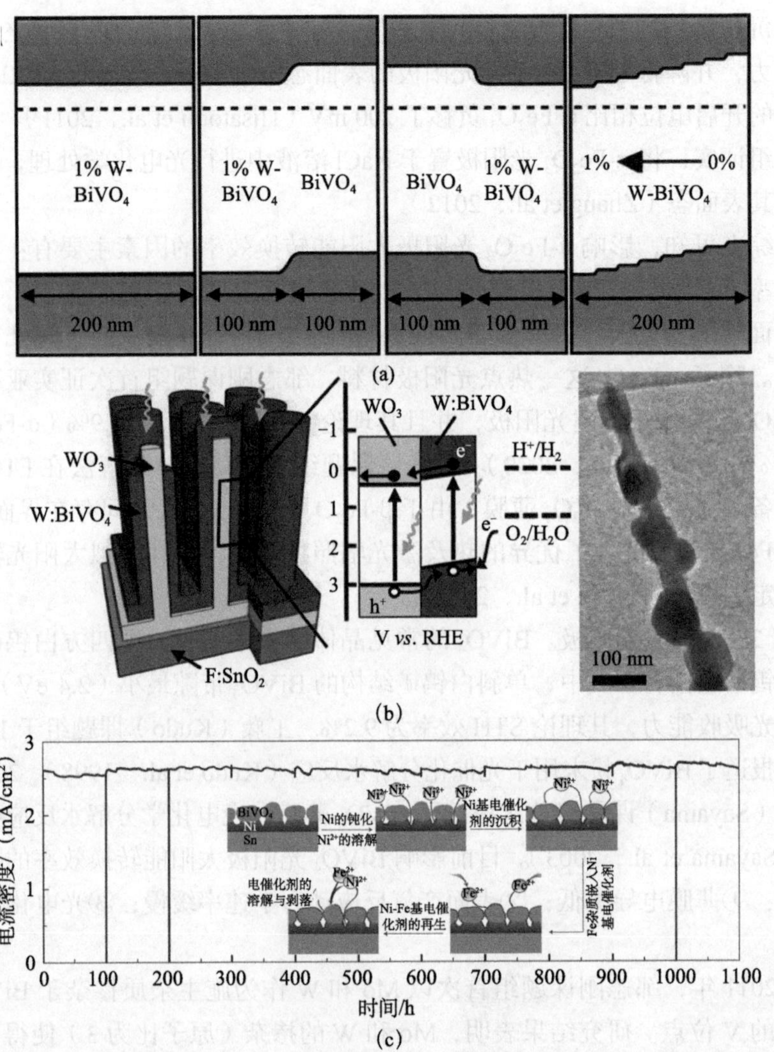

图 1-63 (a) 不同掺杂模式在 BiVO$_4$ 光阳极体相形成的能级排列状况；(b) WO$_3$/BiVO$_4$ 纳米线阵列光阳极的结构示意图、能带示意图及 TEM 表征图；(c) Ni-Fe 基电催化剂修饰的 Sn/Ni/Mo:BiVO$_4$ 光阳极在 1 mol/L 硼酸钾缓冲液（pH = 9）中的电流密度 - 时间曲线，外加偏压为 0.6 V *vs*. RHE，插图为 Ni-Fe 基电催化剂的原位生成和再生机制

构筑异质结也是提升 BiVO$_4$ 光阳极电荷分离效率的重要策略，而将 WO$_3$ 与 BiVO$_4$ 组合形成Ⅱ型异质结最常见。郑晓琳课题组在 WO$_3$ 纳米线表面引入 BiVO$_4$ 构建了核壳结构的 WO$_3$/BiVO$_4$ 异质结光阳极；得益于二者之间的能带排列，光生电子可由 BiVO$_4$ 向内注入 WO$_3$ 纳米线，继而快速输运至导电

衬底，而 $BiVO_4$ 的空穴则向外传输至固液界面，由此该异质结光阳极实现了光生电子和空穴的高效分离与迁移［图 1-63（b）］（Rao et al.，2014）。然而，WO_3 作为导电支架或衬层，物化性质变化如何影响 $WO_3/BiVO_4$ 异质结光阳极性能输出仍有待阐明。邹志刚课题组通过构建结晶性与厚度变化的 WO_3 衬层及相应的 $WO_3/BiVO_4$ 异质结光阳极，分析了 WO_3 衬层在 $WO_3/BiVO_4$ 异质结光阳极中的作用机制：① WO_3 衬层的覆盖度要足够高，以此确保 FTO 衬底与 $BiVO_4$ 及电解液的完全隔离；WO_3 衬层对 FTO 的完全覆盖能最大限度地确保 $BiVO_4$ 薄膜的无应力成核与生长，与此同时，能避免水氧化中间产物与 FTO 直接接触引发的逆反应。② WO_3 衬层的表面缺陷密度要尽量低，以此确保其与 $BiVO_4$ 薄膜接触后形成高质量的Ⅱ型异质结。③ WO_3 衬层的体相电导率要尽量高，以此确保电子在传输至导电衬底 FTO 的过程中电阻损失最小。④ WO_3 衬层的引入有利于形成高质量、低表界面缺陷密度的 $BiVO_4$ 薄膜，这能促进表面水氧化反应的电荷转移过程（Feng et al.，2021）。

与 α-Fe_2O_3 光阳极类似，负载 OER 电催化剂是提升 $BiVO_4$ 光阳极产氧反应动力学速率的有效手段。理想条件下，用于光阳极的 OER 电催化剂需满足以下几个条件：①其本征 OER 活性要高，产氧过电位低；②其在光阳极表面的负载不影响光阳极对入射光的捕获；③其与光阳极接触后能形成合适的能级排列，以高效提取光阳极的光生空穴；④具备优良的空穴传导能力，以降低电荷传输损失；⑤廉价易得，稳定性好，易于在光阳极负载或成膜。诺塞拉课题组开发的 Co-Pi、硼酸钴（Co-Bi）、硼酸镍（Ni-Bi）基 OER 电催化剂因廉价易得、易于制备、本征活性高、可自修复等优点，被广泛用作 $BiVO_4$ 光阳极的 OER 电催化剂（Kanan & Nocera，2008；Esswein et al.，2011）。崔景申课题组发现，电辅助光沉积法制备的 FeOOH 基 OER 电催化剂对 $BiVO_4$ 光阳极性能的提升远高于同等条件下的 Co-Pi，而 FeOOH 的本征 OER 活性低于 Co-Pi。由此，崔景申等认为，OER 电催化剂与光阳极的界面质量、能级分布等对最终复合光阳极的性能有重要影响（Seabold & Choi，2012）。后续，崔景申课题组构建了 $BiVO_4$/FeOOH/NiOOH 复合光阳极，一方面优化了 $BiVO_4$ 光阳极与 OER 电催化剂之间的界面质量，另一方面确保了电极表面高的 OER 活性。最终该光阳极在 0.6 V *vs.* RHE 处的光电流密度达到 2.73 mA/cm^2，ABPE 达到 1.75%，是当时的世界纪录（Kim & Choi，2014）。也有研究者认为，OER 电催化剂并没有加快光阳极的表面产氧反应动力学速率，而是拓宽了光阳极的空间电荷层厚度，使其体相电荷分离效率得到了提升（Barroso et al.，2012）。邹志刚课题组构建了 $BiVO_4/AgO_x/NiO_x$ 复合光阳极，通过电化学阻抗、光辅助开尔文探针显微镜、开路电压等手段证实，OER 电催化剂不仅能加快 $BiVO_4$ 光阳极的产氧反应动力学速率，而且能增强 $BiVO_4$ 光阳极的能带弯曲，继而提升其体相电荷分离效率（Hu et al.，2018）。

相比于 α-Fe_2O_3 光阳极，$BiVO_4$ 光阳极的光电化学稳定性较差，需以合适的手段抑制其表面腐蚀或钝化反应。表面负载 OER 电催化剂能够大幅度提升 $BiVO_4$ 光阳极的光电化学稳定性。堂免一成课题组采用颗粒转移法制备了 Sn/Ni/Mo:$BiVO_4$ 光阳极，一方面，Mo:$BiVO_4$ 颗粒经由高温烧结（800 ℃）展现了优异的抗光电化学腐蚀能力；另一方面，电极中的 Ni 可原位释放 Ni^{2+}，并随后沉积在 Mo:$BiVO_4$ 颗粒表面形成高效的 OER 电催化剂。最终该光阳极在 0.6 V *vs.* RHE 的偏压及大于 2.5 mA/cm^2 的光电流密度下可稳定工作 1100 h，是当时的世界纪录［图 1-63（c）］（Kuang et al.，2017）。崔景申课题组则从电解液组分入手，认为 $BiVO_4$ 光阳极中 V 的溶解流失是造成其光电化学稳定性不佳的原因。基于此，他们使用 V^{5+} 饱和的硼酸钾缓冲溶液（pH = 9.3）作为支持电解液，使得 $BiVO_4$/FeOOH/NiOOH 复合光阳极在约 3.2 mA/cm^2 的光电流密度下可稳定工作 450 h（Lee & Choi，2018）。

2. 光阴极材料的研究现状

目前，用于光阴极的材料主要有 Si、金属氧化物半导体、Ⅲ-Ⅴ族半导体、铜基硫族化合物和Ⅱ-Ⅵ族半导体。

1）Si 基光阴极材料

因地球表面储量丰富、带隙小、能级位置适宜，p-Si 是构筑光阴极最受欢迎的材料。路易斯课题组于 1979 年首次报道了用于水分解制氢的 p-Si 基光阴极，在 514.5 nm 波长的激光（氩离子激光器）照射下 ABPE 达到 2.4%（Lewis et al.，1979）。目前影响 p-Si 基光阴极太阳能转换效率的因素主要有：①表面产氢反应动力学速率缓慢；②光电化学稳定性差；③开启电位不佳。

1980 年，路易斯课题组在 p-Si 表面担载 Pt 作为析氢反应电催化剂，该 p-Si/Pt 光阴极在 632.8 nm 波长的激光（氦氖激光器）照射下 ABPE 值可达 5%（Lewis et al.，1980）。除了贵金属基 HER 电催化剂，非贵金属材料（如 MoS_2、Co-P）（Kempler et al.，2018）等也可用于提升 Si 基光阴极的 HER 活性以及光电化学稳定性。例如，哈拉米洛等通过磁控溅射金属 Mo 而后硫化的手段，在平面结构的 pn^+-Si 基光阴极表面制备了 MoS_2 薄层，该光阴极在 0 V *vs.* RHE 处的光电流密度达到了−17 mA/cm^2，且可在 0.5 mol/L H_2SO_4 电解液中稳定运行 100 h［图 1-64（a）］（Benck et al.，2014）。由于大多数 Si

基光阴极只适合在酸性电解液中工作，为了和其他光阳极组合形成 p/n 双光子光电化学水分解池，有必要开发适用于中性乃至碱性电解液的 Si 基光阴极。戴宏杰课题组以电子束蒸发技术在 p-Si 表面先后沉积了 15 nm 的 Ti 和 5 nm 的 Ni，该 p-Si/Ti/Ni 光阴极在 pH = 9.5 的硼酸钾缓冲溶液及−10 mA/cm^2 的光电流密度下可稳定约 12 h；在 1 mol/L KOH 电解液及−10 mA/cm^2 的光电流密度下，该光阴极连续运行 12 h 后损失了约 100 mV 的光电压（Feng et al., 2015）。胡喜乐课题组则采用磁控溅射技术在无定形 Si/AZO①/TiO_2 光阴极

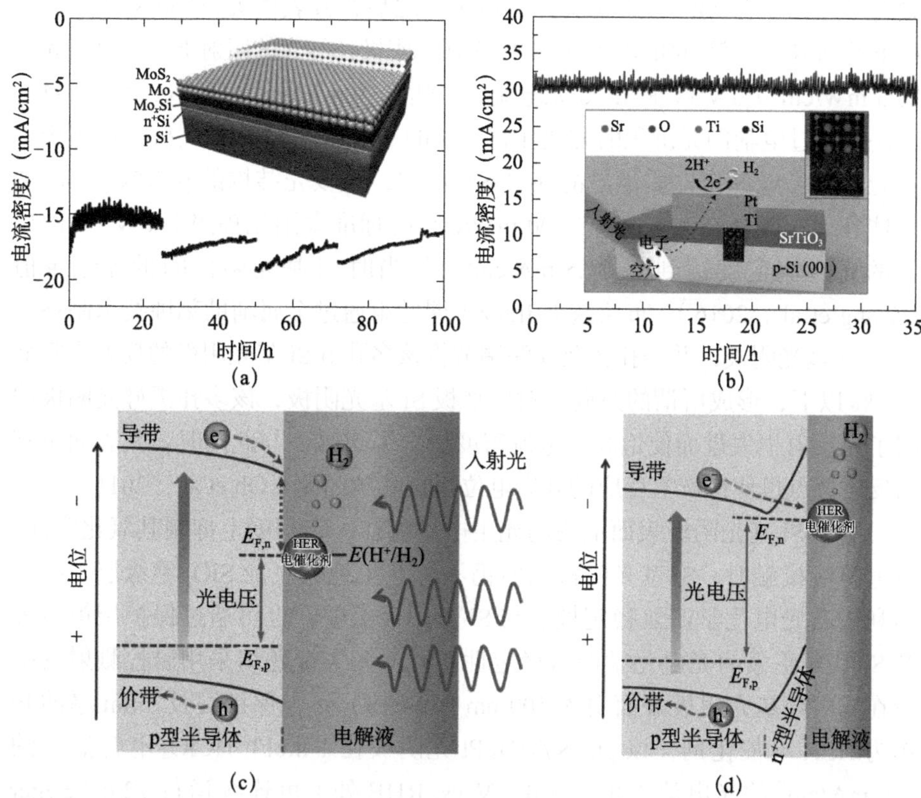

图 1-64　(a) pn$^+$-Si/MoS_2 光阴极在 0.5 mol/L H_2SO_4 电解液中的电流密度－时间曲线，外加偏压为 0 V *vs*. RHE，插图为电极构造示意图；(b) p-Si/$SrTiO_3$/Ti/Pt 光阴极在 0.5 mol/L H_2SO_4 电解液中的电流密度－时间曲线，外加偏压为 0 V *vs*. 银 / 氯化银电极，插图为电极构造示意图；导带位置高（负）的 p 型半导体与 (c) 电解液直接接触及 (d) 表面引入 n$^+$ 发射层形成 pn$^+$-Si 掩埋结的光电压产生状况

其中，$E_{F,n}$：电子准费米能级；$E_{F,p}$：空穴准费米能级

① Al 掺杂 ZnO（AZO）。

表面负载了 Mo_2C 薄层，该无定形 Si/AZO/TiO_2/Mo_2C 光阴极在 0.1 mol/L H_2SO_4 和 1 mol/L KOH 电解液中均能实现−11 mA/cm^2 的光电流密度（0 V *vs.* RHE 处）；在 1 mol/L KOH 电解液及−10 mA/cm^2 的光电流密度下该光阴极可稳定约 1 h（Morales-Guio et al.，2015）。

如前所述，微纳阵列结构的电极拥有光捕获能力强、少数载流子传输距离短、比表面积大等优势。由于 Si 是一种间接带隙半导体，光吸收效率低，因此也有研究者致力于构建微纳阵列结构的 Si 基光阴极。乔根多夫等利用光刻技术在 Si 片上刻蚀形成 Si 微米柱阵列，之后在其表面沉积 Mo_3S_4 团簇作为 HER 电催化剂。该 p-Si/Mo_3S_4 微米柱阵列光阴极在红光的照射下（$\lambda > 620$ nm，28.3 mW/cm^2）0 V *vs.* RHE 处光电流密度达到−9 mA/cm^2（Hou et al.，2011）。金松等通过电辅助光沉积的方法在竹笋状的 Si 纳米柱阵列光阴极表面沉积了透光良好、电催化性质优异的 $NiCoSe_x$ 薄层，该核壳结构的 p-Si/$NiCoSe_x$ 纳米柱阵列光阴极在 AM 1.5 G 100 mW/cm^2 的标准太阳光辐照下 0 V *vs.* RHE 处的光电流密度达到了−37.5 mA/cm^2，是当时 Si 基光阴极的世界纪录值（Zhang et al.，2016）。布兰兹（Branz）课题组通过金属辅助刻蚀技术制备了多孔 Si 基光阴极，并经孔径及孔深调节将该多孔 p-Si 基光阴极的反射率降低至 2% 以下，形成所谓的黑硅。对比平板 Si 基光阴极，该多孔黑硅光阴极抑制了光反射损失进而使饱和光电流密度提升了 20%，与此同时，其表面丰富的反应位点也使得产氢反应的开启电位正移了 70 mV（Oh et al.，2011）。

虽然 Si 基光阴极表面汇集的光生电子能在一定程度上抑制其氧化腐蚀，与电解液接触时，Si 基光阴极表面仍不可避免地会生成 SiO_2 绝缘层，继而不利于其光电化学产氢稳定性。与 Si 光阳极类似，也可通过保护层策略提升 Si 基光阴极的光电化学产氢稳定性。乔根多夫课题组采用磁控溅射的手段在 pn^+-Si 基光阴极表面引入 100 nm 厚的 TiO_2 作为保护层及 5 nm 厚的 Pt 作为 HER 电催化剂，该 pn^+-Si/TiO_2/Pt 光阴极在 1 mol/L $HClO_4$ 电解液及约−20 mA/cm^2 的光电流密度下（0.3 V *vs.* RHE 处）可连续运行 72 h（Seger et al.，2013）。后续该课题组采用 ALD 技术在 pn^+-Si 基光阴极表面引入 100 nm 厚的 TiO_2（并经 400 ℃真空热处理）作为保护层及 Pt 纳米颗粒作为 HER 电催化剂，在红光照射下（$\lambda > 635$ nm，38.6 mW/cm^2），该 pn^+-Si/ALD-TiO_2/Pt 光阴极在 1 mol/L $HClO_4$ 电解液及约−21 mA/cm^2 的光电流密度下（0.3 V *vs.* RHE 处）可稳定运行 14 d（336 h）不衰减，该光阴极运行 14 d 后性能的逐步衰减被证实主要来源于 Pt 纳米颗粒的流失（Seger et al.，2013）。沈明荣课题组在 pn^+-Si 金字塔光阴极表面先后引入 Pt 纳米颗粒作为 HER 电催化剂及 ALD-TiO_2 作为保护层，得益于 Pt 纳米颗粒在电极表面的均匀分布、Pt 纳米颗粒与 Si 的直接接触、ALD-TiO_2 对 Pt 纳米颗粒的黏附与保护作用，以及 ALD-TiO_2 对 Si 表面的钝化与保护作用，该 pn^+-Si/Pt/ALD-TiO_2 光阴极的 ABPE 值达到 10.8%，且在 1 mol/L $HClO_4$ 电解液及−10 mA/cm^2 的光电流密度下可稳定运行 7 d（168 h）不衰减（Fan et al.，2017）。季力课题组以分子束外延技术在 p-Si 表面引入 4 个晶胞厚度的 $SrTiO_3$ 外延薄膜，而后以光刻外加电子束蒸发手段引入网状图案的 Ti 与 Pt 纳米薄膜。由于 Si 与 $SrTiO_3$ 外延薄膜之间界面质量高、晶格匹配度高、导带位置匹配，以及 $SrTiO_3$ 的单晶特质与稳定性，该 MIS 结构的 p-Si/$SrTiO_3$/Ti/Pt 光阴极 ABPE 值可达 4.9%，在0.5 mol/L H_2SO_4 电解液及约−30 mA/cm^2 的光电流密度下（0.2 V *vs.* RHE 处）可稳定运行 35 h 不衰减［图 1-64（b）］（Ji et al.，2015）。贝彻（Boettcher）课题组以喷雾热裂解法在织构化的 pn^+-Si/Ti 光阴极表面制备了 F:SnO_2（F 掺杂 SnO_2）及 TiO_2 薄膜作为保护层，电子束蒸发技术制备了 2 nm 厚的 Ir 薄膜作为 HER 电催化剂；在 1 mol/L KOH 电解液中，该 pn^+-Si/Ti/F:SnO_2/TiO_2/Ir 光阴极 ABPE 值可达 10.9%，0 V *vs.* RHE 处光电流密度可达−35 mA/cm^2，且可稳定运行 24 h（Kast et al.，2014）。

造成 Si 基光阴极开启电位不佳（过负）的原因主要有两个：一是 Si 基光阴极的产氢反应动力学速率缓慢，这可以通过负载高效 HER 电催化剂的途径予以解决；二是 Si 基光阴极的光电压值过小。理论上，光电极的光电压值取决于其内建电位，实际工作时光电压的产生过程与电荷分离效率密切相关。因此，为了在 Si 基光阴极中获得最大化的光电压值，需要对其内建电位进行优化并尽可能抑制载流子在电极体相及表界面处的复合。路易斯课题组指出，由于 p-Si 的费米能级与 H^+/H_2 的标准电极电位差值小，p-Si/ 电解液结光阴极的内建电位小、空间电荷区厚度薄，不利于光电压的产生以及光生电荷的高效分离。针对上述问题，路易斯课题组提出在 p-Si 表面引入 n^+-Si 发射层以形成 pn^+-Si 掩埋结，可大幅提升 p-Si 基光阴极的内建电位及空间电荷区厚度，继而获得优化的光电压值及 ABPE 效率［图 1-64（c）、图 1-64（d）］（Boettcher et al.，2011；Feng et al.，2020）。除了形成 p-n 同质结，构建 MIS 结、异质结等高质量固体结，以及表面修饰铁电薄膜或偶极分子，该方法也可用来提升 Si 基光阴极的电荷分离效率及光电压值。

2）金属氧化物半导体光阴极材料

一些金属氧化物半导体，如二元氧化物 Cu_2O（2.0 eV），三元氧化物 $CaFe_2O_4$（1.9 eV）、$CuNb_3O_8$（1.5 eV）、$CuFeO_2$（1.5 eV）和 $LaFeO_3$（2.0～

2.6 eV），因结构中存在金属空位而表现为 p 型半导体。同时，它们储量丰富、价格低廉、易于制备，是一类极具潜力的光阴极材料。然而，这些 p 型氧化物半导体的光吸收系数较小、载流子迁移率低、寿命短，最终构建光阴极时太阳能转换效率往往不高（Matsumoto et al.，1987；Viswanathan et al.，2014；Prévot et al.，2015；Joshi & Maggard，2012）。

Cu_2O 是 p 型氧化物半导体的典型代表，其理论 STH 效率为 18%，理论光电流密度为 14.7 mA/cm^2。1999 年，钟（de Jongh）课题组首次报道了 Cu_2O 多晶光阴极可实现氧气和甲基紫精的光电化学还原（de Jongh et al.，1999）。用于光电化学产氢反应时，Cu_2O 光阴极的光生电子容易还原自身晶格中的 Cu^+，导致其发生不可逆的腐蚀降解，光电化学稳定性差。格拉兹尔课题组在提升 Cu_2O 光阴极的稳定性方面做了大量的工作。2011 年，该课题组以 ALD 技术在 Cu_2O 表面先后沉积了 AZO 缓冲层与 TiO_2 保护膜，并以电沉积法引入 Pt 纳米颗粒作为 HER 电催化剂。由于 Cu_2O 吸光层、AZO 缓冲层、TiO_2 保护膜、H^+/H_2 的标准电极电位之间形成的级联能级排列能有效促进光生电子从 Cu_2O 至电解液的定向传输与转移，以及 Pt 优异的 HER 电催化效率，该 FTO/Au/Cu_2O/AZO/TiO_2/Pt 光阴极在标准太阳光辐照下 0 V *vs.* RHE 处光电流密度可达−7.6 mA/cm^2，且 1 h 稳定性测试后仍有可观的光电流密度［图 1-65（a）］（Paracchino et al.，2011）。2012 年，该课题组确认了 Cu_2O 吸光层与 AZO 缓冲层之间形成了高质量的 p-n 结，而对 ALD 工艺优化获得半晶态的 TiO_2 保护膜后 Cu_2O 光阴极的光电化学稳定性得到大幅提升，10 h 稳定性测试后光电流密度仍有初始值的 62%（Paracchino et al.，2012）。基于上述工作，2014 年，胡喜乐课题组以 MoS_{2+x} 作为 HER 电催化剂，最终 FTO/Au/Cu_2O/AZO/TiO_2/MoS_{2+x} 光阴极在 pH = 4 的邻苯二甲酸氢钾缓冲溶液及−4.5 mA/cm^2 的光电流密度下可连续运行 10 h，几乎无衰减［图 1-65（b）］（Morales-Guio et al.，2014）。

2015 年，德洛奈（Delaunay）课题组以 ALD-Ga_2O_3 作为 Cu_2O 吸光层与 ALD-TiO_2 保护层之间的缓冲层；由于 Cu_2O 与 Ga_2O_3 之间导带匹配度高，形成了高质量的 p-n 结，该 Cu/Cu_2O/Ga_2O_3/TiO_2 纳米线光阴极经表面沉积 Pt 基 HER 电催化剂后其开启电位可达 1.02 V *vs.* RHE，0 V *vs.* RHE 处光电流密度可达−2.95 mA/cm^2，且在 pH = 4.26 的磷酸钾缓冲溶液及约−3.3 mA/cm^2 的光电流密度下可稳定运行 2 h 不衰减（Li et al.，2015）。与上述工作类似，2018 年，格拉兹尔课题组采用阳极氧化法制备了 Cu_2O 纳米线阵列，后通过 ALD 技术先后引入 Ga_2O_3 层和 TiO_2 保护层；该 Cu_2O/Ga_2O_3/TiO_2 纳米线阵列光阴极

经表面沉积 RuO_x 基 HER 电催化剂后在 pH = 5 的磷酸钠缓冲溶液及 0.5 V *vs*. RHE 偏压下连续工作 100 h，光电流密度从−6 mA/cm² 仅衰减至−5.5 mA/cm²，是目前的世界纪录［图 1-65（c）、图 1-65（d）］（Pan et al.，2018）。相比于 Cu_2O 光阴极，其他氧化物基光阴极的太阳能转换效率及光电化学稳定性均有待进一步优化。前述针对 Cu_2O 光阴极的电极构建及保护层策略，将为提升其他氧化物基光阴极的性能及稳定性提供有价值的参考。

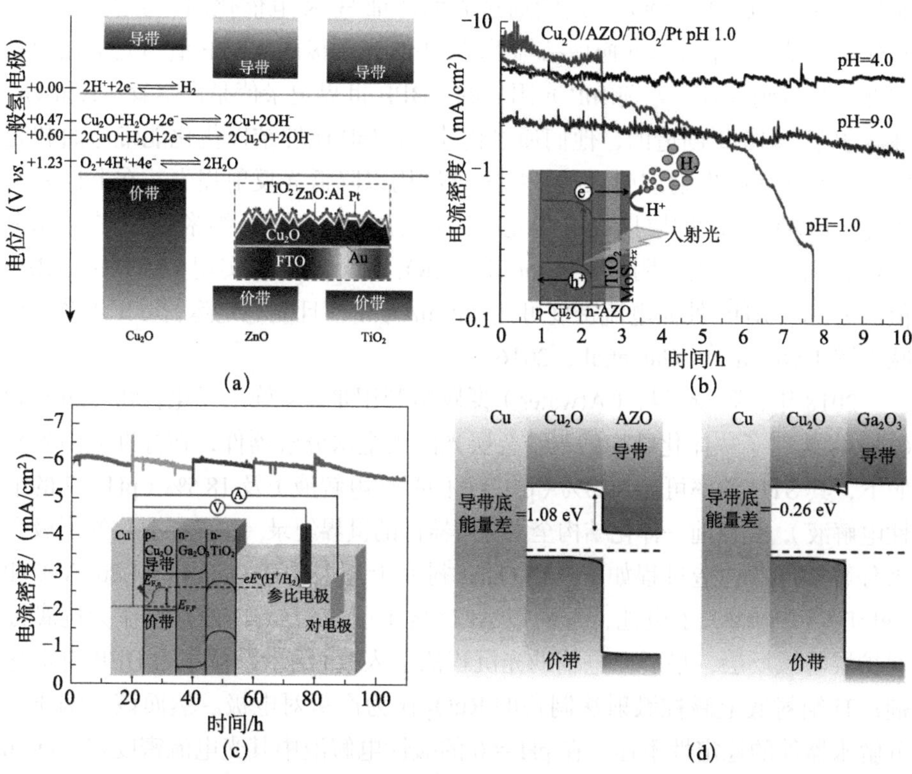

图 1-65　(a) FTO/Au/Cu_2O/AZO/TiO_2/Pt 光阴极的能级排列状况，插图为电极构造示意图；(b) FTO/Au/Cu_2O/AZO/TiO_2/MoS_{2+x} 光阴极在不同 pH 值电解液中的电流密度－时间曲线，外加偏压为 0 V *vs*. RHE，插图为电极构造示意图；(c) 在 pH = 5 的磷酸钠缓冲溶液中 Cu/Cu_2O/Ga_2O_3/TiO_2/RuO_x 纳米线阵列光阴极的电流密度－时间曲线，外加偏压为 0.5 V *vs*. RHE，插图为电极构造示意图；(d) Cu_2O/AZO 及 Cu_2O/Ga_2O_3 异质结的能带对齐状况

其中，$E_{F,n}$：电子准费米能级；$E_{F,p}$：空穴准费米能级

3）Ⅲ-Ⅴ族半导体光阴极材料

由于Ⅲ-Ⅴ族半导体材料在水溶液中稳定性欠佳，在氧化性条件下更易发生腐蚀反应，因此，将Ⅲ-Ⅴ族半导体设计成表面汇集光生电子的光阴极而非表面汇集光生空穴的光阳极，更有利于其在光电催化水分解领域的应用（Malizia et al.，2014；Standing et al.，2015；Gu et al.，2016）。

1980 年，赫勒（Heller）等首次报道了 p-InP 单晶可用于光电化学分解水制氢（Heller et al., 1980）。后续通过表面修饰 HER 电催化剂、引入保护层、构建微纳结构、制备 p-n 掩埋结等手段，InP 光阴极的 ABPE 值与光电化学稳定性逐步得到提升。当前 InP 光阴极的 ABPE 世界纪录值是由亨森（Hensen）课题组于 2016 年创造的，他们通过构建 pn^+-InP 掩埋结以优化能带弯曲和光电压，在 n^+-InP 发射层引入纳米柱阵列结构以降低光反射损失，在电极表面覆盖 ALD-TiO_2 保护层以提升稳定性，电辅助光沉积 Pt 纳米颗粒作为 HER 电催化剂；最终该光阴极在 1 mol/L $HClO_4$ 电解液中可实现 15.8% 的 ABPE 值，0 V *vs.* RHE 处光电流密度可达−28 mA/cm^2，且能稳定运行 6 h 几乎无衰减［图 1-66（a）］（Gao et al.，2016）。

2018 年，阿特沃特（Atwater）课题组利用Ⅲ-Ⅴ族三元化合物 GaInP 和 GaInAs 构建了一体化结构的双结叠层光阴极全水分解器件，在标准太阳光辐照下，其 STH 效率可达 19.3%（pH = 0 的酸性电解液）及 18.5%（pH = 7 的中性电解液），是目前一体化结构全水分解器件的世界纪录。该双结叠层光阴极全水分解器件的构造过程如下：在 GaAs 衬底上逐层构建 GaInAs（1.26 eV）和 GaInP（1.78 eV）子电池，继而以 ALD 技术引入锐钛矿 TiO_2 膜作为抗腐蚀保护层与减反层，最后以电辅助光沉积法引入 Rh 纳米颗粒作为 HER 电催化剂；Ti 箔衬底上磁控溅射法制备的 RuO_2 作为产氧对电极。然而该无偏压全分解水器件的稳定性不佳，在 pH = 0 的酸性电解液中其光电流密度在 40 min 内即从−15 mA/cm^2 迅速衰减至−5 mA/cm^2；在 pH = 7 的中性电解液中，该器件的稳定性有所提升，连续运行 90 min 其光电流密度从约−11 mA/cm^2 下降至约−8 mA/cm^2［图 1-66（b）］（Cheng et al.，2018）。

InGaN 也是一类典型的Ⅲ-Ⅴ族三元化合物。开发 InGaN 用于光电催化水分解领域有以下几方面的优势：①通过调控 In 与 Ga 的比例，$In_xGa_{1-x}N$（$0 \leqslant x \leqslant 1$）的禁带宽度可在 0.7～3.4 eV 调节，借此可实现对太阳光谱不同波段的吸收；②组分适宜的 InGaN 的能带位置可同时跨越 HER 和 OER 的电极电位，继而在热力学上满足无偏压分解水的要求；③ InGaN 可在 Si 衬底上实现外延生长，便于构建叠层结构或一体化结构的器件；④相比于 GaInP 和

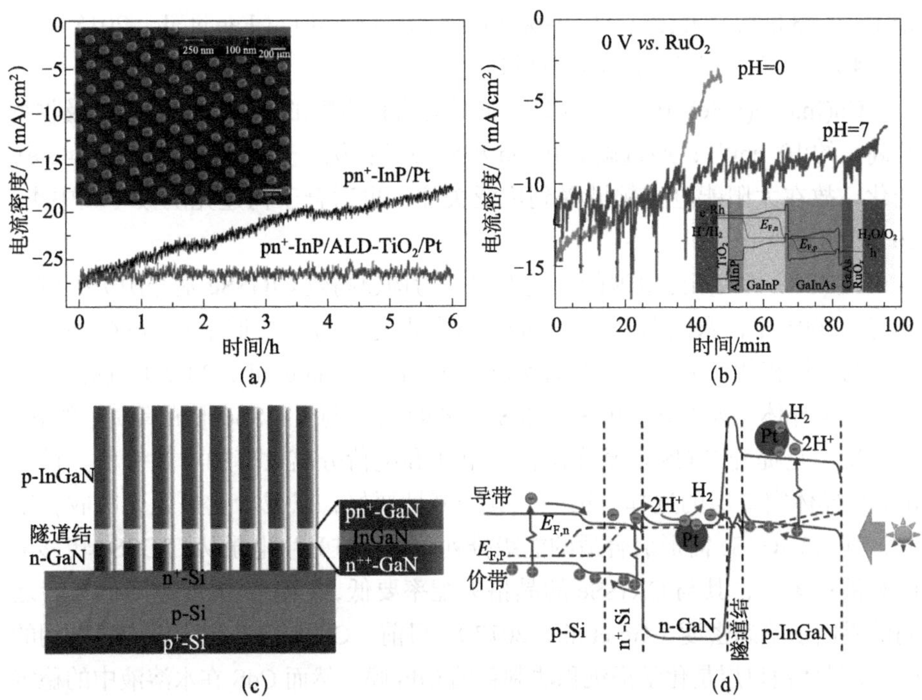

图 1-66 (a) pn$^+$-InP/ALD-TiO$_2$/Pt 光阴极在 1 mol/L HClO$_4$ 电解液中的电流密度－时间曲线，外加偏压为 0 V *vs*. RHE，插图为电极的 SEM 表征图；(b) RuO$_2$-GaAs/GaInAs/GaInP/AlInP/TiO$_2$/Rh 双结叠层光阴极全水分解器件在不同 pH 值电解液中的电流密度－时间曲线，外加偏压为 0 V *vs*. RuO$_2$，插图为电极能级结构示意图；(c) pn$^+$-Si/n-GaN/n^{++}-GaN/InGaN/p^{++}-GaN/p-InGaN 隧道结光阴极的电极构造示意图及其 (d) 能级结构示意图

其中，$E_{F,n}$：电子准费米能级；$E_{F,p}$：空穴准费米能级

GaInAs，InGaN 的抗化学及光电化学腐蚀能力更强，有望构筑稳定性优异的光电极。米泽田课题组通过分子束外延技术在 pn$^+$-Si 晶片表面分段生长了 n-GaN（150 nm）和 p-InGaN（600 nm）纳米线阵列，并在 n-GaN 与 p-InGaN 之间引入 n^{++}-GaN/InGaN/p^{++}-GaN 隧道结，由此构建了 pn$^+$-Si/n-GaN/n^{++}-GaN/InGaN/p^{++}-GaN/p-InGaN 隧道结光阴极。该光阴极中，顶端的 p-InGaN 纳米线阵列吸收波长小于 520 nm 的入射光，产生的光生电子还原质子（电解液）生成氢气，而光生空穴则经由隧道结转移至 n-GaN；底层的 pn$^+$-Si 薄膜捕获穿透 p-InGaN 纳米线阵列且波长小于 1.1 μm 的光子，其光生电子也转移至 n-GaN，部分光生电子与 p-InGaN 转移至此的空穴复合，剩余的光生电子则在 n-GaN 表面还原质子（电解液）生成氢气。最终该隧道结光阴极在 1.3 个太阳光强度下饱和光电流密度高达−40.6 mA/cm^2，ABPE 值可达 8.7%，且能

稳定运行 3 h 几乎无衰减［图 1-66（c）、图 1-66（d）］（Fan et al.，2015）。

4）铜基硫族化合物光阴极材料

$Cu(In,Ga)(S,Se)_2$ (CIGSSe) 是一类具有直接带隙的 p 型半导体，光吸收系数高（约 10^5 cm^{-1}）；通过调节 In/Ga 及 S/Se 的比例，其带隙可在 1.0～2.43 eV 变化，故在太阳能电池领域受到广泛关注，也适合于构建光阴极（关中杰，2016）。

CIGSSe 基太阳能电池的快速发展，为构建高效 CIGSSe 基光阴极提供了参考。2005 年，塞巴斯蒂安（Sebastian）课题组首次报道了 $Cu(In,Ga)Se_2$ 薄膜作为光阴极用于光电化学分解水制氢（Valderrama et al.，2005）。选择合适的 n 型半导体（如 CdS、In_2S_3、ZnS 和 ZnO 等）与 CIGSSe 形成高质量的 p-n 异质结，是提升 CIGSSe 基光阴极光电压和电荷分离效率的主要手段。针对 n 型半导体窗口层的选择，第一，其透光性要好，以减少竞争性光吸收；第二，其与 CIGSSe 需形成合适的能级排列，以促进光生电子从 CIGSSe 转移至电解液；第三，其与 CIGSSe 的晶格失配率要低，以最大限度地降低二者之间的界面缺陷态密度（Ge et al.，2017）。目前，CIGSSe 基光阴极中最常用的 n 型半导体窗口层是化学浴沉积法制备的 CdS 膜。然而 CdS 在水溶液中的稳定性不佳，需以保护层的手段解决 。格拉兹尔课题组采用 ALD 技术在 $CuInS_2$/CdS 异质结光阴极表面引入了致密的 AZO 和 TiO_2 保护层，外加 Pt 纳米颗粒作为 HER 电催化剂，该光阴极在 pH = 5 的缓冲液中及 0 V *vs.* RHE 处的光电流密度可达−2.2 mA/cm^2，且运行 2 h 光电流密度仅衰减了约 20%（Luo et al.，2015）。池田（Ikeda）课题组则通过射频磁控溅射技术在 $CuInS_2$/CdS 光阴极表面沉积了 TiO_2 薄膜作为保护层，后经 Pt 纳米颗粒修饰，该光阴极在 pH = 10 的磷酸钠缓冲液中及 0 V *vs.* RHE 处的光电流密度可达−13.0 mA/cm^2，ABPE 值达 1.82%，且能稳定运行 1 h 光电流密度几乎无衰减［图 1-67（a）］（Zhao et al.，2014）。巩金龙课题组采用超薄 Al_2O_3 层钝化了 CdS 窗口层与 TiO_2 保护层之间的界面缺陷，与此同时优化了 Pt 纳米颗粒 HER 电催化剂的空间分布与尺寸，最终该 $Cu(In,Ga)Se_2/CdS/Al_2O_3/TiO_2/Pt$ 光阴极在 pH = 6.8 的磷酸钠缓冲液中及 0 V *vs.* RHE 处的光电流密度可达−26 mA/cm^2，ABPE 值达 6.6%，且连续运行 8 h 后光电流密度仅衰减了 4.5%；而在 1 mol/L $HClO_4$ 电解液中，该光阴极的 ABPE 值可达 9.3%，连续运行 2 h 后光电流密度仅衰减了 6%［图 1-67（b）］（Chen et al.，2018）。

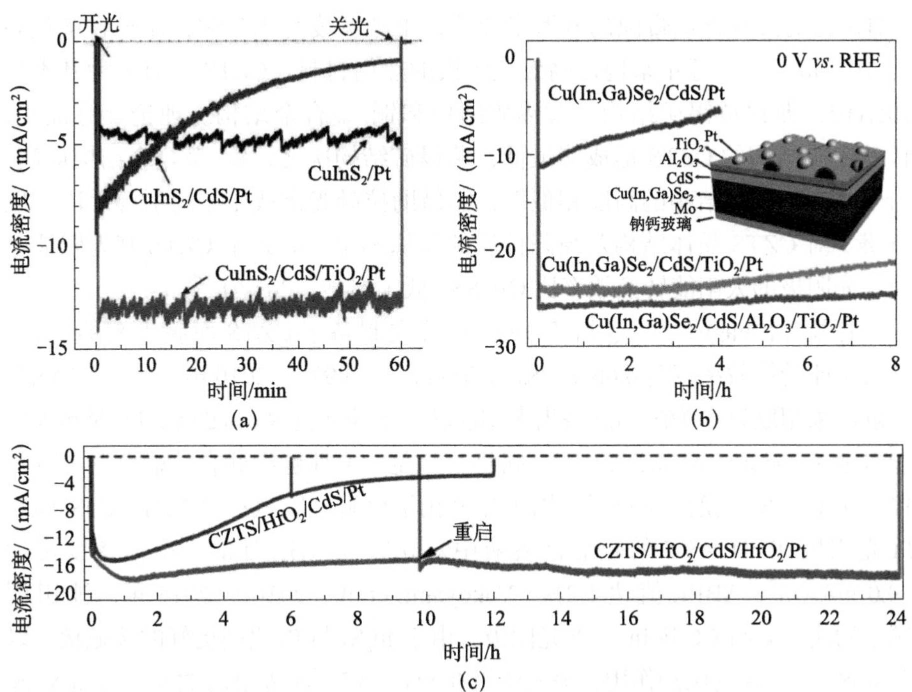

图 1-67 (a) $CuInS_2/CdS/TiO_2/Pt$、$CuInS_2/CdS/Pt$、$CuInS_2/Pt$ 光阴极在 pH = 10 的磷酸钠缓冲液中的电流密度－时间曲线，外加偏压为 0 V *vs*. RHE；(b) $Cu(In,Ga)Se_2/CdS/Al_2O_3/TiO_2/Pt$、$Cu(In,Ga)Se_2/CdS/TiO_2/Pt$、$Cu(In,Ga)Se_2/CdS/Pt$ 光阴极在 pH = 6.8 的磷酸钠缓冲液中的电流密度－时间曲线，外加偏压为 0 V *vs*. RHE，插图为电极构造示意图；(c) $CZTS/HfO_2/CdS/HfO_2/Pt$、$CZTS/HfO_2/CdS/Pt$ 光阴极在 pH = 6.5 的磷酸钠缓冲液中的电流密度－时间曲线，外加偏压为 0 V *vs*. RHE

当前针对 CIGSSe 基光阴极的研究有以下几个方面：①继续优化电极构型及保护层，以提升长时间运行稳定性；②开发新型 n 型半导体窗口材料，以取代有毒的 CdS；③提高结晶性，抑制杂相、缺陷的生成，以提升太阳能转换效率；④开发价格低廉的 HER 电催化剂，以取代 Pt 等贵金属材料；⑤构筑微纳结构，以降低光反射损失及缩短少数载流子传输距离；⑥优化 In/Ga 及 S/Se 的比例，以使其与其他光阳极或太阳能电池组合形成叠层全水分解器件后工作电流最大。实际上，光阴极的 ABPE 值越大并不意味着其与光阳极或太阳能电池组合形成全水分解器件的性能越高。

虽然 CIGSSe 基光阴极的性能和稳定性进展迅速，但其所含的昂贵金属元素 In、Ga 及有毒元素 Se 限制了此类光阴极的大规模应用。相比之下，Cu_2ZnSnS_4 (CZTS) 的原料来源广泛、无毒，适合大规模开发应用。与此同时，

CZTS 也是具有直接带隙的 p 型半导体，带隙宽度为 1.5 eV，光吸收系数高（$> 10^4$ cm^{-1}），是近年来构建高效光阴极的热门材料。CZTS 共有三种基本晶体结构，即锌黄锡矿结构、黄锡矿结构和铜-金合金结构。理论计算证实，锌黄锡矿结构的 CZTS 形成能最低，黄锡矿结构次之，铜-金合金结构最高。由于 CZTS 三种结构的高度相似性，需借助拉曼光谱或中子衍射等手段予以分辨。对 CZTS 晶体结构及缺陷性质的认识不足，导致了 CZTS 基太阳能电池及光阴极的太阳能转换效率比 CIGSSe 低（温鑫，2017）。

片桐（Katagiri）课题组于 1997 年首次制备了 CZTS 薄膜太阳能电池，其太阳能转换效率仅为0.66%（Katagiri et al.，1997）。2010年，米兹（Mitzi）课题组采用联氨作为溶剂制备出太阳能转换效率高达 9.6% 的 CZTS 薄膜太阳能电池（Todorov et al.，2010）。同年，堂免一成课题组联合片桐课题组首次报道了 CZTS 薄膜作为光阴极用于光电化学分解水制氢；该 CZTS/CdS/TiO_2/Pt 光阴极在 pH = 9.5 的硫酸钠溶液中及 0 V *vs.* RHE 处的光电流密度可达 −9.0 mA/cm^2，ABPE 值达 1.2%（Yokoyama et al.，2010）。2015 年，池田课题组构建了 CZTS/CdS/In_2S_3/Pt 光阴极，由于 In_2S_3 与 Pt 之间更好的接触状态以及 In_2S_3 对 CdS 的保护作用，该光阴极在 pH = 6.5 的磷酸钠缓冲液中及 0 V *vs.* RHE 处的光电流密度可达−9.3 mA/cm^2，ABPE 值达 1.63%，且连续运行 3 h 后光电流密度无明显衰减（Jiang et al.，2015）。2017 年，邹志刚课题组通过 Ge 固溶和高硫分压硫化的策略制得了高质量的 Ge-CZTS 薄膜，据此构建的 Ge-CZTS/CdS/In_2S_3/Pt 光阴极在 pH = 6.5 的磷酸钠缓冲液中及 0 V *vs.* RHE 处的光电流密度可达−11.1 mA/cm^2，ABPE 值达 1.7%，且连续运行 2 h 后光电流密度仅衰减了 9%（Wen et al.，2017）。2021 年，江丰课题组采用 ALD-HfO_2 薄膜钝化了 CdS 窗口层与 CZTS 光吸收层之间的界面缺陷，与此同时，在 CdS 窗口层表面沉积 ALD-HfO_2 薄膜作为保护层，最终该 CZTS/HfO_2/CdS/HfO_2/Pt 光阴极在 pH = 6.5 的磷酸钠缓冲液中及 0 V *vs.* RHE 处的光电流密度可达−18 mA/cm^2，ABPE 值达 5.57%，且连续运行 24 h 后光电流密度仅衰减了 3%；而在 pH = 3 的磷酸钠缓冲液中，该光阴极在 0 V *vs.* RHE 处光电流密度可达−28 mA/cm^2，ABPE 值高达 7.27%，是目前 CZTS 基光阴极的世界纪录值，其连续运行 3 h 后光电流密度衰减了 14% [图 1-67（c）]（Huang et al.，2021）。

显而易见，用于提升 CIGSSe 基光阴极性能和稳定性的策略在很大程度上也有望用于优化 CZTS 基光阴极。除了 CIGSSe 和 CZTS 这两类明星材料，铜基硫族化合物还包括诸如 Cu_3BiS_3、Cu_2BaSnS_4、Cu_2SnS_3、$Cu_2Sn_xGe_{1-x}S_3$ 等材料。前述的针对 CIGSSe 和 CZTS 基光阴极的优化策略将为这些新型铜基硫化物光阴极的性能及稳定性提升提供有价值的参考。例如，2021 年江丰课题组构建了 Cu_3BiS_3/CdS/ALD-TiO_2/Pt 光阴极，得益于 Cu_3BiS_3 吸光层与 CdS 窗口层之间形成的高质量 p-n 异质结，以及 ALD-TiO_2 对 Cu_3BiS_3 和 CdS 的钝化与保护作用，该光阴极在 pH = 6.5 的磷酸钠缓冲液中及 0 V *vs.* RHE 处的光电流密度可达−7 mA/cm^2，ABPE 值达 1.7%，是目前 Cu_3BiS_3 基光阴极的世界纪录值，且连续运行 10 h 后光电流密度仅衰减了 10%（Huang et al.，2021）。

5）Ⅱ-Ⅵ族半导体光阴极材料

如前所述，典型的Ⅱ-Ⅵ族半导体材料 CdS 用作光阳极时稳定性不佳，易发生氧化性腐蚀反应。虽然元素掺杂等手段较难使 CdS 的导电类型从 n 型转变为 p 型，然而可以预见，制备 p 型 CdS 继而用于构建表面汇集光生电子的光阴极，将有利于拓展 CdS 半导体在光电催化水分解领域的应用。肖旭东课题组通过 Cu 掺杂的手段获得了具有大比表面积、2.37 eV 直接带隙宽度的 p 型 CdS 薄膜；在其表面引入 Pt 纳米颗粒作为 HER 电催化剂后，该 Cu:CdS/Pt 光阴极于无牺牲剂存在的中性电解液中产氢光电流密度在 0 V *vs.* RHE 处达到了−0.6 mA/cm^2（Huang et al.，2014）。后续该课题组通过物理气相沉积技术制备了一系列 Cu 掺杂的 p 型 $CdS_{1-x}Se_x$ 薄膜，这些薄膜的带隙可在 1.69～2.40 eV 调节；基于这些薄膜构建的 Cu:$CdS_{1-x}Se_x$/Pt 光阴极中 Cu:$CdS_{0.7}Se_{0.3}$/Pt 的性能最高，在−0.12 V *vs.* RHE 处光电流密度可达−0.82 mA/cm^2，相比 Cu:CdS/Pt 光阴极提高了 37%（Ye et al.，2020）。

大桥（Ohashi）课题组于 1977 年首次报道了用于水分解制氢的 p-CdTe 单晶光阴极。相比于 CdS，CdTe 在诸多光电领域，特别是太阳能电池领域，应用前景广阔。CdTe 太阳能电池的快速发展，也为构建高效的 CdTe 基分解水光阴极提供了可靠途径（Ohashi et al.，1977）。堂免一成课题组通过近空间升华法在 FTO/Au/Cu 衬底上制备了 p-CdTe 多晶薄膜，后经 $CdCl_2$ 处理、化学浴沉积法制备 CdS 窗口层、真空蒸镀 2 nm 厚的 Pt 作为 HER 电催化剂，该 FTO/Au/Cu/p-CdTe/CdS/Pt 光阴极在 pH = 8 的磷酸钾缓冲液中 ABPE 值可达 3.7%，0 V *vs.* RHE 处光电流密度可达−22 mA/cm^2，且可持续运行约 70 min（Su et al.，2017）。

三、光电催化产氢材料研究目前面临的重大科学问题及前沿方向

在碳达峰、碳中和背景下，通过太阳能光电催化分解水制备绿氢是可行

的技术路线之一。目前，利用 InGaP（1.9 eV）/ GaAs（1.4 eV）/GaInNAs(Sb)（1.0 eV）叠层太阳能电池在聚焦太阳光下驱动两组串联的质子交换膜电解池，已能实现大于 30% 的 STH 效率。然而由于电催化剂的流失以及质子交换膜的破坏，该分解水系统的运行稳定性不佳（Jia et al.，2016）。将太阳能电池浸入电解液形成光阳极或光阴极结构的全水分解器件，也获得了较高的 STH 效率。例如，以Ⅲ-Ⅴ族化合物构建的双结叠层光阴极全水分解器件 RuO_2/GaAs（衬底）/GaInAs（1.26 eV）/ GaInP（1.78 eV）/AlInP/TiO_2/Rh，其 STH 效率可达 19.3%（Cheng et al.，2018）；双结叠层光阳极全水分解器件 CoP/GaAs（1.42 eV）/ InGaP（1.84 eV）/TiO_2/Ni 的 STH 效率达到了 10.0%（Sun et al.，2016）。除了材料成本高、制作工艺复杂，太阳能电池浸入电解液形成的全水分解器件的稳定性也有待提升。以非光伏级半导体构建光阴极或光阳极，继而组建 PV/n、PV/p、p/n 结构的双光子光电催化水分解池，其 STH 效率进展也比较缓慢（Ben-Naim et al.，2020）。究其原因，在很大程度上是非光伏级半导体形成的光阴极或光阳极性能欠佳造成的。例如，以非晶 Si 钝化的 n 型单晶 Si 作为光阴极（外加 ALD-TiO_2 作为保护层，Pt 作为 HER 电催化剂），$BiVO_4$/FeOOH/NiOOH 作为光阳极，巩金龙课题组构建的 p(PV)/n 双光子分解水器件 STH 效率为 3.7%（Liu et al.，2021）。堂免一成课题组和江丰课题组分别开发了 $CuIn_{0.5}Ga_{0.5}Se_2$/CdS/Pt 及 Cu_2ZnSnS_4/HfO_2/CdS/HfO_2/Pt 光阴极，与 $BiVO_4$/$NiFeO_x$ 光阳极串联后 STH 效率分别为 3.7% 与 3.17%（Kobayashi et al.，2018；Huang et al.，2021）。格拉兹尔课题组利用 Au/Cu_2O/Ga_2O_3/TiO_2/RuO_x 光阴极与 $BiVO_4$/$NiFeO_x$ 光阳极构建了 p(PV)/n 双光子光电催化水分解池，其 STH 效率约为 3%（Pan et al.，2018）。使用更复杂的器件结构，如 PV/PV/n、PV+PV/n、PV/n/n 等，太阳能电池串联非光伏级半导体光阳极或光阴极组成的光电催化水分解池 STH 效率可得到进一步提升但仍未突破 10%（Higashi et al.，2019）。基于当前的研究进展，为推动光电催化制氢技术尽快走向实用化，需重点考虑以下几个研究方向。

（1）宽光谱响应、高效载流子传输的半导体材料探索。半导体材料是光电催化分解水研究的核心，目前仍缺乏理想的兼具宽光谱响应与高效载流子传输的稳定半导体材料。因此，需要结合理论（高通量计算、机器学习等方法）和实验（组合化学等方法）探索合适的半导体材料。

（2）高效电荷分离的新策略与新方法。光生电荷分离是光电催化反应的关键。结合已有的手段，如异质结、异相结、MIS 结、晶面效应、梯度掺杂以及表面修饰铁电薄膜或偶极分子等，继续开发新策略与新方法，以优化内建电位及空间电荷区厚度，优化光吸收材料、保护层、电催化剂的界面特性，有望实现光生电荷的定向迁移与高效分离。

（3）表面修饰的新材料与新工艺。开发本征活性高、与光电催化材料匹配的电催化剂，以及耐光电化学腐蚀且具优异电荷传输特性的保护层，是提升光电催化材料反应动力学速率及使役稳定性的关键。理解光电极、电催化剂及保护层在使役过程中的失活、失效机制，并依此优化电催化剂和保护层的材料选择与沉积工艺。

（4）载流子动力学及反应中间产物的原位表征技术。利用现代光谱技术原位监测光生载流子的寿命与水分解过程的动力学信息，有助于理解光电催化反应的限制因素。借助原位表征技术，探测光电催化过程中反应底物分子的吸附状态、反应历程等信息，阐明光生载流子的表界面转移和能量转换机制以及材料的损伤和失效机制，建立宏观光电催化特征与微观分子反应机制的联系。

（5）光电催化水分解制氢的规模化探索。作为一项绿氢制备技术，光电催化水分解的研究目前仍处于实验室阶段，缺乏大面积器件的设计与搭建。相关的探索包括光电极的规模化制备工艺、气体产物的分离、器件的长期使役特性、小面积器件的模块集成化等。

（6）高值化学品的制备。通过水分解反应获得氧气与从空气中提取氧气相比，在规模化生产与应用方面无成本优势。而利用光电催化在阳极侧生产高值氧化产物，如氯气、溴、次氯酸、过氧化氢等，有望同时获得氢气与经济效益较高的化学品。

四、研究总结与展望

光电催化分解水制氢及其材料研究是新能源材料领域的研究热点之一。亟待解决的关键科学问题为开发高效宽光谱吸收和利用的光电催化材料、发展光生电荷高效分离的策略、设计高效稳定的电催化剂。自 1972 年 TiO_2 光电催化分解水研究以来，光电催化材料的光响应范围已从紫外光拓展到可见光波段，发展了元素掺杂、表面态钝化、构建异质结、MIS 结等促进光生电荷分离的策略，开发了过渡金属层状羟基化合物、硫化物、磷化物、碳化物等一系列高效稳定的电催化剂，引入保护层概念提升了易腐蚀光电催化材料的运行稳定性。对光电催化过程的认识从宏观的光电化学特性逐步推进到时空分辨的载流子产生、传输过程和表面反应的分子历程。例如，近年发展的原位瞬态吸收光谱、光辅助的电化学阻抗谱、调制光电压（光电流）谱等表征技术可对载流子从产生、复合、传输到表面反应的历程进行全面的分析；光辅助的扫描开尔文探针、电化学势敏感的原子力显微镜对光电极产生驱动力的认识深入微纳尺度；原位红外技术等能够指认表面反应的中间产物。但目前光电催化技术仍难以实现规模化应用，因此，未来应结合材料信息学的进展来开发新型光电催化材料，发展各种原位光谱技术和原位表征技术揭示光电催化过程中的限制因素并提出改善策略，设计可规模化应用的示范装置，结合其他研究领域的需求，拓展光电催化的应用范围。

五、学科发展政策建议与措施

目前，光电催化分解水制氢已满足规模化、产业化的效率要求，而提升器件稳定性、降低开发成本是推动本领域从基础研究向产业应用发展的关键。建议整合全国优势资源和人才力量，对以下方向进行重点支持：①通过整合光电催化材料的关键参数（能带位置、光电特性、稳定性等）建立数据库，利用机器学习分析数据，建立高效预测、设计和筛选材料的理论方法；②开发兼具宽光谱响应及高效载流子分离特性的材料体系；③发展表界面修饰的新理念、新手段，提升电荷分离效率、反应动力学速率及器件稳定性；④发展具有时间和空间分辨的原位表征技术，在原子 / 分子至器件尺度上揭示光电催化的能量转换、电荷输运与转移、表界面反应及材料 / 器件失效机制；⑤研制光电催化集成系统，发展高值精细化学品的光电催化制备新路径。

冯建勇（南京大学）、李朝升（南京大学）

本节参考文献

关中杰 . 2016. p 型 $Cu(In,Ga)S_2$ 及 Cu_2ZnSnS_4 光阴极的溶液法制备及其太阳能分解水性能研究 . 南京：南京大学博士学位论文 .

温鑫 . 2017. 微纳结构调控 Cu_2ZnSnS_4 光阴极载流子输运及其光解水性能研究 . 南京：南京大学博士学位论文 .

Abdi F F, Han L H, Smets A H M, et al. 2013. Efficient solar water splitting by enhanced charge separation in a bismuth vanadate-silicon tandem photoelectrode. Nature Communications, 4: 2195.

Abe R, Higashi M, Domen K. 2010. Facile fabrication of an efficient oxynitride TaON photoanode for overall water splitting into H_2 and O_2 under visible light irradiation. Journal of the American Chemical Society, 132: 11828-11829.

Bae D, Seger B, Vesborg P C K, et al. 2017. Strategies for stable water splitting via protected photoelectrodes. Chemical Society Reviews, 46: 1933-1954.

Banerjee S, Mohapatra S K, Misra M. 2009. Synthesis of TaON nanotube arrays by sonoelectrochemical anodization followed by nitridation: a novel catalyst for photoelectrochemical hydrogen generation from water. Chemical Communications, 46: 7137-7139.

Barroso M, Mesa C A, Pendlebury S R, et al. 2012. Dynamics of photogenerated holes in surface modified α-Fe_2O_3 photoanodes for solar water splitting. Proceedings of the National Academy of Sciences of the United States of America, 109: 15640-15645.

Becquerel E. 1839. Recherches Sur les effets de la radiation chimique de la lumiere solaire, au moyen des courants electriques. Comptes Rendus de l'Académie des Sciences, 9: 145-149.

Benck J D, Lee S C, Fong K D, et al. 2014. Designing active and stable silicon photocathodes for solar hydrogen production using molybdenum sulfide nanomaterials. Advanced Energy Materials, 4: 1400739.

Ben-Naim M, Britto R J, Aldridge C W, et al. 2020. Addressing the stability gap in photoelectrochemistry: molybdenum disulfide protective catalysts for tandem Ⅲ-Ⅴ unassisted solar water splitting. ACS Energy Letters, 5: 2631-2640.

Boettcher S W, Warren E L, Putnam M C, et al. 2011. Photoelectrochemical hydrogen evolution using Si microwire arrays. Journal of the American Chemical Society, 133: 1216-1219.

Brattain W H, Garrett C G B. 1955. Experiments on the interface between germanium and an electrolyte. Bell System Technical Journal, 34: 129-176.

Brillet J, Yum J, Comuz M, et al. 2012. Highly efficient water splitting by a dual-absorber tandem cell. Nature Photonics, 6: 824-828.

Chen L, Yang J H, Klaus S, et al. 2015. p-Type transparent conducting oxide/n-type semiconductor heterojunctions for efficient and stable solar water oxidation. Journal of the American Chemical Society, 137: 9595-9603.

Chen M X, Liu Y, Li C C, et al. 2018. Spatial control of cocatalysts and elimination of interfacial defects towards efficient and robust CIGS photocathodes for solar water splitting. Energy & Environmental Science, 11: 2025-2034.

Chen Y W, Prange J D, Duehnen S, et al. 2011. Atomic layer-deposited tunnel oxide stabilizes silicon photoanodes for water oxidation. Nature Materials, 10: 539-544.

Cheng W H, Richter M H, May M M, et al. 2018. Monolithic photoelectrochemical device for direct water splitting with 19% efficiency. ACS Energy Letters, 3: 1795-1800.

Chun W J, Ishikawa A, Fujisawa H, et al. 2003. Conduction and valence band positions of Ta_2O_5, TaON, and Ta_3N_5 by UPS and electrochemical methods. Journal of Physical Chemistry B, 107: 1798-1803.

Cong Y Q, Park H S, Wang S J, et al. 2012. Synthesis of Ta_3N_5 nanotube arrays modified with electrocatalysts for photoelectrochemical water oxidation. Journal of Physical Chemistry C, 116: 14541-14550.

de Jongh P E, Vanmaekelbergh D, Kelly J J. 1999. Cu_2O: a catalyst for the photochemical decomposition of water? Chemical Communications, 12: 1069-1070.

Digdaya I A, Adhyaksa G W P, Trześniewski B J, et al. 2017. Interfacial engineering of metal-insulator-semiconductor junctions for efficient and stable photoelectrochemical water oxidation. Nature Communications, 8: 15968.

Ebaid M, Priante D, Liu G Y, et al. 2017. Unbiased photocatalytic hydrogen generation from pure water on stable Ir-treated $In_{0.33}Ga_{0.67}N$ nanorods. Nano Energy, 37: 158-167.

Esswein A J, Surendranath Y, Reece S Y, et al. 2011. Highly active cobalt phosphate and borate based oxygen evolving catalysts operating in neutral and natural waters. Energy & Environmental Science, 4: 499-504.

Fan R L, Dong W, Fang L, et al. 2017. More than 10% efficiency and one-week stability of Si photocathodes for water splitting by manipulating the loading of the Pt catalyst and TiO_2 protective layer. Journal of Materials Chemistry A, 5: 18744-18751.

Fan S Z, AlOtaibi B, Woo S Y, et al. 2015. High efficiency solar-to-hydrogen conversion on a monolithically integrated InGaN/GaN/Si adaptive tunnel junction photocathode. Nano Letters, 15: 2721-2726.

Fang T, Huang H T, Feng J Y, et al. 2019. Reactive inorganic vapor deposition of perovskite oxynitride films for solar energy conversion. Research, 2019: 9282674.

Fang Y X, Hou Y D, Fu X Z, et al. 2022. Semiconducting polymers for oxygen evolution reaction under light illumination. Chemical Reviews, 122: 4204-4256.

Feng J, Gong M, Kenney M J, et al. 2015. Nickel-coated silicon photocathode for water splitting in alkaline electrolytes. Nano Research, 8: 1577-1583.

Feng J Y, Cao D P, Wang Z Q, et al. 2014. Ge-mediated modification in Ta_3N_5 photoelectrodes with enhanced charge transport for solar water splitting. Chemistry: A European Journal, 20: 16384-16390.

Feng J Y, Huang H T, Fang T, et al. 2019. Defect engineering in semiconductors: manipulating nonstoichiometric defects and understanding their impact in oxynitrides for solar energy conversion. Advanced Functional Materials, 29: 1808389.

Feng J Y, Huang H T, Guo W X, et al. 2021. Evaluating the promotional effects of WO_3 underlayers in $BiVO_4$ water splitting photoanodes. Chemical Engineering Journal, 417: 128095.

Feng J Y, Huang H T, Yan S C, et al. 2020. Non-oxide semiconductors for artificial photosynthesis: progress on photoelectrochemical water splitting and carbon dioxide reduction. Nano Today, 30: 100830.

Feng J Y, Luo W J, Fang T, et al. 2014. Highly photo-responsive $LaTiO_2N$ photoanodes by improvement of charge carrier transport among film particles. Advanced Functional Materials, 24: 3535-3542.

Feng X J, LaTempa T J, Basham J I, et al. 2010. Ta_3N_5 nanotube arrays for visible light water photoelectrolysis. Nano Letters, 10: 948-952.

Formal F L, Grätzel M, Sivula K. 2010. Controlling photoactivity in ultrathin hematite films for solar water-splitting. Advanced Functional Materials, 20: 1099-1107.

Formal F L, Tétreault N, Cornuz M, et al. 2011. Passivating surface states on water splitting hematite photoanodes with alumina overlayers. Chemical Science, 2: 737-743.

Fujishima A, Honda K. 1972. Electrochemical photolysis of water at a semiconductor electrode. Nature, 238: 37-38.

Gao L, Cui Y C, Vervuurt R H J, et al. 2016. High-efficiency InP-based photocathode for hydrogen production by interface energetics design and photon management. Advanced Functional Materials, 26: 679-686.

Ge J, Koirala P, Grice C R, et al. 2017. Oxygenated CdS buffer layers enabling high open-circuit voltages in earth-abundant Cu_2BaSnS_4 thin-film solar cells. Advanced Energy Materials, 7: 1601803.

Gerischer H. 1966. Electrochemical behavior of semiconductors under illumination. Journal of the Electrochemical Society, 113: 1174-1182.

Grätzel M. 2001. Photoelectrochemical cells. Nature, 414: 338-344.

Gu J, Yan Y, Young J L. et al. 2016. Water reduction by a p-$GaInP_2$ photoelectrode stabilized by an amorphous TiO_2 coating and a molecular cobalt catalyst. Nature Materials, 15: 456-460.

Guo B D, Batool A, Xie G C, et al. 2018. Facile integration between Si and catalyst for high-performance photoanodes by a multifunctional bridging layer. Nano Letters, 18: 1516-1521.

Hajibabaei H, Little D J, Pandey A, et al. 2019. Direct deposition of crystalline Ta_3N_5 thin films on FTO for PEC water splitting. ACS Applied Materials & Interfaces, 11: 15457-15466.

Hajibabaei H, Zandi O, Hamann T W. 2016. Tantalum nitride films integrated with transparent conductive oxide substrates via atomic layer deposition for photoelectrochemical water splitting. Chemical Science, 7: 6760-6767.

Hardee K L, Bard A J. 1977. Semiconductor electrodes: X. Photoelectrochemical behavior of several polycrystalline metal oxide electrodes in aqueous solutions. Journal of the Electrochemical Society, 124: 215-224.

He Y M, Thorne J E, Wu C H, et al. 2016. What limits the performance of Ta_3N_5 for solar water splitting? Chem, 1: 640-655.

Heller A, Miller B, Lewerenz H J, et al. 1980. An efficient photocathode for semiconductor liquid junction cells: 9.4% solar conversion efficiency with p-InP/VCl_3-VCl_2-HCl/C. Journal of the American Chemical Society, 102: 6555-6556.

Higashi M, Domen K, Abe R. 2012. Highly stable water splitting on oxynitride TaON photoanode system under visible light irradiation. Journal of the American Chemical Society, 134: 6968-6971.

Higashi M, Domen K, Abe R. 2013. Fabrication of an efficient $BaTaO_2N$ photoanode harvesting a wide range of visible light for water splitting. Journal of the American Chemical Society, 135: 10238-10241.

Higashi M, Kato Y, Iwase Y, et al. 2021. RhO_x cocatalyst for efficient water oxidation over TaON photoanodes in wide pH range under visible-light irradiation. Journal of Photochemistry & Photobiology, A: Chemistry, 419: 113463.

Higashi T, Nishiyama H, Suzuki Y, et al. 2019. Transparent Ta_3N_5 photoanodes for efficient oxygen evolution toward the development of tandem cells. Angewandte Chemie International Edition, 58: 2300-2304.

Hisatomi T, Brillet J, Cornuz M, et al. 2012. A Ga_2O_3 underlayer as an isomorphic template for ultrathin hematite films toward efficient photoelectrochemical water splitting. Faraday Discussions, 155: 223-232.

Hisatomi T, Dotan H, Stefik M, et al. 2012. Enhancement in the performance of ultrathin hematite photoanode for water splitting by an oxide underlayer. Advanced Materials, 24: 2699-2702.

Hisatomi T, Formal F L, Cornuz M, et al. 2011. Cathodic shift in onset potential of solar oxygen evolution on hematite by 13-group oxide overlayers. Energy & Environmental Science, 4: 2512-2515.

Hou J G, Wang Z, Yang C, et al. 2013. Cobalt-bilayer catalyst decorated Ta_3N_5 nanorod arrays as integrated electrodes for photoelectrochemical water oxidation. Energy & Environmental Science, 6: 3322-3330.

Hou Y D, Abrams B L, Vesborg P C K, et al. 2011. Bioinspired molecular co-catalysts bonded to a silicon photocathode for solar hydrogen evolution. Nature Materials, 10: 434-438.

Hu S, Shaner M R, Beardslee J A, et al. 2014. Amorphous TiO_2 coatings stabilize Si, GaAs, and

GaP photoanodes for efficient water oxidation. Science, 344: 1005-1009.

Hu Y F, Wu Y Q, Feng J Y, et al. 2018. Rational design of electrocatalysts for simultaneously promoting bulk charge separation and surface charge transfer in solar water splitting photoelectrodes. Journal of Materials Chemistry A, 6: 2568-2576.

Huang D W, Li L T, Wang K, et al. 2021. Wittichenite semiconductor of Cu_3BiS_3 films for efficient hydrogen evolution from solar driven photoelectrochemical water splitting. Nature Communications, 12: 3795.

Huang D W, Wang K, Li L T, et al. 2021. 3.17% efficient Cu_2ZnSnS_4-$BiVO_4$ integrated tandem cell for standalone overall solar water splitting. Energy & Environmental Science, 14: 1480-1489.

Huang Q, Li Q, Xiao X D. 2014. Hydrogen evolution from Pt nanoparticles covered p-type CdS:Cu photocathode in scavenger-free electrolyte. Journal of Physical Chemistry C, 118: 2306-2311.

Ishikawa A, Takata T, Kondo J N, et al. 2004. Electrochemical behavior of thin Ta_3N_5 semiconductor film. Journal of Physical Chemistry B, 108: 11049-11053.

Ji L, McDaniel M D, Wang S J, et al. 2015. A silicon-based photocathode for water reduction with an epitaxial $SrTiO_3$ protection layer and a nanostructured catalyst. Nature Nanotechnology, 10: 84-90.

Jia J Y, Seitz L C, Benck J D, et al. 2016. Solar water splitting by photovoltaic-electrolysis with a solar-to-hydrogen efficiency over 30%. Nature Communications, 7: 13237.

Jiang F, Gunawan, Harada T, et al. 2015. Pt/In_2S_3/CdS/Cu_2ZnSnS_4 thin film as an efficient and stable photocathode for water reduction under sunlight radiation. Journal of the American Chemical Society, 137: 13691-13697.

Joshi U A, Maggard P A. 2012. $CuNb_3O_8$: a p-type semiconducting metal oxide photoelectrode. Journal of Physical Chemistry Letters, 3:1577-1581.

Kanan M W, Nocera D G. 2008. *In situ* formation of an oxygen-evolving catalyst in neutral water containing phosphate and Co^{2+}. Science, 321: 1072-1075.

Kast M G, Enman L J, Gurnon N J, et al. 2014. Solution-deposited F:SnO_2/TiO_2 as a base-stable protective layer and antireflective coating for microtextured buried-junction H_2-evolving Si photocathodes. ACS Applied Materials & Interfaces, 6: 22830-22837.

Katagiri H, Sasaguchi N, Hando S, et al. 1997. Preparation and evaluation of Cu_2ZnSnS_4 thin films by sulfurization of E-B evaporated precursors. Solar Energy Materials & Solar Cells, 49: 407-414.

Kempler P A, Gonzalez M A, Papadantonakis K M, et al. 2018. Hydrogen evolution with minimal parasitic light absorption by dense Co-P catalyst films on structured p-Si photocathodes. ACS

Energy Letters, 3: 612-617.

Kenney M J, Gong M, Li Y G, et al. 2013. High-performance silicon photoanodes passivated with ultrathin nickel films for water oxidation. Science, 342: 836-840.

Khaselev O, Turner J A. 1998. A monolithic photovoltaic-photoelectrochemical device for hydrogen production via water splitting. Science, 280: 425-427.

Kim T W, Choi K S. 2014. Nanoporous $BiVO_4$ photoanodes with dual-layer oxygen evolution catalysts for solar water splitting. Science, 343: 990-994.

Kobayashi H, Sato N, Orita M, et al. 2018. Development of highly efficient $CuIn_{0.5}Ga_{0.5}Se_2$-based photocathode and application to overall solar driven water splitting. Energy & Environmental Science, 11: 3003-3009.

Kuang Y B, Jia Q X, Ma G J, et al. 2017. Ultrastable low-bias water splitting photoanodes via photocorrosion inhibition and *in situ* catalyst regeneration. Nature Energy, 2: 16191.

Kudo A, Ueda K, Kato H, et al. 1998. Photocatalytic O_2 evolution under visible light irradiation on $BiVO_4$ in aqueous $AgNO_3$ solution. Catalysis Letters, 53: 229-230.

Lee D K, Choi K S. 2018. Enhancing long-term photostability of $BiVO_4$ photoanodes for solar water splitting by tuning electrolyte composition. Nature Energy, 3: 53-60.

Lewis N S, Nocera D G. 2006. Powering the planet: chemical challenges in solar energy utilization. Proceedings of the National Academy of Sciences of the United States of America, 103: 15729-15735.

Li C L, Hisatomi T, Watanabe O, et al. 2015. Positive onset potential and stability of Cu_2O-based photocathodes in water splitting by atomic layer deposition of a Ga_2O_3 buffer layer. Energy & Environmental Science, 8: 1493-1500.

Li M X, Luo W J, Cao D P, et al. 2013. A cocatalyst-loaded Ta_3N_5 photoanode with a high solar photocurrent for water splitting upon facile removal of the surface layer. Angewandte Chemie International Edition, 52: 11016-11020.

Li Y B, Takata T, Cha D, et al. 2013. Vertically aligned Ta_3N_5 nanorod arrays for solar-driven photoelectrochemical water splitting. Advanced Materials, 25: 125-131.

Li Y B, Zhang L, Torres-Pardo A, et al. 2013. Cobalt phosphate-modified barium-doped tantalum nitride nanorod photoanode with 1.5% solar energy conversion efficiency. Nature Communications, 2013, 4: 2566.

Li Y, Zhang N S, Liu C H, et al. 2021. Metastable-phase β-Fe_2O_3 photoanodes for solar water splitting with durability exceeding 100 h. Chinese Journal of Catalysis, 42: 1992-1998.

Li Z S, Feng J Y, Yan S C, et al. 2015. Solar fuel production: strategies and new opportunities with nanostructures. Nano Today, 10: 468-486.

Li Z S, Luo W J, Zhang M L, et al. 2013. Photoelectrochemical cells for solar hydrogen production: current state of promising photoelectrodes, methods to improve their properties, and outlook. Energy & Environmental Science, 6: 347-370.

Liao M J, Feng J Y, Luo W J, et al. 2012. Co_3O_4 nanoparticles as robust water oxidation catalysts towards remarkably enhanced photostability of a Ta_3N_5 photoanode. Advanced Functional Materials, 22: 3066-3074.

Lichterman M F, Carim A I, McDowell M T, et al. 2014. Stabilization of n-cadmium telluride photoanodes for water oxidation to O_2(g) in aqueous alkaline electrolytes using amorphous TiO_2 films formed by atomic-layer deposition. Energy & Environmental Science, 7: 3334-3337.

Liu B, Wang S J, Feng S J, et al. 2021. Double-side Si photoelectrode enabled by chemical passivation for photoelectrochemical hydrogen and oxygen evolution reactions. Advanced Functional Materials, 31: 2007222.

Liu G J, Shi J Y, Zhang F X, et al. 2014. A tantalum nitride photoanode modified with a hole-storage layer for highly stable solar water splitting. Angewandte Chemie International Edition, 53: 7295-7299.

Liu G J, Ye S, Yan P L, et al. 2016. Enabling an integrated tantalum nitride photoanode to approach the theoretical photocurrent limit for solar water splitting. Energy & Environmental Science, 9: 1327-1334.

Luo J S, Tilley S D, Steier L, et al. 2015. Solution transformation of Cu_2O into $CuInS_2$ for solar water splitting. Nano Letters, 15: 1395-1402.

Luo W J, Yang Z S, Li Z S, et al. 2011. Solar hydrogen generation from seawater with a modified $BiVO_4$ photoanode. Energy & Environmental Science, 4: 4046-4051.

Malizia M, Seger B, Chorkendorff I. et al. 2014. Formation of a p-n heterojunction on GaP photocathodes for H_2 production providing an open-circuit voltage of 710 mV. Journal of Materials Chemistry A, 2: 6847-6853.

Matsumoto Y, Omae M, Sugiyama K, et al. 1987. New photocathode materials for hydrogen evolution: calcium iron oxide ($CaFe_2O_4$) and strontium iron oxide ($Sr_7Fe_{10}O_{22}$). Journal Physical Chemistry, 91: 577-581.

Mei B, Pedersen T, Malacrida P, et al. 2015. Crystalline TiO_2: a generic and effective electron-conducting protection layer for photoanodes and -cathodes. Journal of Physical Chemistry C, 119: 15019-15027.

Morales-Guio C G, Thorwarth K, Niesen B, et al. 2015. Solar hydrogen production by amorphous silicon photocathodes coated with a magnetron sputter deposited Mo_2C catalyst. Journal of the American Chemical Society, 137: 7035-7038.

Morales-Guio C G, Tilley S D, Vrubel H, et al. 2014. Hydrogen evolution from a copper(I) oxide photocathode coated with an amorphous molybdenum sulphide catalyst. Nature Communications, 5: 3059.

Moreno-Hernandez I A, Brunschwig B S, Lewis N S. 2018. Tin oxide as a protective heterojunction with silicon for efficient photoelectrochemical water oxidation in strongly acidic or alkaline electrolytes. Advanced Energy Materials, 8: 1801155.

Oh J H, Deutsch T G, Yuan H C, et al. 2011. Nanoporous black silicon photocathode for H_2 production by photoelectrochemical water splitting. Energy & Environmental Science, 4: 1690-1694.

Ohashi K, Mccann J, Bockris J O'M. 1977. Stable photoelectrochemical cells for the splitting of water. Nature, 266: 610-611.

Pan L F, Kim J H, Mayer M T, et al. 2018. Boosting the performance of Cu_2O photocathodes for unassisted solar water splitting devices. Nature Catalysis, 1: 412-420.

Paracchino A, Laporte V, Sivula K, et al. 2011. Highly active oxide photocathode for photoelectrochemical water reduction. Nature Materials, 10: 456-461.

Paracchino A, Mathews N, Hisatomi T, et al. 2012. Ultrathin films on copper (Ⅰ) oxide water splitting photocathodes: a study on performance and stability. Energy & Environmental Science, 5: 8673-8681.

Pijpers J J H, Winkler M T, Surendranath Y, et al. 2011. Light-induced water oxidation at silicon electrodes functionalized with a cobalt oxygen-evolving catalyst. Proceedings of the National Academy of Sciences of the United States of America, 108: 10056-10061.

Pinaud B A, Vesborg P C K, Jaramillo T F. 2012. Effect of film morphology and thickness on charge transport in Ta_3N_5/Ta photoanodes for solar water splitting. Journal of Physical Chemistry C, 116: 15918-15924.

Prévot M S, Guijarro N, Sivula K. 2015. Enhancing the performance of a robust sol-gel-processed p-type delafossite $CuFeO_2$ photocathode for solar water reduction. ChemSusChem, 8: 1359-1367.

Rao P M, Cai L L, Liu C, et al. 2014. Simultaneously efficient light absorption and charge separation in WO_3/$BiVO_4$ core/shell nanowire photoanode for photoelectrochemical water oxidation. Nano Letters, 14: 1099-1105.

Sayama K, Nomura A, Zou Z G, et al. 2003. Photoelectrochemical decomposition of water on nanocrystalline $BiVO_4$ film electrodes under visible light. Chemical Communications, 23: 2908-2909.

Scheuermann A G, Lawrence J P, Kemp K W, et al. 2016. Design principles for maximizing photovoltage in metal-oxide-protected water-splitting photoanodes. Nature Materials, 15: 99-105.

Seabold J A, Choi K S. 2012. Efficient and stable photo-oxidation of water by a bismuth vanadate photoanode coupled with an iron oxyhydroxide oxygen evolution catalyst. Journal of the American Chemical Society, 134: 2186-2192.

Seger B, Pedersen T, Laursen A B, et al. 2013. Using TiO_2 as a conductive protective layer for photocathodic H_2 evolution. Journal of the American Chemical Society, 135: 1057-1064.

Seger B, Tilley D S, Pedersen T, et al. 2013. Silicon protected with atomic layer deposited TiO_2: durability studies of photocathodic H_2 evolution. RSC Advances, 3: 25902-25907.

Shaner M R, Hu S, Sun K, et al. 2015. Stabilization of Si microwire arrays for solar-driven H_2O oxidation to O_2(g) in 1.0 M KOH(aq) using conformal coatings of amorphous TiO_2. Energy & Environmental Science, 8: 203-207.

Sivula K, Formal F L, Grätzel M. 2009. WO_3-Fe_2O_3 photoanodes for water splitting: a host scaffold, guest absorber approach. Chemistry of Materials, 21: 2862-2867.

Sivula K, Formal F L, Grätzel M. 2011. Solar water splitting: progress using hematite (α-Fe_2O_3) photoelectrodes. ChemSusChem, 4: 432-449.

Sivula K, van de Krol R. 2016. Semiconducting materials for photoelectrochemical energy conversion. Nature Review Materials, 1: 15010.

Standing A, Assali S, Gao L, et al. 2015. Efficient water reduction with gallium phosphide nanowires. Nature Communications, 6: 7824.

Steinmiller E M P, Choi K S. 2009. Photochemical deposition of cobalt-based oxygen evolving catalyst on a semiconductor photoanode for solar oxygen production. Proceedings of the National Academy of Sciences of the United States of America, 106: 20633-20636.

Su J, Minegishi T, Domen K. 2017. Efficient hydrogen evolution from water using CdTe photocathodes under simulated sunlight. Journal of Materials Chemistry A, 5: 13154-13160.

Sun K, Liu R, Chen Y K, et al. 2016. A stabilized, intrinsically safe, 10% efficient, solar-driven water-splitting cell incorporating earth-abundant electrocatalysts with steady-state pH gradients and product separation enabled by a bipolar membrane. Advanced Energy Materials, 6: 1600379.

Sun K, McDowell M T, Nielander A C, et al. 2015. Stable solar-driven water oxidation to O_2(g) by Ni-oxide-coated silicon photoanodes. Journal of Physical Chemistry Letters, 6: 592-598.

Sun K, Saadi F H, Lichterman M F, et al. 2015. Stable solar-driven oxidation of water by semiconducting photoanodes protected by transparent catalytic nickel oxide films. Proceedings of the National Academy of Sciences of the United States of America, 112: 3612-3617.

Tilley S D, Cornuz M, Sivula K, et al. 2010. Light-induced water splitting with hematite: improved nanostructure and iridium oxide catalysis. Angewandte Chemie International Edition, 49: 6405-6408.

Todorov T K, Reuter K B, Mitzi D B. 2010. High-efficiency solar cell with earth-abundant liquid-processed absorber. Advanced Materials, 22: E156- E159.

Valderrama R C, Sebastian P J, Pantoja Enriquez J, et al. 2005. Photoelectrochemical characterization of CIGS thin films for hydrogen production. Solar Energy Materials & Solar Cells, 88: 145-155.

Viswanathan B, Subramanian V, Lee J S. 2014. Materials and Processes for Solar Fuel Production. New York: Springer-Verlag.

Walter M G, Warren E L, McKone J R, et al. 2010. Solar water splitting cells. Chemical Reviews, 110: 6446-6473.

Wang L, Dionigi F, Nguyen N T, et al. 2015. Tantalum nitride nanorod arrays: introducing Ni-Fe layered double hydroxides as a cocatalyst strongly stabilizing photoanodes in water splitting. Chemistry of Materials, 27: 2360-2366.

Wang L, Nguyen N T, Huang X J, et al. 2017. Hematite photoanodes: synergetic enhancement of light harvesting and charge management by sandwiched with $Fe_2TiO_5/Fe_2O_3/Pt$ structures. Advanced Functional Materials, 27: 1703527.

Wang Z L, Mao X, Chen P, et al. 2019. Understanding the roles of oxygen vacancies in hematite-based photoelectrochemical processes. Angewandte Chemie International Edition, 58: 1030-1034.

Warren S C, Voïtchovsky K, Dotan H, et al. 2013. Identifying champion nanostructures for solar water-splitting. Nature Materials, 12: 842-849.

Wen X, Luo W J, Guan Z J, et al. 2017. Boosting efficiency and stability of a Cu_2ZnSnS_4 photocathode by alloying Ge and increasing sulfur pressure simultaneously. Nano Energy, 41: 18-26.

Xu G Z, Xu Z, Shi Z, et al. 2017. Silicon photoanodes partially covered by $Ni@Ni(OH)_2$ core-shell particles for photoelectrochemical water oxidation. ChemSusChem, 10: 2897-2903.

Yang J H, Cooper J K, Toma F M, et al. 2016. A multifunctional biphasic water splitting catalyst tailored for integration with high-performance semiconductor photoanodes. Nature Materials, 16: 335-341.

Yao T T, Chen R T, Li J J, et al. 2016. Manipulating the interfacial energetics of n-type silicon photoanode for efficient water oxidation. Journal of the American Chemical Society, 138: 13664-13672.

Ye Z, Hu Z F, Yang L X, et al. 2020. Stable p-type $Cu:CdS_{1-x}Se_x/Pt$ thin-film photocathodes with fully tunable bandgap for scavenger-free photoelectrochemical water splitting. Solar RRL, 4: 1900567.

Yokoyama D, Hashiguchi H, Maeda K, et al. 2011. Ta_3N_5 photoanodes for water splitting prepared by sputtering. Thin Solid Films, 519: 2087-2092.

Yokoyama D, Minegishi T, Jimbo K, et al. 2010. H_2 evolution from water on modified Cu_2ZnSnS_4 photoelectrode under solar light. Applied Physics Express, 3: 101202.

Yu X W, Yang P, Chen S, et al. 2017. NiFe alloy protected silicon photoanode for efficient water splitting. Advanced Energy Materials, 7: 1601805.

Yu Y H, Sun C L, Yin X, et al. 2018. Metastable intermediates in amorphous titanium oxide: a hidden role leading to ultra-stable photoanode protection. Nano Letters, 18: 5335-5342.

Zhang H X, Ding Q, He D H, et al. 2016. A p-Si/NiCoSe$_x$ core/shell nanopillar array photocathode for enhanced photoelectrochemical hydrogen production. Energy & Environmental Science, 9: 3113-3119.

Zhang M L, Luo W J, Zhang N S, et al. 2012. A facile strategy to passivate surface states on the undoped hematite photoanode for water splitting. Electrochemistry Communications, 23: 41-43.

Zhang N S, Wang X, Feng J Y, et al. 2020. Paving the road toward the use of β-Fe_2O_3 in solar water splitting: Raman identification, phase transformation and strategies for phase stabilization. National Science Review, 7: 1059-1067.

Zhao J, Minegishi T, Zhang L, et al. 2014. Enhancement of solar hydrogen evolution from water by surface modification with CdS and TiO_2 on porous $CuInS_2$ photocathodes prepared by an electrodeposition-sulfurization method. Angewandte Chemie International Edition, 53: 11808-11812.

Zhen C, Wang L Z, Liu G, et al. 2013. Template-free synthesis of Ta_3N_5 nanorod arrays for efficient photoelectrochemical water splitting. Chemical Communications, 49: 3019-3021.

Zhong D K, Cornuz M, Sivula K, et al. 2011. Photo-assisted electrodeposition of cobalt-phosphate (Co-Pi) catalyst on hematite photoanodes for solar water oxidation. Energy & Environmental Science, 4: 1759-1764.

Zhong M, Hisatomi T, Sasaki Y, et al. 2017. Highly active GaN-stabilized Ta_3N_5 thin-film photoanode for solar water oxidation. Angewandte Chemie International Edition, 56: 4739-4743.

Zhong Y J, Li Z S, Zhao X, et al. 2016. Enhanced water-splitting performance of perovskite $SrTaO_2N$ photoanode film through ameliorating interparticle charge transport. Advanced Functional Materials, 26: 7156-7163.

Zhou X H, Liu R, Sun K, et al. 2015. Interface engineering of the photoelectrochemical performance of Ni-oxide-coated n-Si photoanodes by atomic-layer deposition of ultrathin films of cobalt oxide. Energy & Environmental Science, 8: 2644-2649.

Zhou X H, Liu R, Sun K, et al. 2016. 570 mV photovoltage, stabilized n-Si/CoO_x heterojunction photoanodes fabricated using atomic layer deposition. Energy & Environmental Science, 9: 892-897.

Zhu T, Chong M N. 2015. Prospects of metal-insulator-semiconductor (MIS) nanojunction structures for enhanced hydrogen evolution in photoelectrochemical cells: a review. Nano Energy, 12: 347-373.

Zou Z G, Ye J H, Sayama K, et al. 2001. Direct splitting of water under visible light irradiation with an oxide semiconductor photocatalyst. Nature, 414: 625-627.

第五节　光催化 CO_2 还原材料

一、光催化 CO_2 还原材料研究概述

随着全球工业化进程的日益加快，人们对能源的需求与日俱增。目前全球范围内的能源结构仍然以化石燃料为主，在加剧了此类不可再生能源快速消耗的同时，也使得 CO_2 的排放量急剧增加。开发新型清洁能源并减少大气中的 CO_2 含量，对解决全球能源危机和环境问题具有重要意义。光催化 CO_2 还原为碳氢燃料是一种理想的解决能源短缺并减小温室效应影响的新技术（Tu et al.，2014；Zhang et al.，2018）。

光催化反应是指利用光能驱动光催化材料产生光生电荷，降低反应活化能，将光能转化为化学能的现象。其反应过程主要分为三部分：①光照时，半导体材料吸收能量等于或者大于其禁带宽度 E_g 的入射光，从而激发价带上的电子，使其跃迁至导带成为激发态电子；同时，价带上会产生相应数量的空穴，形成光生电子–空穴对。②光生电子–空穴对在内建电场的作用下发生分离，并迁移至半导体材料表面活性反应位点，此过程需尽量降低光生电子与空穴的复合（包括体相复合和表面复合）。③当迁移至半导体材料的表面后，由于光生空穴、电子分别具有强的氧化与还原能力，进而能够在半导体材料的表面诱发吸附的 H_2O 和 CO_2 发生氧化还原反应。其中，空穴氧化分解 H_2O，产生 O_2 和 H^+。光生电子与生成的 H^+ 和 CO_2 反应，生成 CH_4 或 CO 等碳氢燃料，反应机理示意图如图 1-68 所示。

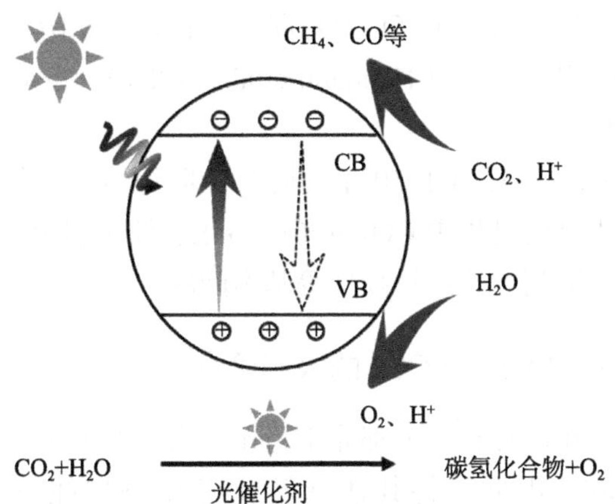

图 1-68 光催化 CO_2 还原为碳氢燃料示意图

光催化 CO_2 还原反应是一个多电子参与的反应。参与反应的电子数不同，其还原产物也不同，如 CO、HCOOH、HCHO、CH_3OH 和 CH_4 等，可能涉及的反应如下。

$$CO_2 + 2H^+ + 2e^- \longrightarrow CO + H_2O \quad E^0 = -0.53\ V\ vs.\ RHE,\ pH=7 \quad (1\text{-}16)$$

$$CO_2 + 2H^+ + 2e^- \longrightarrow HCOOH \quad E^0 = -0.61\ V\ vs.\ RHE,\ pH=7 \quad (1\text{-}17)$$

$$CO_2 + 4H^+ + 4e^- \longrightarrow HCHO + H_2O \quad E^0 = -0.48\ V\ vs.\ RHE,\ pH=7 \quad (1\text{-}18)$$

$$CO_2 + 6H^+ + 6e^- \longrightarrow CH_3OH + H_2O \quad E^0 = -0.38\ V\ vs.\ RHE,\ pH=7 \quad (1\text{-}19)$$

$$CO_2 + 8H^+ + 8e^- \longrightarrow CH_4 + 2H_2O \quad E^0 = -0.24\ V\ vs.\ RHE,\ pH=7 \quad (1\text{-}20)$$

$$CO_2 + e^- \longrightarrow CO_2^- \quad E^0 = -1.90\ V\ vs.\ RHE,\ pH=7 \quad (1\text{-}21)$$

其中，E^0 为相对于可逆氢电极（RHE，pH = 7）还原反应进行所需的最低电位，单位为伏（V）。

二、光催化 CO_2 还原材料研究的历史及现状

光催化 CO_2 还原材料的研究源于光催化的兴起。1972 年，藤岛昭和本多健一在紫外光照射条件下，利用单晶 TiO_2 成功将 H_2O 分解成 H_2 和 O_2（Fujishima & Honda，1972）。这一现象引起人们的极大兴趣，促使光催化成为研究热点。1978 年，哈尔曼（Halmann）首次报道了光催化 CO_2 还原的研究，他以 p 型半导体 GaP 为光电极，在水溶液中将 CO_2 还原为 CH_3OH（Halmann，1978）。1979 年，井上（Inoue）等开发了基于 TiO_2、GaP、CdS、ZnO 等光催化材料，并对光催化 CO_2 还原的反应机理进行了阐述（Inoue et al.,

1979）。随后，哈尔曼等又以 $SrTiO_3$ 为光催化材料，在水溶液中将 CO_2 还原为 CH_3OH、HCOOH 和 HCHO（Halmann et al.，1983）。自此，光催化 CO_2 还原的研究逐渐兴起。

光催化 CO_2 还原虽然经过了 50 年的研究，但仍然存在很多问题，面临巨大挑战，如光催化反应过程中 CO_2 分子的活化、催化剂中光生载流子的分离与迁移、光催化 CO_2 还原产物选择性等基本问题。

（一）光催化还原反应中 CO_2 的吸附活化

线性的 CO_2 分子在通常状态下都非常稳定，由于 CO_2 中所含有的 C ═ O 键能约为 750 kJ/mol，因此，将其还原转化必须添加能量（Chang et al.，2016）。目前，大多数研究认为，CO_2 光还原反应的第一步通常是 CO_2 得单电子形成 $CO_2^{\cdot-}$ 的中间物种（$CO_2 + e^- \longrightarrow CO_2^{\cdot-}$，−1.9V *vs.* NHE）（Qiao et al.，2014）。较高的负还原电位在热力学上对绝大多数半导体而言都不利于 CO_2 的光还原。因此，在光催化 CO_2 还原反应中，CO_2 得单电子的预活化被视为整个催化反应的速率控制步骤。为了克服 CO_2 光催化还原过程中较高的反应能垒，调控催化剂表面上 CO_2 的吸附和预活化对整个催化反应具有重要意义。催化剂表面的缺陷结构被视为一种有效的 CO_2 吸附活化的活性位点。诱导产生的空位缺陷可以有效增加 CO_2 的吸附，并且作为电子捕获中心，有利于光生载流子的分离以及 CO_2 的活化。以典型的 TiO_2 光催化剂为例，已有大量实验和理论证明，催化剂中的缺陷位点可以有效促进 CO_2 的吸附和活化（Qiao et al.，2014；Xi et al.，2012；Xie et al.，2011；Zhao et al.，2015；Sun et al.，2018；Lee et al.，2011）。例如 CO_2 不能在化学计量比的 TiO_2 晶体表面发生解离，但在其中引入表面氧空位缺陷之后，CO_2 中的 C ═ O 键能够和氧空位产生新的稳定吸附结构，并使之裂解生成 CO（Lee et al.，2011）。STM 可以直观地表征 CO_2 在氧空位上的吸附解离过程。如图 1-69 所示，通过施加负电位，可以使电子从催化剂表面转移到吸附在氧空位上的 CO_2 分子上，这种吸附和连续电荷转移可以使 CO_2 分子活化解离，生成 CO。

在光催化 CO_2 还原反应过程中，除生成典型的 C1 产物，如 CH_4、CO 之外，通过调制催化剂表面缺陷活性位点，还能制备一些多碳产物。如石原（Ishihara）团队通过制备含有氧空位的超细 $WO_3 \cdot H_2O$ 纳米管（H-WO_3），可以有效将 CO_2 还原为 CH_3COOH（Sun et al.，2018），其 CH_3COOH 平均产率为 9.4 μmol/（g·h），选择性高达 85%。通过密度泛函理论计算以及原位红外

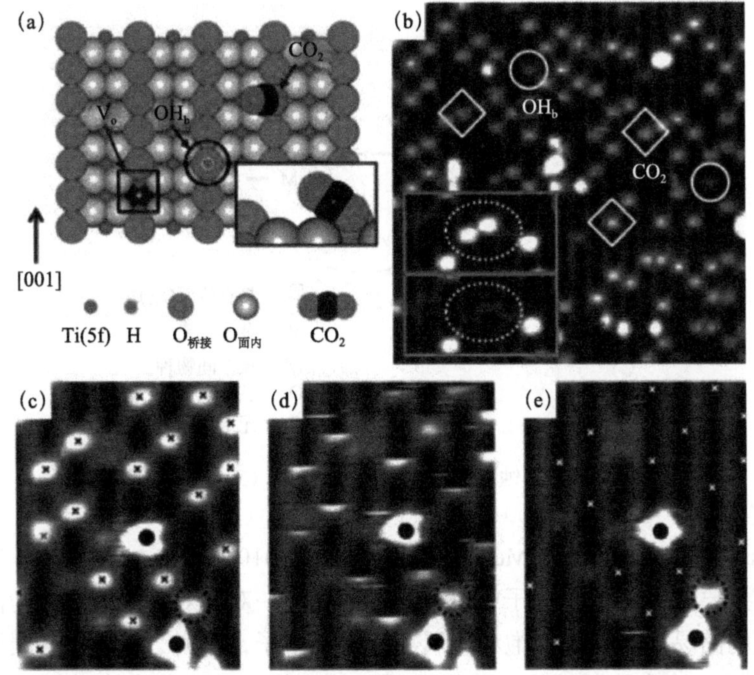

图 1-69　(a) CO_2 吸附在氧空位处的示意图；(b) CO_2 在 55 K 下吸附在 $TiO_2(110)$ 表面的 STM 图，OH_b 中的下角标 b 用来标识 OH 类型为桥羟基；(c)~(e) CO_2 分别在 $TiO_2(110)$ 施加脉冲电压前、施加脉冲电压过程中以及施加脉冲电压后的 STM 图（Xie et al., 2011）

观察到，吸附在氧空位处的 CO_2 中 C ═ O 键被拉长，活化后的 CO_2 更有利于 *COOH 物种的生成。

氧空位除活化 CO_2 并通过光催化转化生成碳氢产物之外，闫世成课题组提出光致缺陷辅助 CO_2 直接劈裂产生 C 和 O_2，为 CO_2 的资源化提供了另一种思路（Wang et al., 2018）。其反应机理为：利用光腐蚀反应，通过光生空穴腐蚀氧化物半导体中的氧原子，产生氧缺陷，进而捕获和活化 CO_2，因 CO_2 分子中的 O 原子被催化剂晶格氧空位捕获而有效抑制了逆反应过程，并因光生空穴的持续性氧化而释放 O_2，光生电子则将 CO_2 还原为单质 C（图 1-70）。

除单独的氧空位对 CO_2 分子有活化作用之外，有报道称，由空位缺陷形成不饱和金属离子作为路易斯酸位点，连同一些材料中存在的路易斯碱位点如羟基基团，二者可以形成受阻的路易斯酸碱对，可用于 CO_2 活化

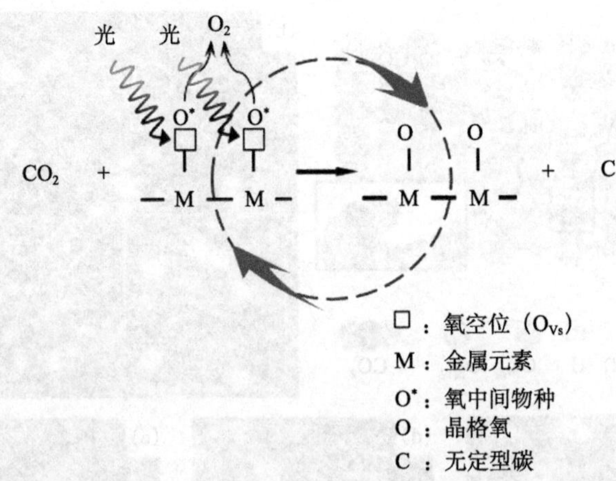

图 1-70 基于光腐蚀反应的光催化直接分解 CO_2 反应机理示意图

（Ghoussoub et al.，2016；Ménard & Stephan，2010；Mömming et al.，2009）。其原理主要是由于 CO_2 分子中每个氧原子含有一对孤对电子，可以为由氧空位形成的路易斯酸位点不饱和金属离子提供电子，同时，CO_2 中的碳原子可以从材料中的路易斯碱位点获得电子，从而使 CO_2 活化。这种结构可以为 CO_2 的活化裂解提供一种新策略。例如，奥兹（G. A. Ozin）团队从理论和实验上详细论证了 $In_2O_{3-x}(OH)_y$ 中临近氧空位的不饱和 In 位点作为路易斯酸位点，In-OH 作为路易斯碱位点，二者形成的受阻路易斯酸碱对可以有效地活化 CO_2 并将其还原为 CO（Ghuman et al.，2016，2015；Hoch et al.，2016）。此外，邹志刚团队（Wang et al.，2018）提出利用光生空穴氧化氢氧化物材料中的晶格羟基生成氧空位的方法，在光催化材料 $CoGeO_2(OH)_2$ 中通过真空辐照引入氧空位，使之形成不饱和的金属离子作为路易斯酸位点，材料自身羟基作为路易斯碱位点，形成受阻路易斯酸碱对，加速了 CO_2 活化，实现了高效的光催化 CO_2 还原。同时，由于催化剂本身含有的羟基可以直接作为质子源参与催化反应，无须再添加额外的质子源（如 H_2O 或 H_2），降低了光催化 CO_2 还原反应的活化能，进一步实现了羟基作为质子源并可再生的“自呼吸”式光催化 CO_2 还原。其反应机理如图 1-71 所示。

除催化剂表面的氧空位可以有效活化 CO_2 之外，固体碱也常用来对 CO_2 分子进行活化。CO_2 分子是一种典型的酸性分子，固体碱则呈现碱性。例如，碱金属氧化物中的氧离子（O_2^-）就是典型的碱性位点，两者之间存在很强的化学吸附作用（Copperthwaite et al.，1988；Espinal et al.，2012；Wang et al.，

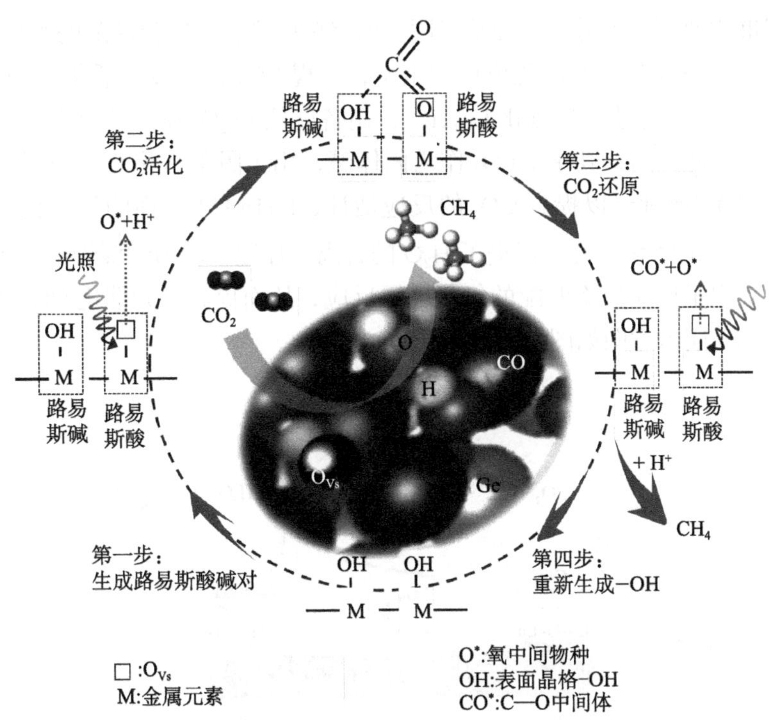

图 1-71　受阻路易斯酸碱对活化 CO_2 光催化反应机理示意图

2011）。这种强的化学吸附作用能够自发地将稳定的 CO_2 分子转化成低能的碳酸盐物种（CO_3^{2-}），同时 CO_2 分子的直线型结构相应地发生弯曲，C ═ O 双键将会变长。甚至，在暗反应条件下能够实现其中一个 C ═ O 双键的断裂。当然，CO_2 的吸附活化取决于氧化物表面的化学性质。常见的几种 CO_2 吸附模式如图 1-72 所示（Tu et al.，2014）。事实上，固体碱的强吸附作用主要应用于 CO_2 的碳捕集与封存（carbon capture and storage，CCS）和热催化领域。直到 21 世纪初，固体碱的化学吸附作用才开始应用于光催化 CO_2 还原领域。

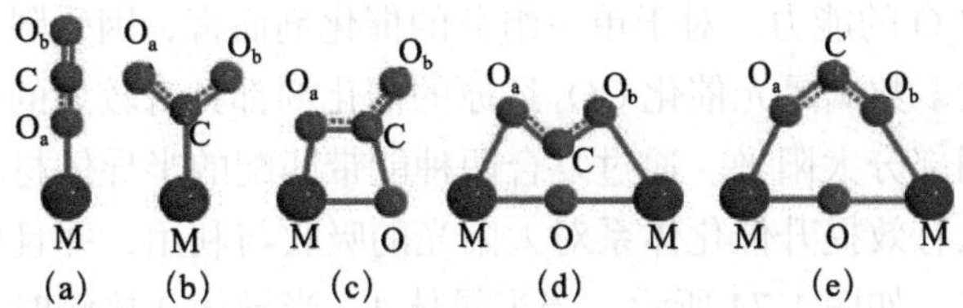

图 1-72　催化剂表面常见的几种 CO_2 吸附模式（Lu et al.，2017）

M 为金属元素

闫世成课题组报道了一种基于固体碱修饰的富含氧空位缺陷的光催化剂，提出了构筑空间分离的活化位点的策略，实现 CO_2 和 H_2O 双活化（Lu et al., 2017）。具体是通过一步氮化 La_2TiO_5，制备原位 La_2O_3 修饰的富含氧空位的 $LaTiO_2N$ 光催化剂，其中 La_2O_3 作为固体碱，用于吸附活化 CO_2 分子，生成低能的 CO_3^{2-} 物种，以提高 CO_2 的反应活性，$LaTiO_2N$ 表面的氧空位缺陷作为 H_2O 分子的活化中心，促进了 H_2O 的解离，加速了动力学缓慢的质子释放过程。实现了反应速率匹配的氧化还原反应，从而提高了光催化 CO_2 还原的性能，其反应示意图如图 1-73 所示。

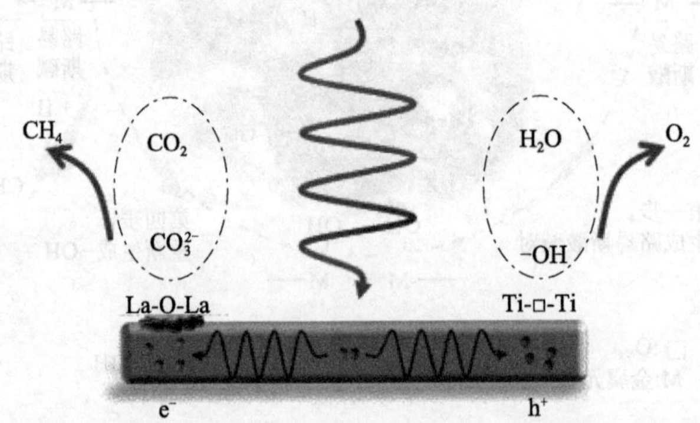

图 1-73 CO_2 和 H_2O 双活化提高 $LaTiO_2N$ 光催化 CO_2 还原示意图

□：氧空位（O_{Vs}）

（二）光催化反应中电荷的分离与输运

根据光催化 CO_2 还原基本原理，催化剂中的光生载流子的高效分离和输运对催化性能的提升有重要意义。通过对催化剂的改性，比如目前发展的异质结内建电场、合理构建 Z 型机制（Z-scheme）体系、晶面工程、形貌调控等方法，均可有效促进光生载流子的分离和输运。

光催化 CO_2 还原需要半导体材料的能带结构同时满足其导带还原 CO_2 以及价带氧化 H_2O 的能力。对于单一组分的催化剂而言，因受限于材料本身的能带分布，大多数满足光催化 CO_2 还原的催化剂都具有较大的带隙，因而只能吸收与利用部分太阳光。通过复合两种能带匹配的半导体材料，合理构筑异质结，可以有效提升催化体系对太阳光的吸收与利用，并且促进光生载流子的高效分离。如图 1-74 所示，当半导体 1、半导体 2 接触时，接触界面处的电荷将重新分布，直到费米能级相等为止，二者的能带发生弯曲，并在界面处形成空间电荷区。光激发半导体 1、半导体 2 时，光生电子和空穴分别位于它们的导带与价带上，空间电荷区中的电场驱使电子从半导体 1 的导带迁移到半导体 2 的导带上参与还原反应，同时空穴从半导体 2 的价带迁移到半导体 1 的价带上参与氧化反应。异质结光催化材料实现了氧化还原位点的分离，可以降低光生载流子复合的可能性。基于异质结构的催化原理，包括 TiO_2/CuO（Xi et al.，2011）、TiO/ZnO（Ong et al.，2015）、C_3N_4/ 石墨烯（Li et al.，2011）、Cu_2O/SiC（Bai et al.，2015）等异质结催化体系都已经成功合成并应用于光催化反应中。

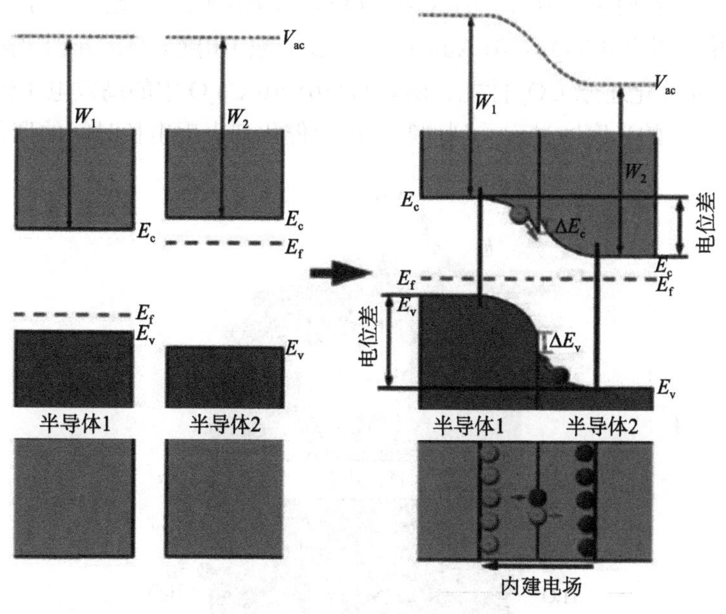

图 1-74　半导体－半导体异质结的形成的能带示意图（Li et al.，2011）

在上述的半导体异质结中，由于电荷的重新分布，光生电子-空穴对在分离后分别处于一个相对较正的导带和较负的价带上，它们的氧化还原反应的驱动力将会减弱。换句话说，这些异质结构的设计是为了提高电荷的分离效率而牺牲了电荷的氧化还原能力。近年来，通过模拟自然光合作用发展的一种 Z 型复合催化系统可以避免这种情况的发生。Z 型复合结构通常由具有交错排列的带结构的两个半导体组成。两种半导体的导带和价带不满足单独全反应的氧化还原电位要求，但它们可以分别进行单独的还原或氧化半反应。在 Z 型复合催化系统中，来自具有较正导带的半导体 1 的光生电子将与具有较负价带的来自半导体 2 的空穴重新结合。此外，残留在半导体 2 的导带中的光生电子与残留在半导体 1 的价带中的空穴可分别用于还原和氧化半反应，构成体氧化还原反应（Zhou et al.，2018）。显然，该体系中电子与空穴的空间分离将大大有助于减少电荷的复合。更重要的是，Z 型复合催化系统具有其他体系难以实现的优势：剩余的电子和空穴很好地保持了其能量水平，从而使反应动力不会受到影响。此外，在这种设计中，使用的半导体通常为具有用于可见光吸收的窄带隙半导体。闫世成课题组报道，通过逐步合成路线将 Au-Cu_2O 核壳结构沉积到 $BiVO_4$ 截角八面体的 {010} 晶面上，从而构建了晶面依赖的 $BiVO_4$-Au-Cu_2O 三元 Z 型结构（Zhou et al.，2018）。如图 1-75 所示，合成的 $BiVO_4$(010)-Au-Cu_2O 可以激活无光催化活性的 $BiVO_4$ 和 Cu_2O，并且相比 $BiVO_4$(110)-Au-Cu_2O 和 $BiVO_4$(010)-Cu_2O，分别提高了 3 倍和 5 倍的光催化还原 CO_2 性能，$BiVO_4$(110)-Au-Cu_2O 中的高效电子传输通道使得 Cu_2O 中的光腐蚀过程受到抑制，因而催化剂表现出长时间的催化活性。

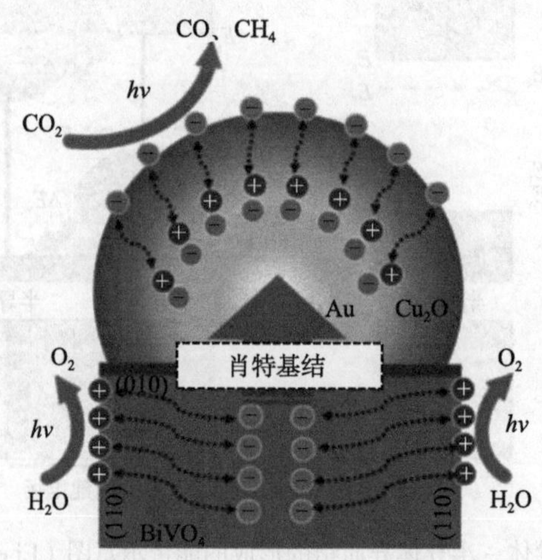

图 1-75　$BiVO_4$-Au-Cu_2O Z 型结构中载流子迁移示意图

光催化反应是个表面反应，而半导体催化剂的表面物理化学性质是由其表面原子结构决定的。同一半导体单晶的不同晶面可能由于其原子配位的不同而表现出表面能、电子能带结构、催化反应活性等性质的差异（Ran et al.，2018）。基于半导体晶面进行表面修饰、特定晶面暴露、助催化剂选择性沉积等，可以有效地改善材料的催化性能。此外，不同晶面之间存在的晶面结效应，可以有效促进催化剂中光生载流子的分离。

（三）光催化 CO_2 还原的产物选择性与反应路径

C 在 CO_2 中呈现最高价态，因此，根据电子转移数目的不同，CO_2 还原具有很多不同价态的潜在产物，如气相中的 CO、CH_4 及一些高级烃类，以及液相中的 $HCOOH$、CH_3OH。在实际反应过程中，这种产物多样性的特点也得到了很好的验证。例如，井上（Inoue et al.，1979）等以 TiO_2 为光催化剂，在 CO_2 饱和水溶液中可将 CO_2 还原为 $HCOOH$、CH_3OH、$HCHO$ 和少量的 CH_4。近年来，光催化 CO_2 还原的材料体系及性能已得到较好的发展，但对其产物选择性机理的认识还有待提高（Ran et al.，2018；Zheng et al.，2017；Wu et al.，2017）。产物选择性机理的揭示对光催化 CO_2 还原的发展具有非常重要的意义，它不仅有助于新型光催化材料的开发，而且能够为设计新颖的反应路径提供指导思想，以实现产物的定向调控。

催化剂是催化反应的重要媒介，不同催化剂的 CO_2 还原产物选择性往往不同。以气相 CO_2 还原为例，TiO_2（Liu et al.，2012）和 g-C_3N_4（Jiang et al.，2018）的主要还原产物为 CO，次要产物为 CH_4。以 Zn_2GeO_4（Yan et al.，2013）、$ZnGa_2O_4$（Liu et al.，2010）和 WO_3（Chen et al.，2012）为光催化剂时，CH_4 为主要还原产物。这有可能是由催化剂表面不同的化学性质决定的。光电化学电池体系中半导体光催化剂材料对还原 CO_2 产物的选择性同样起到至关重要的作用。石谷等（Tamaki et al.，2012）利用 Ru(Ⅱ) 的复合物（图 1-76）作为光敏剂，合成了一种超分子的光催化材料。该高分子材料在可见光下光电催化还原 CO_2 表现出了极高的 $HCOOH$ 产物选择性。这一还原反应由光敏剂光化学还原和催化材料分子内电子传递协同完成。哈默斯（Hamers）等（Zhang et al.，2014）采用溶剂化电子提高了光电化学体系中还

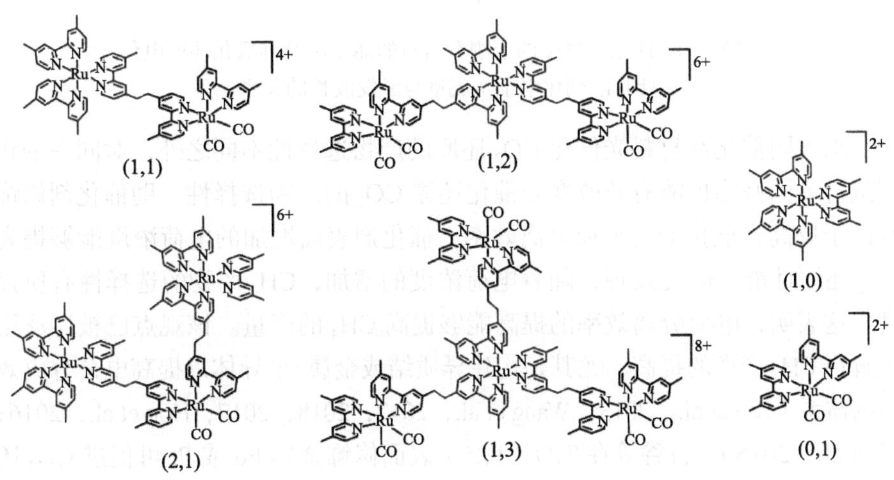

图 1-76 Ru(Ⅱ) 复合物的化学结构示意图

原 CO_2 产 CO 的选择性。他们利用具有 H 原子终端的金刚石作为光催化材料（图 1-77），在紫外光照条件下，生成的电子直接传递至水中，形成溶剂化电子并还原水中的 CO_2 分子。

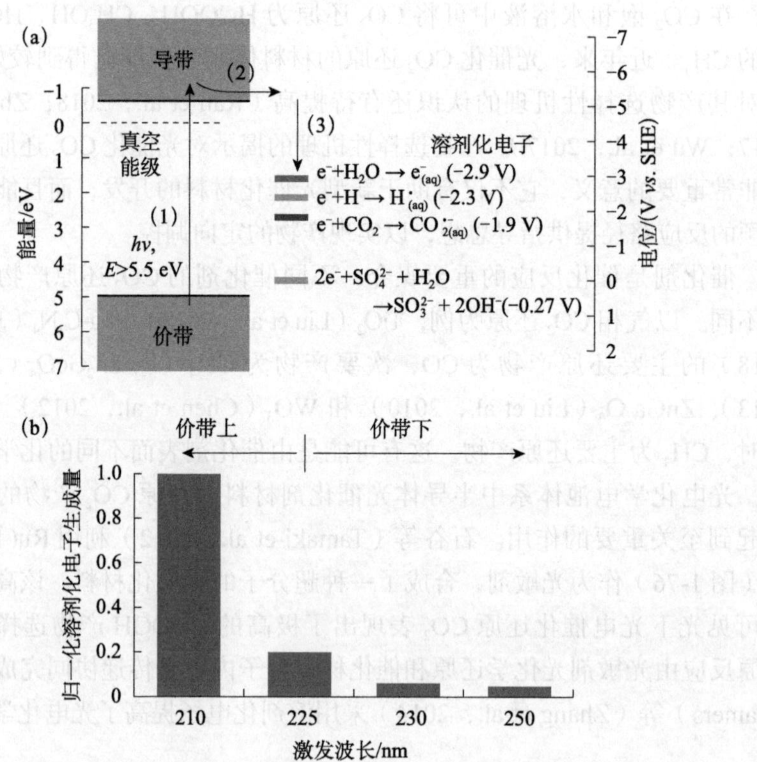

图 1-77 （a）H 原子终端的金刚石材料的能带图及其氧化还原电位，（b）溶剂化电子生成量与激发波长的关系

除不同催化剂材料光催化 CO_2 还原的产物选择性不同之外，对同一催化剂进行表面修饰也能有效改变光催化还原 CO_2 的产物选择性。助催化剂修饰有助于提高光催化剂的电荷分离效率，催化剂表面增加的电荷浓度能够提高 CO_2 还原性能。研究发现，随着电荷浓度的增加，CH_4 产物的选择性有所提高。这表明，电荷分离效率的提高能够提高 CH_4 的产量。该观点已被广泛用于解释 CH_4 产率的提高，尤其是构建异质结或金属-半导体以提高电荷分离效率的体系（Xie et al.，2013；Wang et al.，2012，2018，2017；Hou et al.，2016；Li et al.，2016）。石谷等在 TiO_2（P25）表面修饰金属 Pd 或 Pt 可促进 CO_2 还原为 CH_4，同样地，在 g-C_3N_4 表面修饰金属助催化剂后，如 Cu（Tahir et al.，

2017）、Pt（Ong et al.，2015）、Pt-Cu 合金（Lang et al.，2017）等，其产物选择性也发生了改变，由原来的 CO 产物转变为 CH_4 产物。但是需要指出的是，不同的助催化剂通常具有不同的还原产物，这意味着，除了电子浓度，催化剂表面的化学性质对其产物的选择性有着重要影响。有文献报道，TiO_2 在修饰 Au、Rh 或 Ru 时，CO_2 的还原产物主要为 CH_3COOH（Ishitani et al.，1993）。

闫世成课题组通过在 Ta_3N_5、$LaTiO_2N$ 以及 $Ta_3N_5/LaTiO_2N$ 异质结上的光催化实验，结合理论计算，发现催化剂表面化学性质比反应动力学更能影响催化剂产物的选择性（Lu et al.，2018）。周勇课题组采用包裹 Au 纳米颗粒的碳球为模板，通过调节钛酸四丁酯水解来控制 TiO_2 壳的厚度，合成 Au@TiO_2 蛋黄形空心球，并应用于光催化 CO_2 还原（Tu et al.，2015）。纯 TiO_2 空心球只能光还原 CO_2 为 CH_4，而 Au@TiO_2 空心球可高效地光还原 CO_2 为 CH_4 和 C_2H_6。结果表明，Au@TiO_2 空心球 中 Au 纳米颗粒的等离子共振效应产生强烈的非均匀局域电磁场，该电磁场不仅能促进光生电子-空穴对的产生和分离，而且能促进涉及多个 e^-/H^+ 转移的高级碳物种的形成，从而提高光催化 CO_2 还原性能及产 C_2H_6 的选择性。基于上述研究结果可以推测，CO_2 还原的产物选择性高度依赖于催化剂的表面物理化学性质。

通过对催化剂进行晶面调控可以改变催化剂表面化学性质，因此同样可以改变催化剂的选择性。闫世成课题组（Yan et al.，2013）通过一种低温简易的方法合成了不同长宽比的 Zn_2GeO_4 纳米棒，并通过光催化 CO_2 还原的实验证明，暴露{110}晶面比例最大的 Zn_2GeO_4 样品材料光催化产 CO 的选择性最高。相似的晶面效应在 TiO_2 和 $BiVO_4$ 材料均有报道（Mao et al.，2012；Liu et al.，2009）。刘岗等（Jiao et al.，2012）以 PO_4^{3-}/F^- 形貌控制剂合成了暴露{101}晶面的中空锐钛矿 TiO_2，与暴露相同晶面的固体锐钛矿 TiO_2 相比，中空锐钛矿 TiO_2 材料对还原 CO_2 产 CH_4 的选择性更高。

除此之外，调控同一半导体材料的形貌、晶型或者掺杂改性也可以改变光催化 CO_2 还原的产物选择性。黄柏标等（Cheng et al.，2012）通过阴离子交换的方法合成了中空状的 Bi_2WO_6 微球，相比于实心状固态 Bi_2WO_6，其中空状的微结构形貌可以提高对 CO_2 的吸附能力，有利于提升材料在可见光下光催化 CO_2 还原产甲醇的性能。Cu-Mn 和 Ni-N 共掺的 TiO_2 可将 CO_2 还原为 CH_3OH（Luo et al.，2011；Hou et al.，2011），In 掺杂的 TiO_2 则能够将 CO_2 还原为 CH_4（Tahir & Amin，2015）。

光催化 CO_2 还原反应的产物选择性还与反应物的活化方式有关。黄柏标课题组（Liu et al., 2013）在金属有机框架光催化材料中加入 Cu^{2+}，可以有效提高其对 CO_2 的吸附活化能力，降低 CO_2 的还原势垒。相比于不加 Cu^{2+} 的材料，加入 Cu^{2+} 的样品对产物甲醇的选择性提升了 6 倍。这种产物选择性的差异可直接在气/液反应条件中体现。如上所述，以 TiO_2 为催化剂时，气相反应中 CO_2 的还原产物为 CO（主要产物）和 CH_4（Xie et al., 2013），明显不同于液相环境下的还原产物（如 HCOOH、CH_3OH 和 HCHO 等）（Tu et al., 2014）。科契（Kočí）等（Kočí et al., 2009）研究发现，在 NaOH 溶液中，TiO_2 可将 CO_2 还原为 CH_4（主要产物）和 CH_3OH。王野等（Xie et al., 2013）研究发现，TiO_2 表面 MgO 修饰有助于进一步提高气相体系中的 CO 产物选择性。经对比发现，在 NaOH 溶液中，CO_2 被活化为 HCO_3^-，而 MgO 能够将 CO_2 活化为 CO_3^{2-}。

在半导体光催化 CO_2 还原的气相体系中，除光催化材料本身的性质影响光催化 CO_2 还原活性和产物选择性之外，其他的因素，比如反应物投加比、光强以及牺牲剂等均可影响这一反应的产物选择性。迪米特里耶维奇（Dimitrijevic）等（Dimitrijevic et al., 2011）利用电子顺磁技术研究了在 TiO_2 紫外光催化 CO_2 还原产 CH_4 反应过程中 H_2O 和 CO_2 的多重作用。实验结果表明，H_2O 分子在反应中主要有以下三方面作用：①稳定光生电荷，抑制光生电子和空穴的结合；②作为电子给体，与光生空穴反应产生 $^{\cdot}OH$ 自由基；③作为电子接受体，与光生电子反应产生 H 原子。以 CO_3^{2-}或 HCO_3^-形式存在的溶解性 CO_2，可被视为空穴捕获剂，并与 H_2O 分子与空穴反应相竞争，产生 $CO_3^{\cdot-}$自由基。池上等（Ikeue et al., 2001）发现，H_2O/CO_2 投加比较大时，催化材料表面吸附的—OH 基团增加，提高了光生电子的还原和中间产物的质子化，提升了 CH_4 产率；当 H_2O 的投加量较少时，则催化剂表面较少的—OH 基团有利于气相体系选择性生成 CH_3OH，而抑制 CH_4 的形成。安东尼·斯坦丁（Anthony Standing）等（Zhang et al., 2009）的研究也证明，在 Pt/TiO_2 纳米管光催化 CO_2 还原的过程中，提高 H_2O/CO_2 的投加比能促进 CH_4 的生成。但该研究还发现，在以 Pt/TiO_2 纳米颗粒为催化剂时，H_2O/CO_2 的投加比对产物选择性影响不大。因此可以得出结论，半导体光催化 CO_2 还原气相体系中的反应活性及产物的选择性不是由某个单一因素所决定的，而是由多种因素相互作用决定的。

实验研究表明，在气相中加入 H_2、H_2S 等可以提高反应体系中产物的产率和选择性。加入油酸三乙醇胺（Wang et al., 2015）、异丙醇（Kaneco et al.，1998）等极性溶剂作为牺牲剂可以提高 CO_2 的还原效率。堂免一成等（Lo et al.，2007）研究了加入不同还原剂（H_2、H_2+H_2O、H_2O）对 TiO_2 和 ZrO_2 光催化 CO_2 还原反应产物选择性的影响。结果表明，同时添加 H_2 和 H_2O，对 TiO_2 光催化 CO_2 还原产 CH_4 的效率最佳；ZrO_2 则在 H_2 气氛下对产物 CO 的选择性最好。叶金花课题组（Li et al.，2015）研究了以 H_2O 作为还原剂，采用 Ru/$NaTaO_3$ 和 Pt/$NaTaO_3$ 材料光催化 CO_2 还原的反应活性及产物选择性。研究表明，反应过程中加入 H_2 可以明显提高光催化还原 CO_2 的反应活性，并有效提升 Ru/$NaTaO_3$ 对 CH_4 产物以及 Pt/$NaTaO_3$ 对 CO 产物的选择性。通过同位素示踪法证明，H_2 在反应过程中仅作为电子给体，将反应中间产物 O^* 转化为 H_2O 分子，而并非直接还原 CO_2。此外，该课题组还证明了以 $N_2H_4 \cdot H_2O$ 代替 H_2O 分子，也可以提高反应活性及产物选择性（Kang et al.，2015），这是因为反应过程中 $N_2H_4 \cdot H_2O$ 不仅可以作为还原剂，而且可以作为牺牲剂来提供额外电子。

理论计算发现，不同的 CO_2 还原产物还会受到不同的反应路径影响。以 TiO_2 为模型的理论计算显示，在 CO_2 还原的过程中，可能存在两种不同的反应路径（Habisreutinger et al.，2013；Ji & Luo，2016），一种是快速脱氧路径，亦称碳烯路径：$CO_2 \longrightarrow CO \longrightarrow C^* \longrightarrow CH_2 \longrightarrow CH_4$；另一种是快速加氢路径，亦称甲醛路径：$CO_2 \longrightarrow HCOOH \longrightarrow HCHO \longrightarrow CH_3OH \longrightarrow CH_4$。反应路径的揭示也有助于理解 CO_2 还原产物选择性的影响因素。

（四）新型人工光合作用系统

由于光催化 CO_2 还原反应的动力学比较缓慢，人们正在积极探索新的人工光合成系统，希望实现热力学和动力学更有利的反应路径。目前该研究已取得一定的进展。加州大学杨培东课题组（Sakimoto et al.，2016）创新性地将具有非光合作用的热醋穆尔氏菌（*Moorella thermoacetica*）与无机半导体 CdS 纳米粒子进行组合，构建了细菌杂化的无机半导体人工光合系统。如图 1-78 所示，在光照条件下，该系统中的 CdS 能够提供光生电子-空穴对。但与传统人工光合成系统不同的是，光激发 CdS 产生的电子将会注入细菌体内，由细菌利用自身的新陈代谢机制完成 CO_2 的还原工作。该系统能够将 CO_2 转化为 CH_3COOH。此外，哈佛大学的斯利弗（Sliver）和诺塞拉等（Liu et al.，2016）开发了一种基于富养罗尔斯通氏菌（*Ralstonia eutropha*）杂化 Co-P 水分解电催化剂的人工光合作用系统。其工作原理是，Co-Pi 阳极与 Co-P 阴极分别氧化和还原 H_2O，生成 O_2 和 H_2。富养罗尔斯通氏菌能够以 H_2 为电子供体，

将 CO_2 转化为生物质或生物燃料，如聚三羟基丁酸酯和 C3～C5 醇类。同样，扬氏梭菌（*Clostridium ljungdahlii*）和木醋杆菌（*Acetobacterium woodii*）等细菌均可通过厌氧乙酰辅酶 A（acetyl-CoA）提供电子，并将 CO_2 还原为有机酸或醇（Ragsdale，1997）。当该类系统与太阳能转化器件耦合时，其能量转化效率远高于自然界中光合作用的太阳能转换效率。

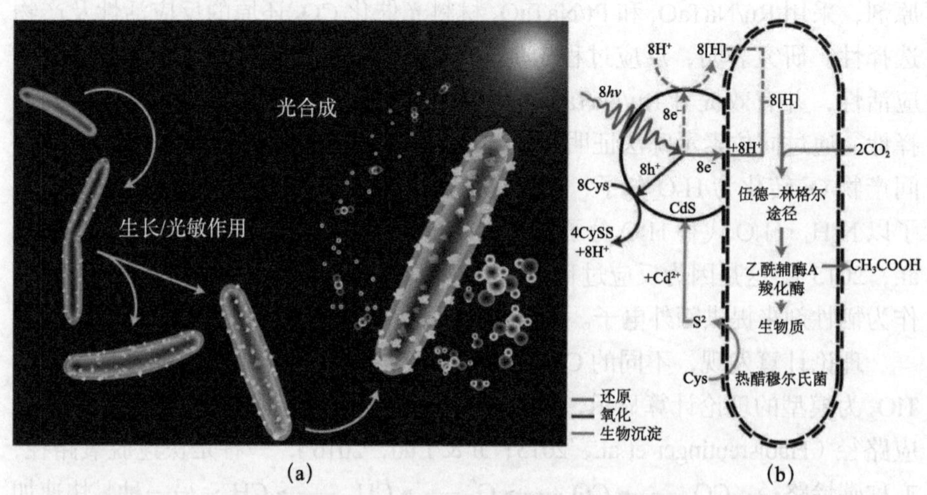

图 1-78　细菌 / 无机半导体杂化的人工光合作用系统还原 CO_2 示意图

（a）热醋穆尔氏菌与无机半导体 CdS 杂化系统。图中箭头方向表示该杂化体系的光合作用过程，包含细胞捕捉 CdS 颗粒，并在光合作用生长过程中将 CO_2 转化为乙酸。（b）热醋穆尔氏菌与无机半导体 CdS 杂化系统产物形成路径图。两种可能的路径：一种是细胞外形成途径（如点画线所示）；一种是直接电子传输到细胞的生成途径（实线所示路径）。Cys：半胱氨酸；CySS：胱氨酸

三、光催化 CO_2 还原材料目前面临的重大科学问题及前沿方向

目前，人们对光催化 CO_2 还原材料的认知与研究取得了长足进步，但是其效率仍然较低，远未达到工业化生产条件。若能有效地解决以下这些问题，光催化 CO_2 还原的应用将上一个台阶。

（1）半导体材料只有吸收相应波长的太阳光才能激发电子，产生光生电子-空穴对，从而驱动光催化反应的进行。因此，只有尽可能地缩小催化剂的光学带隙，才有可能实现高的太阳光利用率。要驱动光催化反应的进行，催化剂的导价带位置须横跨反应所需的氧化还原电位。

（2）由反应机理可知，氧化还原反应的驱动力为光生电子-空穴对，它们在从材料体相迁移至表面活性位点以及参与反应的过程中，存在着严重的

电荷复合（体相复合和表面复合），导致实际参与反应的光生载流子数量减少，大大降低了光催化反应的效率。

（3）CO_2 活化是化学反应中最具挑战性的过程之一，因为 CO_2 具有高度稳定性和惰性，C═O 键的破坏以及 C—H 键的形成需要消耗很高的能量（Halmann，1978），此过程也涉及多个电子和相应数量的质子的参与，从而造成 CO_2 的催化还原需要较高的反应势垒，导致反应动力学缓慢。同时，CO_2 的还原还涉及 H_2O 的氧化，而这一步恰恰是 H_2O 分解的速率控制步骤。虽然引入 O 空位能够活化 CO_2 或 H_2O，但如何同时活化 CO_2 和 H_2O，以匹配氧化和还原两个半反应之间的反应速率，减少空穴累积和光生电子-空穴对复合仍然是一大难点。

（4）光催化 CO_2 还原反应涉及多电子转移，产物较多，但对于气相反应而言，CH_4 和 CO 却是常见产物。催化剂不同，其还原产物有可能也不同。虽然可通过实验仪器检测到不同的产物，但其具体反应机理尚不明确，这也是光催化 CO_2 还原反应研究中的一大障碍。

四、研究总结与展望

CO_2 是造成温室效应的主要气体之一，也是非常重要的碳源。基于半导体光催化技术，可直接利用太阳能，将 CO_2 和 H_2O 转换为高附加值的碳氢燃料，实现太阳能到化学能的有效转化与存储，从而实现碳资源的循环利用，这不仅为优化能源结构、解决由 CO_2 排放引起的环境负效应提供了一种理想途径，更是实现社会可持续发展的有力保障。总体来说，通过近几十年的研究，光催化 CO_2 还原得到了十足的发展：光催化材料的太阳能转换效率逐步提高，改善光催化反应效率的手段趋于明确化，对光催化机理的认识逐步深入、表征手段快速发展，基于新奇物理机制的光催化材料逐渐兴起。但是，为达到实用化目标，CO_2 还原的研究仍有待质的飞跃发展。首先面临的巨大问题是如何大幅度地提高光催化 CO_2 还原性能，除了需要依赖新材料的开发，如何实现光催化材料带隙与太阳光谱匹配、如何实现光催化材料的导价带位置与反应物电极电位匹配、如何降低光生电子-空穴对复合提高量子效率、如何提高光催化材料的稳定性等问题仍是这一领域必须要解决的关键科学问题。此外，现有的表征技术已不能完全满足在催化机理方面认识的需求，因此，需要借助于一些先进的原位表征手段，从光催化物理本质出发，揭示影响光催化反应过程的关键因素所在。深化对光催化反应机制的认识，由宏观、定性的描述到微观、定量的研究，对光吸收、光生载流子的激发和输运过程以及界面动力学过程进行综合研究，阐明能量传递和转换的机制，以指导如何高效地发挥现有光催化材料的催化活性和开发高量子效率的光催化材料。突破现有理论框架，积极推进光化学学科与其他学科的交叉融合，实现高效稳定的光驱动 CO_2 转化为可再生的碳基能源的新途径。

五、学科发展政策建议与措施

（一）大力发展原位光谱学表征手段

CO_2 还原反应路径复杂，产物选择性低且与产氢反应相竞争。其反应路径的复杂性带来了中间物种的多样性，亟须发展时间分辨的原位光谱学手段捕捉中间态物种，以促进对 CO_2 还原催化反应机制的认识，进而指导催化材料的设计。

（二）发展高效材料设计方法

目前，由于缺乏材料设计的指导性理论方法，对于 CO_2 还原材料的开发仍然采用传统的试错法，从而制约了这一领域的发展。基于机器学习的高效材料设计方法已经在材料设计领域内初步显示出了优越性，有助于发现能够精确描述材料催化特性的描述符，帮助人们更加深入地理解催化机理。

闫世成（南京大学）

本节参考文献

Bai S, Jiang J, Zhang Q, et al. 2015. Steering charge kinetics in photocatalysis: intersection of materials syntheses, characterization techniques and theoretical simulations. Chemical Society Reviews, 44: 2893-2939.

Chang X X, Wang T, Gong J L. 2016. CO_2 photo-reduction: insights into CO_2 activation and reaction on surfaces of photocatalysts. Energy & Environmental Science, 9: 2177-2196.

Chen X Y, Zhou Y, Liu Q, et al. 2012. Ultrathin, single-crystal WO_3 nanosheets by two-dimensional oriented attachment toward enhanced photocatalystic reduction of CO_2 into hydrocarbon fuels under visible light. ACS Applied Materials & Interfaces, 4: 3372-3377.

Cheng H F, Huang B B, Liu Y Y, et al. 2012. An anion exchange approach to Bi_2WO_6 hollow microspheres with efficient visible light photocatalytic reduction of CO_2 to methanol. Chemical Communications, 48: 9729-9731.

Copperthwaite R G, Davies P R, Morris M A, et al. 1988. The reactive chemisorption of carbon dioxide at magnesium and copper surfaces at low temperature. Catalysis Letters, 1: 11-19.

Dimitrijevic N M, Vijayan B K, Poluektov O G, et al. 2011. Role of water and carbonates in photocatalytic transformation of CO_2 to CH_4 on titania. Journal of the American Chemical Society, 133: 3964-3971.

Espinal L, Wong-Ng W, Kaduk J A, et al. 2012. Time dependent CO_2 sorption hysteresis by octahedral molecular sieves with manganese oxide framework. Journal of the American Chemical Society, 134: 7944-7951.

Fujishima A, Honda K. 1972. Electrochemical photolysis of water at a semiconductor electrode. Nature, 238: 37-38.

Ghoussoub M, Yadav S, Ghuman K K, et al. 2016. Metadynamics-biased *ab initio* molecular dynamics study of heterogeneous CO_2 reduction via surface frustrated Lewis pairs. ACS Catalysis, 6: 7109-7117.

Ghuman K K, Hoch L B, Szymanski P, et al. 2016. Photoexcited surface frustrated Lewis pairs for heterogeneous photocatalytic CO_2 reduction. Journal of the American Chemical Society, 138: 1206-1214.

Ghuman K K, Wood T E, Hoch L B, et al. 2015. Illuminating CO_2 reduction on frustrated Lewis pair surfaces: investigating the role of surface hydroxides and oxygen vacancies on nanocrystalline $In_2O_{(3-x)}(OH)_y$. Physical Chemistry Chemical Physics, 17: 14623-14635.

Habisreutinger S N, Schmidt-Mende L, Stolarczyk J K. 2013. Photocatalytic reduction of CO_2 on TiO_2 and other semiconductors. Angewandte Chemie International Edition, 52: 7372-7408.

Halmann M. 1978. Photoelectrochemical reduction of aqueous carbon dioxide on p-type gallium phosphide in liquid junction solar cells. Nature, 275: 115-116.

Halmann M, Ulman M, Aurian-Blajeni B. 1983. Photochemical solar collector for the photoassisted reduction of aqueous carbon dioxide. Solar Energy, 31: 429-431.

Hoch L B, Szymanski P, Ghuman K K, et al. 2016. Carrier dynamics and the role of surface defects: designing a photocatalyst for gas-phase CO_2 reduction. Proceedings of the National Academy of Sciences USA, 113: E8011-E8020.

Hou J G, Cao S Y, Wu Y Z, et al. 2016. Perovskite-based nanocubes with simultaneously improved visible-light absorption and charge separation enabling efficient photocatalytic CO_2 reduction. Nano Energy, 30: 59-68.

Hou W B, Hung W H, Pavaskar P, et al. 2011. Photocatalytic conversion of CO_2 to hydrocarbon fuels via plasmon-enhanced absorption and metallic interband transitions. ACS Catalysis, 1:

929-936.

Ikeue K, Yamashita H, Anpo M, et al. 2001. Photocatalytic reduction of CO_2 with H_2O on Ti-β zeolite photocatalysts: effect of the hydrophobic and hydrophilic properties. Journal of Physical Chemistry B, 105: 8350-8355.

Inoue T, Fujishima A, Konishi S, et al. 1979. Photoelectrocatalytic reduction of carbon dioxide in aqueous suspensions of semiconductor powders. Nature, 277: 637-638.

Ishitani O, Inoue C, Suzuki Y, et al. 1993. Photocatalytic reduction of carbon dioxide to methane and acetic acid by an aqueous suspension of metal-deposited TiO_2. Journal of Photochemistry & Photobiology A: Chemistry, 72: 269-271.

Ji Y F, Luo Y. 2016. New mechanism for photocatalytic reduction of CO_2 on the anatase TiO_2 (101) surface: the essential role of oxygen vacancy. Journal of the American Chemical Society, 138: 15896-15902.

Jiang Z F, Wan W M, Li H M, et al. 2018. A hierarchical Z-scheme α-Fe_2O_3/g-C_3N_4 hybrid for enhanced photocatalytic CO_2 reduction. Advanced Materials, 30: 1706108.

Jiao W, Wang L Z, Liu G, et al. 2012. Hollow anatase TiO_2 single crystals and mesocrystals with dominant {101} facets for improved photocatalysis activity and tuned reaction preference. ACS Catalysis, 2: 1854-1859.

Kaneco S, Shimizu Y, Ohta K, et al. 1998. Photocatalytic reduction of high pressure carbon dioxide using TiO_2 powders with a positive hole scavenger. Journal of Photochemistry & Photobiology A: Chemistry, 115: 223-226.

Kang Q, Wang T, Li P, et al. 2015. Photocatalytic reduction of carbon dioxide by hydrous hydrazine over Au-Cu alloy nanoparticles supported on $SrTiO_3$/TiO_2 coaxial nanotube arrays. Angewandte Chemie International Edition, 54: 841-845.

Kočí K, Obalová L, Matějová L, et al. 2009. Effect of TiO_2 particle size on the photocatalytic reduction of CO_2. Applied Catalysis B: Environmental, 89: 494-502.

Lang Q Q, Yang Y J, Zhu Y Z, et al. 2017. High-index facet engineering of PtCu cocatalysts for superior photocatalytic reduction of CO_2 to CH_4. Journal of Materials Chemistry A, 5: 6686-6694.

Lee J, Sorescu D C, Deng X Y. 2011. Electron-induced dissociation of CO_2 on TiO_2(110). Journal of the American Chemical Society, 133: 10066-10069.

Li H L, Lei Y G, Huang Y, et al. 2011. Photocatalytic reduction of carbon dioxide to methanol by Cu_2O/SiC nanocrystallite under visible light irradiation. Journal of Natural Gas Chemistry, 20: 145-150.

Li M, Li P, Chang K, et al. 2015. Highly efficient and stable photocatalytic reduction of CO_2 to CH_4 over Ru loaded $NaTaO_3$. Chemical Communications, 51: 7645-7648.

Li M L, Zhang L X, Wu M Y, et al. 2016. Mesostructured CeO_2/*g*-C_3N_4 nanocomposites: remarkably enhanced photocatalytic activity for CO_2 reduction by mutual component activations. Nano Energy, 19: 145-155.

Liu C, Colón B C, Ziesack M, et al. 2016. Water splitting-biosynthetic system with CO_2 reduction efficiencies exceeding photosynthesis. Science, 352: 1210-1213.

Liu G, Yu J C, Lu G Q M, et al. 2011. Crystal facet engineering of semiconductor photocatalysts: motivations, advances and unique properties. Chemical Communications, 47: 6763-6783.

Liu L J, Zhao C Y, Li Y. 2012. Spontaneous dissociation of CO_2 to CO on defective surface of Cu(I)/TiO_{2-x} nanoparticles at room temperature. Journal of Physical Chemistry C, 116: 7904-7912.

Liu L J, Zhao H L, Andino J M, et al. 2012. Photocatalytic CO_2 reduction with H_2O on TiO_2 nanocrystals: comparison of anatase, rutile, and brookite polymorphs and exploration of surface chemistry. ACS Catalysis, 2: 1817-1828.

Liu Q, Zhou Y, Kou J H, et al. 2010. High-yield synthesis of ultralong and ultrathin Zn_2GeO_4 nanoribbons toward improved photocatalytic reduction of CO_2 into renewable hydrocarbon fuel. Journal of the American Chemical Society, 132: 14385-14387.

Liu Y Y, Huang B B, Dai Y, et al. 2009. Selective ethanol formation from photocatalytic reduction of carbon dioxide in water with $BiVO_4$ photocatalyst. Catalysis Communications, 11: 210-213.

Liu Y Y, Yang Y M, Sun Q L, et al. 2013. Chemical adsorption enhanced CO_2 capture and photoreduction over a copper porphyrin based metal organic framework. ACS Applied Materials & Interfaces, 5: 7654-7658.

Lo C C, Hung C H, Yuan C S, et al. 2007. Photoreduction of carbon dioxide with H_2 and H_2O over TiO_2 and ZrO_2 in a circulated photocatalytic reactor. Solar Energy Materials & Solar Cells, 91: 1765-1774.

Lu L, Wang B, Wang S M, et al. 2017. La_2O_3-modified $LaTiO_2N$ photocatalyst with spatially separated active sites achieving enhanced CO_2 reduction. Advanced Functional Materials, 27: 1702447.

Lu L, Wang S M, Zhou C G, et al. 2018. Surface chemistry imposes selective reduction of CO_2 to CO over Ta_3N_5/$LaTiO_2N$ photocatalyst. Journal of Materials Chemistry A, 6: 14838-14846.

Luo D M, Bi Y, Kan W, et al. 2011. Copper and cerium co-doped titanium dioxide on catalytic photo reduction of carbon dioxide with water: experimental and theoretical studies. Journal of Molecular Structure, 994: 325-331.

Mao J, Peng T Y, Zhang X H, et al. 2012. Selective methanol production from photocatalytic reduction of CO_2 on $BiVO_4$ under visible light irradiation. Catalysis Communications, 28:

38-41.

Ménard G, Stephan D W. 2010. Room temperature reduction of CO_2 to methanol by Al-based frustrated Lewis pairs and ammonia borane. Journal of the American Chemical Society, 132: 1796-1797.

Mömming C M, Otten E, Kehr G, et al. 2009. Reversible metal-free carbon dioxide binding by frustrated Lewis pairs. Angewandte Chemie International Edition, 48: 6643-6646.

Ong W J, Tan L L, Chai S P, et al. 2015. Heterojunction engineering of graphitic carbon nitride (g-C_3N_4) via Pt loading with improved daylight-induced photocatalytic reduction of carbon dioxide to methane. Dalton Transactions, 44: 1249-1257.

Ong W J, Tan L L, Chai S P, et al. 2015. Surface charge modification via protonation of graphitic carbon nitride (g-C_3N_4) for electrostatic self-assembly construction of 2D/2D reduced graphene oxide (rGO)/g-C_3N_4 nanostructures toward enhanced photocatalytic reduction of carbon dioxide to methane. Nano Energy, 13: 757-770.

Qiao J L, Liu Y Y, Hong F, et al. 2014. A review of catalysts for the electroreduction of carbon dioxide to produce low-carbon fuels. Chemical Society Reviews, 43: 631-675.

Ragsdale S W. 1997. The Eastern and Western branches of the Wood/Ljungdahl pathway: how the East and West were won? Biofactors, 6: 3-11.

Ran J R, Jaroniec M, Qiao S Z. 2018. Cocatalysts in semiconductor-based photocatalytic CO_2 reduction: achievements, challenges, and opportunities. Advanced Materials, 30: 1704649.

Sakimoto K K, Wong A B, Yang P D. 2016. Self-photosensitization of nonphotosynthetic bacteria for solar-to-chemical production. Science, 351: 74-77.

Sun S M, Watanabe M, Wu J, et al. 2018. Ultrathin $WO_3 \cdot 0.33H_2O$ nanotubes for CO_2 photoreduction to acetate with high selectivity. Journal of the American Chemical Society, 140: 6474-6482.

Tahir B, Tahir M, Amin N A S. 2017. Photo-induced CO_2 reduction by CH_4/H_2O to fuels over Cu-modified g-C_3N_4 nanorods under simulated solar energy. Applied Surface Science, 419: 875-885.

Tahir M, Amin N S. 2015. Indium-doped TiO_2 nanoparticles for photocatalytic CO_2 reduction with H_2O vapors to CH_4. Applied Catalysis B: Environmental, 162: 98-109.

Tamaki Y, Morimoto T, Koike K, et al. 2012. Photocatalytic CO_2 reduction with high turnover frequency and selectivity of formic acid formation using Ru(Ⅱ) multinuclear complexes. Proceedings of the National Academy of Sciences USA, 109: 15673-15678.

Tu W G, Zhou Y, Li H J, et al. 2015. Au@TiO_2 yolk-shell hollow spheres for plasmon-induced photocatalytic reduction of CO_2 to solar fuel via a local electromagnetic field. Nanoscale, 7: 14232-14236.

Tu W G, Zhou Y, Zou Z G. 2014. Photocatalytic conversion of CO_2 into renewable hydrocarbon fuels: state-of-the-art accomplishment, challenges, and prospects. Advanced Materials, 26: 4607-4626.

Wang B, Wang X H, Lu L, et al. 2018. Oxygen-vacancy-activated CO_2 splitting over amorphous oxide semiconductor photocatalyst. ACS Catalysis, 8: 516-525.

Wang S B, Ding Z X, Wang X C. 2015. A stable $ZnCo_2O_4$ cocatalyst for photocatalytic CO_2 reduction. Chemical Communications, 51: 1517-1519.

Wang S B, Guan B Y, Lu Y, et al. 2017. Formation of hierarchical In_2S_3-$CdIn_2S_4$ heterostructured nanotubes for efficient and stable visible light CO_2 reduction. Journal of the American Chemical Society, 139: 17305-17308.

Wang S M, Guan Y, Lu L, et al. 2018. Effective separation and transfer of carriers into the redox sites on Ta_3N_5/Bi photocatalyst for promoting conversion of CO_2 into CH_4. Applied Catalysis B: Environmental, 224: 10-16.

Wang S P, Yan S L, Ma X B, et al. 2011. Recent advances in capture of carbon dioxide using alkali-metal-based oxides. Energy & Environmental Science, 4: 3805-3819.

Wang W N, An W J, Ramalingam B, et al. 2012. Size and structure matter: enhanced CO_2 photoreduction efficiency by size-resolved ultrafine Pt nanoparticles on TiO_2 single crystals. Journal of the American Chemical Society, 134: 11276-11281.

Wu J H, Huang Y, Ye W, et al. 2017. CO_2 reduction: from the electrochemical to photochemical approach. Advanced Science, 4: 1700194.

Xi G C, Ouyang S X, Li P, et al. 2012. Ultrathin $W_{18}O_{49}$ nanowires with diameters below 1 nm: synthesis, near-infrared absorption, photoluminescence, and photochemical reduction of carbon dioxide. Angewandte Chemie International Edition, 51: 2395-2399.

Xi G C, Ouyang S X, Ye J H. 2011. General synthesis of hybrid TiO_2 mesoporous "French fries" toward improved photocatalytic conversion of CO_2 into hydrocarbon fuel: a case of TiO_2/ZnO. Chemistry: A European Journal, 17: 9057-9061.

Xie K, Umezawa N, Zhang N, et al. 2011. Self-doped $SrTiO_{3-\delta}$ photocatalyst with enhanced activity for artificial photosynthesis under visible light. Energy & Environmental Science, 4: 4211-4219 .

Xie S J, Wang Y, Zhang Q H, et al. 2013. Photocatalytic reduction of CO_2 with H_2O: significant enhancement of the activity of Pt-TiO_2 in CH_4 formation by addition of MgO. Chemical Communications, 49: 2451-2453.

Yan S C, Wang J J, Gao H L, et al. 2013. An ion-exchange phase transformation to $ZnGa_2O_4$ nanocube towards efficient solar fuel synthesis. Advanced Functional Materials, 23: 758-763.

Zhang L H, Zhu D, Nathanson G M, et al. 2014. Selective photoelectrochemical reduction of aqueous CO_2 to CO by solvated electrons. Angewandte Chemie International Edition, 126: 9904-9908.

Zhang N, Long R, Gao C, et al. 2018. Recent progress on advanced design for photoelectrochemical reduction of CO_2 to fuels. Science China Materials, 61: 771-805.

Zhang Q H, Han W D, Hong Y J, et al. 2009. Photocatalytic reduction of CO_2 with H_2O on Pt-loaded TiO_2 catalyst. Catalysis Today, 148: 335-340.

Zhao Y F, Chen G B, Bian T, et al. 2015. Defect-rich ultrathin ZnAl-layered double hydroxide nanosheets for efficient photoreduction of CO_2 to CO with water. Advanced Materials, 27: 7824-7831.

Zheng Y, Zhang W Q, Li Y F, et al. 2017. Energy related CO_2 conversion and utilization: advanced materials/nanomaterials, reaction mechanisms and technologies. Nano Energy, 40: 512-539.

Zhou C G, Wang S M, Zhao Z Y, et al. 2018. A facet-dependent Schottky-junction electron shuttle in a $BiVO_4$\{010\}-Au-Cu_2O Z-scheme photocatalyst for efficient charge separation. Advanced Functional Materials, 28: 1801214.

第六节　光伏电池材料

一、光伏电池材料研究概述

当代社会，随着社会经济的快速发展，人们对资源的依赖度不断提升，石油、天然气、煤炭等传统能源物质逐渐消耗，这些能源都属于不可再生资源，而且已经不能满足社会的实际需求了，能源问题的严重性愈发显露。目前，太阳能、地热能、潮汐能、风能等新型能源的开发为解决能源问题提供了有效的策略，其中太阳能在使用过程中不会对环境产生污染，而且因其取之不尽、用之不竭的显著特点备受人们关注。太阳能又称光电池，能通过光—电或光—热—电的途径将太阳能转化成电能。但是，目前大多数太阳能电池存在制造成本昂贵、废弃物污染环境、能量转化效率低和使用稳定性差等缺点，导致其在实际应用中受到极大的限制。因此，开发经济、环境友好型材料用以制备高效且稳定的太阳能电池成为研究的重点。按照材料的不同，太阳能电池可分为薄膜太阳能电池、有机太阳能电池、钙钛矿太阳能电池、硅基太阳能电池、染料敏化太阳能电池等类别，本节选取其中具有代表性的前三种太阳能电池材料阐述其发展历史及现状。

二、光伏电池材料研究的历史及现状

（一）薄膜太阳能电池材料

目前研究较多的薄膜太阳能电池材料有 3 种，即非晶硅薄膜太阳能电池（a-Si）、Ⅲ-Ⅴ族化合物薄膜太阳能电池和Ⅱ-Ⅵ族化合物薄膜太阳能电池，其中非晶硅薄膜太阳能电池的生产比例最大，在薄膜太阳能电池中占 38%。

1. 非晶硅薄膜太阳能电池

非晶硅薄膜太阳能电池一般采用等离子增强型化学气相沉积法制备，所用原料为高纯硅烷，在一定的温度下分解，在玻璃、不锈钢板、陶瓷板、柔性塑料片上沉积薄膜，故通常情况下，非晶硅薄膜太阳能电池的厚度小于 0.5 μm，虽薄薄一层却可有效吸收光子能量，可节省很多硅材料；生产中可以连续在多个真空沉积室完成，沉积分解温度低，易大批量生产。另外，硅烷价格十分便宜，化学工业产量丰富，成本低。

2. Ⅲ-Ⅴ族化合物薄膜太阳能电池

Ⅲ-Ⅴ族的半导体材料具有直接能隙，与太阳光谱匹配较好，典型代表为 GaAs 电池。采用化学气相沉积法在单晶硅基板上析出 GaAs 薄膜，制成薄膜太阳能电池，因其吸光系数高，光电转换率可达 33%（Kojima et al.，2009），且该电池耐反射损伤性佳，对温度变化不敏感，多应用于宇宙空间探测器上，如“神舟三号”飞船、“神舟八号”飞船、“天宫一号”飞船等。该类电池单位价格比多晶硅薄膜太阳能电池高数十倍甚至百倍，因此不是民用主流。

3. Ⅱ-Ⅵ族化合物薄膜太阳能电池

1）碲化镉薄膜太阳能电池

碲化镉能隙值为 1.45 eV，处于理想太阳能电池的能隙范围内且吸光系数高，为获得高光电转化率奠定了基础。加之快速成膜技术的应用，大大缩短了加工时间，提高了生产效率，因而极具应用潜力。近年来，碲化镉薄膜太阳能电池已广泛用于屋顶太阳能电池的制备，但镉污染是制约该薄膜太阳能电池发展的因素。

2）铜铟镓硒薄膜太阳能电池

铜铟镓硒吸光范围广，稳定性优良，尤其是在户外环境下，材料制造成本低，符合太阳能电池材料选材要求；在转化效率方面，在标准光强照射条件下为 14% 左右，采取聚光措施能达到 30% 左右，接近单晶硅太阳能电池的转化效率。其抗辐射损伤能力强，不仅可用于大面积的地面发电，还可应用于太空领域。目前铜铟镓硒薄膜太阳能电池的应用普及度不高，仍需进一步提高光电转换效率及大幅降低成本，这是未来努力的方向。

3）柔性衬底薄膜太阳能电池

在柔性衬底如织布上制备薄膜太阳能电池，可任意卷曲折叠，成本低廉，应用于太阳能背包、太阳能敞篷等，另外可以集成在窗户或屋顶、外墙、汽车、飞机上形成光伏建筑一体化、太阳能汽车，甚至太阳能飞机等，应用前景广阔。目前美国托莱多大学在柔性衬底非晶硅太阳能电池领域的研究处于世界领先地位。

虽然薄膜太阳能电池尚未形成产业化，但材料消耗少、制备能耗低、一个车间内就可完成组件生产的优势势不可挡，相信若干年后随着科学技术的不断进步，薄膜太阳能电池的转化效率将不是问题，规模化生产指日可待，加之光伏建筑一体化等分布式光伏的建设与应用，预计薄膜太阳能电池未来可能成为市场的主流选择。

（二）有机太阳能电池材料

有机太阳能电池材料种类繁多，可大体分为四类：小分子太阳能电池材料、大分子太阳能电池材料、D-A 体系材料和有机无机杂化体系材料。以共轭聚合物为基体代替无机材料制备太阳能电池掀起又一轮太阳能电池的研究热潮。

有机太阳能电池是一类以有机材料为基础的光电转换材料，主要原理是利用有机化合物材料以光伏效应产生电压，形成电流，实现太阳光向电能的转化。有机太阳能电池的突出优势在于使用的原料为聚合物分子，具有成本低廉、工艺简单、可塑性强、便于实现柔性可折叠的透明电极等优点。尤其是在近几十年的发展过程中，有机太阳能电池材料在器件制备以及材料合成等方面得到了充分的应用。相关研究表明，在实验室条件下，研究者通过不断优化制备条件，将有机太阳能电池的转换效率提高到 15%。

（三）钙钛矿太阳能电池材料

近年来，以钙钛矿太阳能电池为代表的新型光伏器件研究受到了广泛关注，得到了快速发展。2009 年世界上首次报道的钙钛矿太阳能电池的转换率仅有 3.8%（Kojima et al.，2009），之后这类电池得到科学界广泛的关注，钙钛矿太阳能电池也因此被《科学》（*Science*）评选为 2013 年十大科学突破之一。有机无机杂化材料 $CH_3NH_3PbX_3$ 为钙钛矿晶型，是一种成本低廉、易成膜、窄带隙、吸光性能好、载流子迁移率高的双极性半导体材料，基于这类材料制备的薄膜太阳能电池被称为钙钛矿太阳能电池。根据最新公布的电池效率数据，钙钛矿太阳能电池认证的最高效率已高达 25.8%（经美国能源部可再生能源实验室认证的效率达到 25.5%）（Min et al.，2021），紧逼多晶硅和硫系异质结太阳能电池，具有极大的商业应用前景。

钙钛矿太阳能电池结构化合物的组成可表示为 ABX_3，A 代表有机阳离子，如甲胺离子（MA^+）、甲脒离子（FA^+）、Cs^+ 等；B 代表金属离子，如 Pb^{2+}、Sn^{2+} 等；X 代表卤素离子，如 Cl^-、Br^-、I^-等。其中卤素离子以共顶的途径相互连成八面体，单位八面体在三维空间内通过无限延伸而形成无机骨架结构（图 1-79）。金属离子位于卤素八面体的中心，有机阳离子层位于层间。无机层和有机层之间存在氢键，并且通过氢键力进行连接，相互交叠而形成稳定的类钙钛矿层状结构，此结构能够提高载流子的传输效率，从而能增加太阳能电池器件的光电转换效率以及改善其环境稳定性。这种独特的结构使其具备良好的非线性光学、磁和传导、电致发光等物理性质。

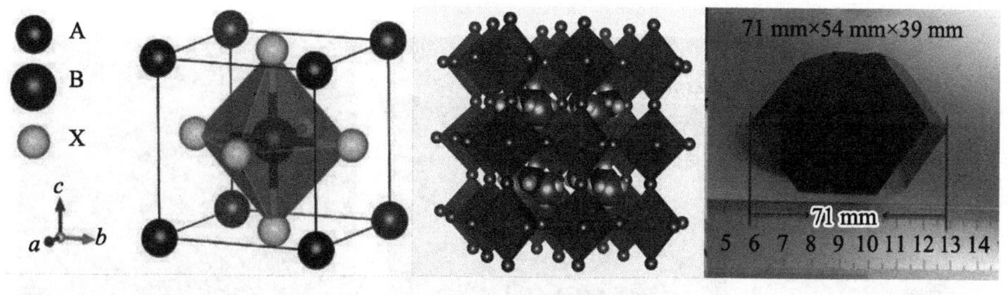

图 1-79　钙钛矿晶体结构

经过近几年的发展，钙钛矿太阳能电池也衍生出多种结构，如图 1-80 所示，分为介孔结构、三重介孔结构、平面结构、叠层结构。其中，介孔结构电池又分为传统介孔结构（n-i-p）与反向介孔结构（p-i-n）；三重介孔结构是以全印刷工艺为主的 C 基无空穴传输层电池结构；平面结构电池也分为传统平面结构（n-i-p）与反向平面结构（p-i-n）；叠层结构电池一般以钙钛矿电池叠加低禁带宽度的 C-Si、CIGS 等为主。图 1-81 和图 1-82 分别显示了传统介孔结构（n-i-p）钙钛矿太阳能电池的微观结构和工作原理。

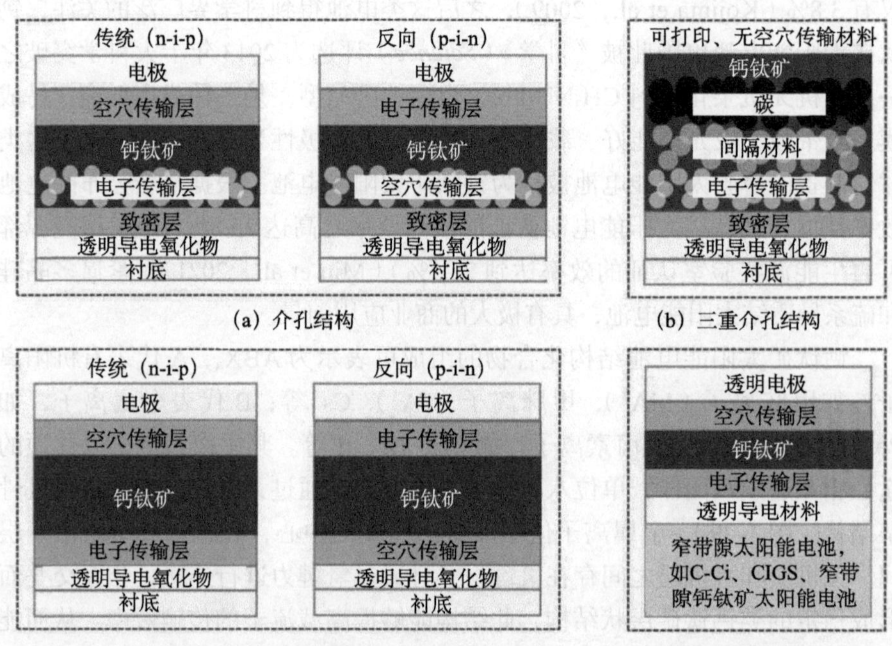

图 1-80 各种钙钛矿太阳能电池结构示意图

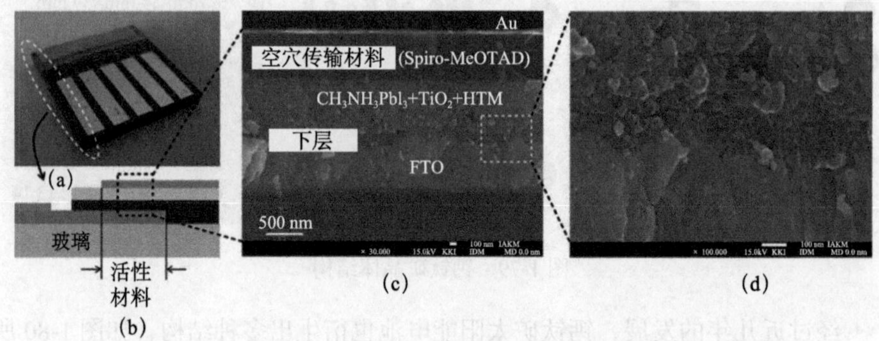

图 1-81 传统介孔结构（n-i-p）钙钛矿太阳能电池微观结构图

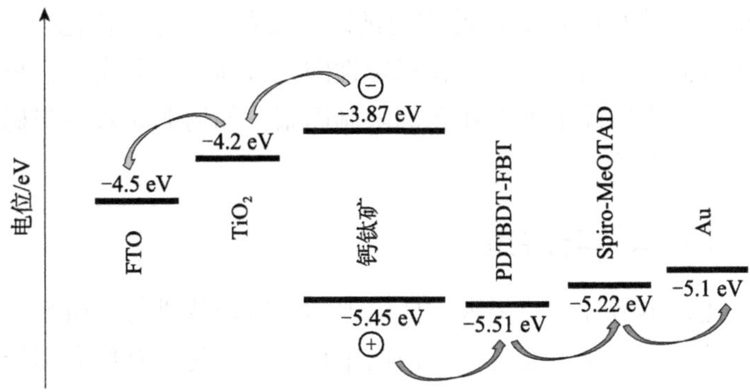

图 1-82 钙钛矿太阳能电池工作原理示意图

三、光伏电池材料研究目前面临的重大科学问题及前沿方向

（一）薄膜太阳能电池

我国薄膜太阳能电池产业起步晚，在最近几年才成规模，技术水平、行业成熟度、供应链等均处于不断探索的过程中。另外，铟和镓的蕴藏量有限且制造工艺复杂、投资成本高等因素对 CIGS 太阳能电池的发展也有一定的影响。砷化镓太阳能电池中砷有毒，对环境安全和生产工人自身的安全都是一个不小的威胁。薄膜太阳能电池生产设备复杂昂贵，降低薄膜太阳能电池成本的关键是实现薄膜产业高端装备国产化，扩大薄膜太阳能电池的产量，以规模化带动成本的降低和效率的提高。

（二）有机太阳能电池

尽管与之前提到的传统硅基太阳能电池相比还有一定的差距，如容量不够大、规模生产转化效率低等，但对有机太阳能电池的研究无论是在性能、机理还是稳定性等许多方面都尚处于初始阶段。因此，进一步借鉴无机太阳能电池的成熟技术及研究思路将会对有机太阳能电池的研究起到推动作用。

（三）钙钛矿太阳能电池

虽然目前钙钛矿太阳能电池的光电转化效率高，发展前景被人看好，但钙钛矿太阳能电池的使用还需要解决面积过小的问题，一般情况下，高效率电池的面积不超过 0.1 cm^2，因此实现钙钛矿太阳能电池产业化的关键在于制造出大面积且高效的转换材料。在稳定性问题方面，在阳光下，材料性能容易衰减是钙钛矿太阳能电池商业化的限制因素，如何兼顾效率与稳定性也是目前面临的一大难题。

四、研究总结与展望

综上所述，本节针对三类太阳能电池的发展现状进行了研究：薄膜太阳能电池材料消耗少、制备能耗低，但尚未形成产业化；有机太阳能电池制作过程简单易实现、生产成本低，但寿命和稳定性还需要提高；以钙钛矿为代表的有机-无机杂化太阳能电池的崛起为研究高效的太阳能电池提供了全新思路。太阳能电池作为一种新兴材料，在解决人类能源危机方面十分具有发展前景。因此，本节主要对各种材料的太阳能电池进行了现状分析，希望对今后太阳能电池材料的发展具有一定的指导意义。

五、学科发展政策建议与措施

（一）机理深入研究

机理的深入研究可指导设计与合成宽吸收和高迁移率的太阳能电池材料。随着器件性能的日益提高，其稳定性研究也将提到日程上来。结合有机材料、无机材料、纳米材料各自的优点，优化器件结构，改善材料性质，以提高太阳能电池的综合性能。

（二）材料设计

从材料的角度讲，目前需要做的是从廉价易得的原料出发，有针对性地设计合成一些化合物，对光诱导电子转移过程和机制进行研究，以指导材料的设计合成。同时，还需要对现有的材料体系进行复合优化，以取得最大效率。

总的来说，价廉、高效、能够大面积制备的太阳能电池材料一直是人们追求的目标。太阳能电池材料若能在分子水平上裁减和设计，简化生产工艺，制备大面积轻盈薄膜，在光电转换性能方面取得进一步的突破，将有可能在生产实践中得到广泛应用，市场前景十分广阔。

周　勇（南京大学）、王镜喆（南京大学昆山创新研究院）

本节参考文献

Kojima A, Teshima K, Shirai Y, et al. 2009. Organometal halide perovskites as visible-light sensitizers for photovoltaic cells. Journal of the American Chemical Society, 131: 6050-6051.

Min H, Lee D, Kim J, et al. 2021. Perovskite solar cells with atomically coherent interlayers on SnO_2 electrodes. Nature, 598: 444-450.

第七节 氢燃料电池

一、氢燃料电池研究概述

（一）产业背景及战略意义

21 世纪，能源与环境问题备受关注。随着全球经济的快速发展，能源的需求量也随之增加，而传统的化石能源如石油等资源日渐枯竭，全球正面临能源危机。与此同时，化石能源在燃烧过程中会产生大量 NO_x、SO_x 等气体，造成环境污染并令全球极端气候愈加频繁。根据联合国环境规划署 2017 年 10 月 31 日在日内瓦发布的《排放差距报告》，各国的减排承诺只达到了实现 2030 年温控目标所需减排水平的 1/3。因此，全球正面临着化石能源短缺、环境问题日益严重等重大问题，节能与环保已成为人类社会可持续发展战略的核心。解决能源消耗与节能环保相互矛盾的问题，关键是亟须开发清洁、可再生的新能源。可重复利用、环境友好且能源转化效率高的新能源技术的研究得到了各国政府大量的政策性扶持和财政支持。

氢能的高热值相比传统能源——汽油、煤炭等有较大优势。人类发展史伴随着能源的利用史，即木材—煤炭—石油（天然气）。可见对能源的利用方式，正在从对高碳能源的使用到对富氢低碳能源的开发转变。同时，氢能作为最清洁的能源之一，能满足人们对减缓气候恶化的需求。对于氢能的开发利用，从形式上主要分为两种：直接燃烧，利用其热值，可直接利用热能，亦可转化为电能使用；通过发电装置直接将氢能转化为电能，其中最重要的应用之一即燃料电池。

燃料电池是一种可以将储存在燃料和氧气中的化学能直接转化为电能的电化学储能装置。普通的内燃机由于需要经历热机过程，受卡诺循环的限制，其能量转化效率为25%～35%；燃料电池不受此限制，因而具有很高的能量转化效率，一般为40%～60%，如果将余热充分利用，甚至可以高达90%。此外，燃料电池在工作时，其反应产物主要为H_2O，几乎不排放NO_x和SO_x，因而不会污染环境，是新一代的高效环保的能源转化材料。即便是采用含碳的燃料，高效燃料电池在工作时排出的CO_2量，也仅相当于同等条件下的传统火力发电厂的60%。因此，燃料电池对解决目前全球所面临的能源短缺和环境污染两大难题都具有极其重要的意义。同时，燃料电池具有能量转化效率高、能量密度高、运行温度低、启动快、环境友好、可持续发电等优点，是未来清洁能源利用的理想选择。

燃料电池种类众多，分类方法也较多，目前常用的是根据电解质类型进行分类，包括PEMFC、碱性燃料电池（alkaline fuel cell，AFC）、磷酸型燃料电池（phosphoric acid fuel cell，PAFC）、熔融碳酸盐燃料电池（molten carbonate fuel cell，MCFC）和固体氧化物燃料电池（solid oxide fuel cell，SOFC）。根据运行温度分类，PEMFC和PAFC的运行温度一般在200 ℃以下，归类为低温燃料电池，而MCFC和SOFC的操作温度可以达到600～1000 ℃，属于高温燃料电池。燃料电池还可以根据燃料的不同分为氢燃料电池、甲烷燃料电池、甲醇燃料电池、乙醇燃料电池和金属燃料电池。采用聚合物质子交换膜作电解质的PEMFC，与其他几种类型的燃料电池相比，具有工作温度低、启动速度快、模块式安装和操作方便等优点，被认为是电动车、潜艇、各种可移动电源、供电电网和固定电源等的最佳替代电源。

20世纪60年代，美国通用电气（General Electric，GE）公司研制出PEMFC，并应用于美国“双子座”（Gemini）航天飞机的辅助电源，但是受限于质子交换膜的寿命，并未在航天领域得到进一步的推广应用。20世纪70年代，美国杜邦公司研制出全氟磺酸系列全氟磺酸膜产品，提高了质子交换膜的热稳定性和耐酸性，从而提高了PEMFC的寿命。同时，随着石墨双极板（bipolar plate）加工技术、气体流道优化以及系统集成等技术的进步，PEMFC的性能进一步提高。1983年，加拿大巴拉德动力系统公司着力发展PEMFC，并取得了突破性进展。国内外对PEMFC的深入研究，使得PEMFC在性能、寿命及成本等方面得到了长足的发展，并且在交通、便携式电源以及分布式发电等领域得到了广泛的应用，逐步推进了PEMFC的商业化。

燃料电池的诸多特性使其广泛应用于电动汽车、航天飞机、潜艇、通信系统、中小规模电站、家用电源，以及其他需要移动电源的场所。汽车行业是燃料电池应用的一个重要领域，以 PEMFC 为动力的电动汽车正处于大规模产业化的前夜，其进一步的发展，有利于减少各国对化石能源的过度依赖，汽车尾气排放的减少也为治理雾霾提供了可能的解决方案。目前全世界的汽车销售量处于快速发展阶段，截至 2018 年，全世界的汽车销售量已经超过 1 万辆，预计 2025 年将超过 11 000 万辆。传统汽车以汽油和柴油为动力源，燃烧产生大量有害气体污染空气，由此带来了汽车燃油短缺和环境污染等问题。因此，PEMFC 汽车的发展有望解决全球的能源短缺和环境污染两大问题。

（二）基本原理

PEMFC 的结构组成如图 1-83 所示。PEMFC 由质子交换膜、催化剂电极层（catalyst electrode layer）、气体扩散层（gas diffusion layer，GDL）和双极板等核心部件组成。气体扩散层、催化剂电极层和质子交换膜通过热压过程制备得到膜电极组件（membrane electrode assembly，MEA）。中间的质子交换膜起到了传导质子、阻止电子传递和隔离阴阳极反应的多重作用；两侧的催化剂电极层是燃料和氧化剂进行电化学反应的场所；气体扩散层的作用主要为支撑催化剂电极层、稳定电极结构、提供气体传输通道及改善水管

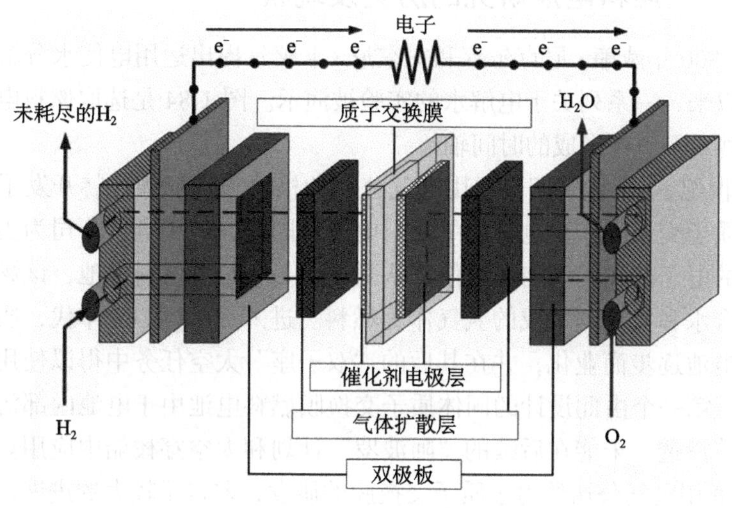

图 1-83　PEMFC 结构图

理；双极板的主要作用则是分隔反应气体，并通过流场将反应气体导入燃料电池中，收集并传导电流，支撑膜电极以及承担整个燃料电池的散热和排水功能。

PEMFC 的工作原理为：燃料（H_2）进入阳极，通过扩散作用到达阳极催化剂表面，在阳极催化剂催化作用下分解形成带正电的质子和带负电的电子，质子通过质子交换膜到达阴极，电子则沿外电路通过负载流向阴极。同时，O_2 通过扩散作用到达阴极催化剂表面，在阴极催化剂催化作用下，电子、质子和 O_2 发生氧化还原反应生成水。电极反应如下：

阳极（氧化反应）：$2H_2 \longrightarrow 4H^+ + 4e^-$ E_a=0 V *vs*. RHE

阴极（还原反应）：$O_2 + 4H^+ + 4e^- \longrightarrow 2H_2O$ E_c=1.23 V *vs*. RHE

总反应：$O_2 + 2H_2 \longrightarrow 2H_2O$ E_{total}=1.23 V *vs*. RHE

质子交换膜是 PEMFC 的电解质，直接影响电池的使用寿命。同时，电催化剂在燃料电池运行条件下会发生奥斯特瓦尔德（Ostwald）熟化作用，降低电池的使用寿命。因此，质子交换膜和电催化剂是影响燃料电池耐久性的主要因素。燃料电池的成本主要由电催化剂（46%）、质子交换膜（11%）、双极板（24%）组成。其中，电催化剂中由于大量贵金属铂的使用，成本占据了燃料电池总成本的近一半。质子交换膜和双极板的高成本也同样增加了燃料电池的总成本。因此，研究开发高性能、高耐久性、低成本的质子交换膜燃料电池新材料是目前该领域的研究热点。

二、氢燃料电池研究的历史及现状

自 1800 年威廉·尼克尔森和安东尼·卡莱尔提出运用电使水分解成氢气和氧气以来，一系列关于电解水的实验被演示。图 1-84 是选取燃料电池发展史上一些重要事件形成的时间轴。

20 世纪 50 年代末期，通用电气公司的格拉布和尼德拉赫开发了一种使用固体质子交换膜作为电解质的燃料电池。同时，通用电气公司为美国海军船舶局的电子部和美国陆军通信部队开发了一款小型燃料电池，该燃料电池通过混合水和氢化锂生成的氢气作为燃料。进入 20 世纪 60 年代，质子交换膜燃料电池逐步商业化，并在其后的“双子座”太空任务中得以使用。但遗憾的是，第一个由此设计的固体质子交换膜燃料电池由于电池内部污染和交换膜气体渗透，未能在后续的“阿波罗”计划和太空穿梭船中应用。其后的 10 年，通用电气公司致力于质子交换膜的研发，取得了较大的进展，质子交换膜燃料电池在军事和航空领域均有应用。

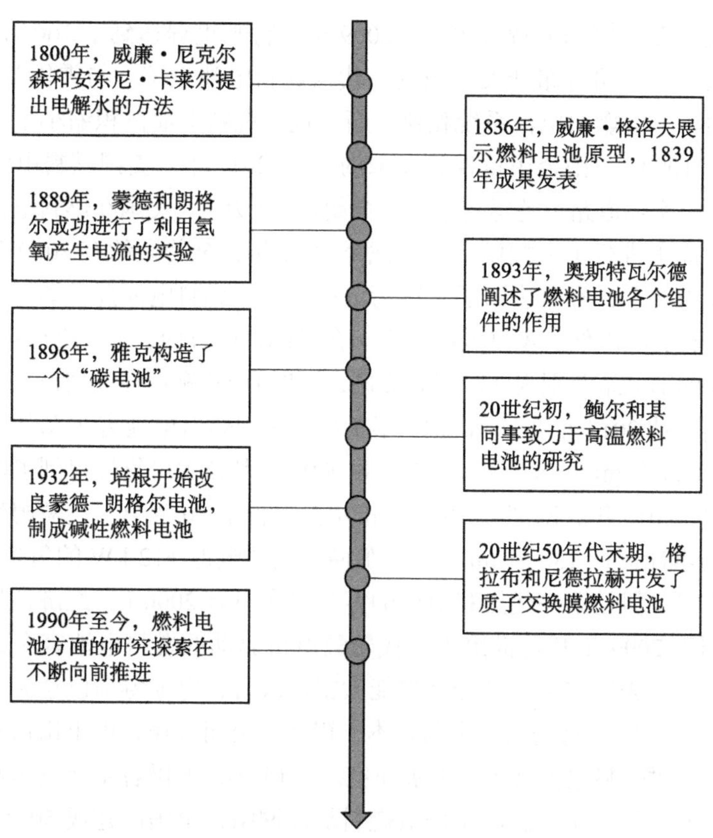

图 1-84　燃料电池发展历程示意图

早在 1966 年，通用汽车公司就开发了世界上第一辆燃料电池公路车雪佛兰 Electrovan，该车使用质子交换膜燃料电池作为动力源，输出功率为 5 kW，行驶里程约为 193 km，最高时速可达 113 km/h。近年来，美国、欧盟、日本、韩国等国家和组织都投入了大量的资金与人力推动燃料电池汽车的研究。通用汽车公司、福特、克莱斯勒、丰田、本田、奔驰等公司都相继研发出燃料电池汽车。

美国是燃料电池研发和示范的主要国家，从 20 世纪 90 年代开始，在美国能源部、交通运输部和国家环境保护局等政府部门的大力支持下，美国诸多知名汽车厂商（如通用汽车公司、福特等）都加大了对燃料电池技术的研发与实验，加拿大也拥有诸多非常著名的燃料电池品牌，其中巴拉德动力系统（Ballard Power System）公司更是燃料电池行业的领头羊。2007 年秋季，美国通用汽车公司启动了车行道（Project Driveway）计划，将 100 辆雪佛兰燃料电池汽车投放到消费者手中，2009 年总行驶里程达到了 160 万 km。同年，通用汽车公司宣布开发全新的一代氢燃料电池系统。与雪佛兰 Equinox 燃料电池车上的燃料电池系统相比，新一代氢燃料电池体积缩小了一半，质量减轻了 100 kg，铂金用量仅为原来的 1/3。2011 年，美国燃料电池混合动力公共汽车实际道路示范运行单车寿命超过 1.1 万 h。美国在燃料电池混合动力叉车方面也进行了大规模示范，截至 2011 年，全美大约有 3000 台燃料电池混合动力叉车，寿命达到了 1.25 万 h 的水平。燃料电池混合动力叉车在室内使用，具有噪声低、零排放的优点。在基础设施方面，截至 2016 年，美国东西海岸的加氢站分别为 34 座与 42 座，同时有 50 座正在计划设计当中。在客车方面，从 2014 年 8 月到 2015 年 7 月，美国燃料电池客车总计运行公里数超过 104.5 万 mi①，运行时间超过 83 000 h，其技术可靠性得到了验证。

2003～2010 年，欧洲在 10 个城市示范运行了 30 辆第一代戴姆勒燃料电池客车，累计运行 130 万 mi。这些车辆采用“电池 +12 kW 的氢燃料电池”的动力形式。但是第一代燃料电池的客车寿命只有 2000 h，经济性较差。戴姆勒集团于 2009 年开始推出第二代轮毂电机驱动的燃料电池客车，主要性能达到了国际先进水平，其经济性能大幅度改善，电池寿命已达到 1.2 万 h。2013 年初，德国宝马公司决定与日本丰田汽车公司合作，由丰田汽车公司向宝马公司提供燃料电池技术并开展研究。2015 年，德国各主要汽车和能源公司与政府共同建立了广泛的全国氢燃料加注网络，全国已建成 50 个加氢站，为全国 5000 辆燃料电池汽车提供加氢服务。

日本在燃料电池技术领域的发展也不甘落后。丰田汽车公司的 2008 版 FCHV-Adv 汽车在实际测试中，实现了在−37 ℃条件下顺利启动的目标，一次加氢行驶里程 830 km，百公里耗氢量为 0.7 kg。2014 年 12 月，丰田汽车公司发布了 Mirai 燃料电池电动汽车。根据丰田汽车公司的官方数据，在参照日本 JC08 燃油模式测试的情况下，其性能表现基本和 1.8 L 汽油车相仿，Mirai 的续航里程达到 650 km，完成单次氢燃料补给仅需约 3 min，10 s 内可以完成百公里加速，最高时速约为 161 km/h，该车完全能够满足日常行车需求。Mirai 的产量在 2017 年达到 3000 辆，成为首款投放市场的量产燃料电池汽车。2018 款 Mirai 燃料电池汽车的性能得到进一步提高，单次充氢后续航将超过 700 km，最大输出功率达到 113 kW。另外，本田汽车公司新开发的 FCX Clarity 燃料电池汽车，其性能可以与 Mirai 相媲美，能够在−30 ℃条件下顺利启动，最大

① 1mi≈1609.344m。

输出功率高达 131 kW，续航将超过 750 km，单次加氢时间为 3～5 min。同年，日本政府宣布关于氢能发展计划，计划到 2020 年保有 4 万辆燃料电池汽车，2025 年达到 20 万辆，2030 年达到 80 万辆，并同时配有 8000 个加氢站。

2017 年，我国的汽车生产量和销售量分别为 2901 万辆和 2888 万辆，连续 8 年汽车销量全球份额排第一位。汽车的使用消耗了大量的石油资源和矿产资源，同时汽车的尾气排放对环境造成了严重污染。因此，我国也在不断加大新能源汽车的投入，燃料电池汽车就是其中重要的一项，并且取得了一定的成绩。

在政府的大力支持下，进入 21 世纪，我国的燃料电池汽车技术研发取得重大进展，初步掌握了整车、动力系统与核心部件的核心技术，基本建立了具有自主知识产权的燃料电池轿车与燃料电池城市客车动力系统技术平台，实现了百辆级动力系统与整车的生产能力。目前，中国燃料电池汽车正处于商业化示范运行考核与应用阶段。2008 年，20 辆我国自主研制的氢燃料电池轿车服务于北京奥运会；2010 年，上海世界博览会上成功运行了近 200 辆具有自主知识产权的燃料电池汽车。上海汽车集团股份有限公司也拥有燃料电池汽车的整车技术，其 2015 年研发的荣威 950 Fuel Cell 插电式燃料电池车以动力蓄电池加氢燃料电池系统作为双动力源，其中以氢燃料电池为主动力源，整车匀速续航里程可达 400 km，并能在−20 ℃的低温环境下启动。2017 年，上汽大通汽车有限公司 FCV80 氢燃料电池轻型客车正式上市，单次加氢仅需要 3 min，续航里程超过 400 km。2017 年 4 月，福田欧辉氢燃料电池客车在上海车展正式上市发售，并斩获批量订单。在技术方面，该款燃料电池客车加氢只需 10 min，续航里程可达 500 km。宇通也一直致力于燃料电池客车的开发，2016 年 5 月，宇通第三代燃料电池客车问世，其加氢时间为 10 min，续航里程超过 600 km。2018 年，宇通推出第四代燃料电池产品，性能再上一级。

我国燃料电池汽车产业的发展与国际相比，存在一定差距，如我国现阶段车用燃料电池的寿命还停留在 3000～5000 h，燃料电池汽车中一些关键的部件如膜、碳纸等，依然大量依靠进口产品。因此，我国的氢能和燃料电池技术及其产业形成还需要长期努力，不断加强技术提升和创新，加快政策标准法规建设和完善。

虽然通过全球氢能科研工作者的努力，车用聚合物电解质膜燃料电池技术取得了显著进展，但燃料电池系统的耐久性和成本还未达到商业化目标。美国能源部制定的《燃料电池技术团队路线图》(*Fuel Cell Technical Team*

Roadmap）中，清晰描绘了车用燃料电池的目标状态图（图 1-85），即开发的运输燃料电池发电系统的能量转换效率需达到 65%，能量密度达到 650 W/L，质量比功率达到 650 W/kg，到 2020 年耐久性达到 5000 h，并最终达到 8000 h，而批量生产成本到 2020 年需控制到 40 美元 /kW，并最终达到 30 美元 /kW。与目标相比，目前制约燃料电池产业化的主要问题是耐久性和成本。因此，开发高性能、低成本的聚合物电解质膜燃料电池新材料及其部件是解决燃料电池系统的耐久性和成本这两大问题的必经之路，也是目前车用燃料电池研究的热点。

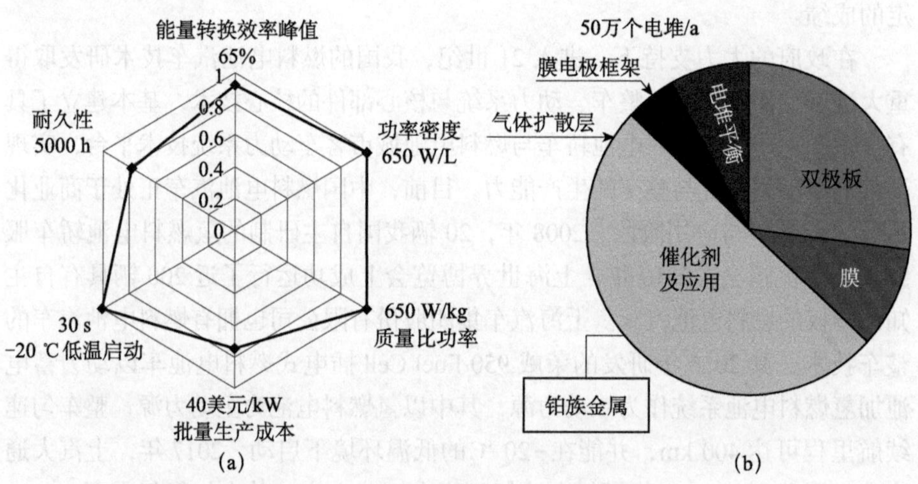

图 1-85　美国能源部提出的燃料电池研究目标及其成本组成

三、氢燃料电池研究目前面临的重大科学问题及前沿方向

（一）催化剂材料

质子交换膜燃料电池中，常用的阴极电催化剂为商业 Pt/C 电催化剂，即 3～5 nm 的 Pt 纳米颗粒担载于高比表面积 C 载体上面（Stamenkovic et al., 2017）。庄信万丰（Johnson Matthey，JM）公司生产的 40 wt% Pt/C 电催化剂在 0.9 V *vs*. RHE 处的氧气还原反应（ORR）质量比活性（mass activity，MA @ 0.9 V *vs*. RHE）为 0.21 A/mg，面积比活性（specific activity，SA）为 0.32 mA/cm^2，远低于美国能源部 2025 年的目标（MA @ 0.9 V *vs*. RHE，0.44 A/mg；SA@0.9 V *vs*. RHE，0.72 mA/cm^2）（表 1-6）。

表 1-6　膜电极组件和催化剂的技术指标

特性	单位	现状	2025 年目标
散热（$Q/\Delta T_i$）	kW/°C	1.45	1.45
膜电极成本	美元 /kW	11.8	10
铂族金属总含量	g/kW	0.125：105（150～250kPa）	≤0.10
循环耐久性	h	4100	8000
性能 @0.8V	mW/cm^2	306	300
性能 @ 额定功率	mW/cm^2	890：1190（150～250 kPa）	1800
鲁棒性（冷启动）	—	未测量	0.7
鲁棒性（热启动）	—	未测量	0.7
鲁棒性（冷启动瞬态工况）	—	未测量	0.7
催化质量活性损失	%	40	≤40%（初始）
0.8 A/cm^2 处性能损失	mV	20	≤30
电催化剂支撑稳定性	%（质量比活性损失）	未测量	≤40
1.5 A/cm^2 处性能损失	mV	>500	≤30
质量比活性	A/mg$_{pgm}$ @900 mV$_{iR}$ 补偿	0.6	0.44
无铂族金属催化剂性能	A/cm^2 @900 mV$_{iR}$ 补偿	0.021	0.044

铂的低储量和高成本也限制了燃料电池的大规模商业化进程。目前 Pt 用量已从 2010 年的 0.8～1.0 g Pt/kW 降至 2020 年的 0.3～0.5 g Pt/kW，并有希望进一步降低，使其催化剂用量达到传统内燃机尾气净化器贵金属的用量水平（<0.05 g Pt/kW），近期目标是 2025 年燃料电池电堆的 Pt 用量降至 0.1 g Pt/kW 左右。铂催化剂除受成本与资源制约外，还存在稳定性问题。通过燃料电池衰减机理分析可知，燃料电池在车辆运行工况下，催化剂会发生衰减，如在动电位作用下会发生 Pt 纳米颗粒的团聚、迁移、流失，在开路、怠速及启停过程中，产生氢气、空气界面引起的高电位导致催化剂碳载体的腐蚀，从而引起催化剂流失。因此，针对目前商用催化剂存在的成本与耐久性问题，研究新型高稳定性、高活性的低 Pt 或非 Pt 催化剂是热点。

常用的降低 Pt 用量、提高催化剂活性和稳定性的方法包括晶体结构调控、过渡金属元素掺杂形成合金、核-壳结构等，并在非贵金属催化剂领域有一定的进展。

1. Pt-M 合金电催化剂

为了降低贵金属 Pt 在燃料电池中的用量，将储量大、低成本的过渡金属 M（M=Ni、Co、Cr、Mn、Fe 等）与 Pt 结合，形成 Pt-M 合金电催化剂，不仅可以降低电催化剂成本，还可以提高燃料电池的 ORR 电催化活性。目前二元或三元铂合金体系，如 PtPd、PtAu、PtAg、PtCu、PtFe、PtNi、PtCo、PtW 和 PtCoMn 等，已被报道具有显著提高的 ORR 活性。研究表明，Pt-M 合金电催化剂的 ORR 活性高于 Pt 电催化剂的主要原因是：过渡金属 M 与 Pt 形成合金后，Pt 的原子结构和电子结构得到优化，Pt-Pt 间距缩短，从而影响吸附物种（反应物、中间产物、产物）在 Pt-M 合金表面的吸附强度，有利于氧的双位解离吸附，有利于 ORR 的发生。过渡金属 M 对 Pt 原子结构和电子结构的影响包括几何效应与电子效应。几何效应指原子半径不同的过渡金属 M 与 Pt 形成合金时，会使 Pt 的晶格收缩或拉伸，改变 Pt-Pt 的原子距离。电子效应指形成 Pt-M 合金后，Pt 得到或失去电子，d 带中心的位置发生变化。Pt-M 合金的几何效应和电子效应相互作用，影响着合金电催化剂的电化学性能，不同的过渡金属 M 与 Pt 形成的合金，几何效应和电子效应不同，ORR 活性不同。例如，对于 Pt_3M（M=Ni、Co、Fe、Ti、V）合金电催化剂，其表面电子结构（d 带中心位置）与 ORR 活性之间的关系如图 1-86 所示。Pt_3M 的 ORR 活性与 d 带中心位置的关系呈现出火山形关系，即适中的 d 带中心

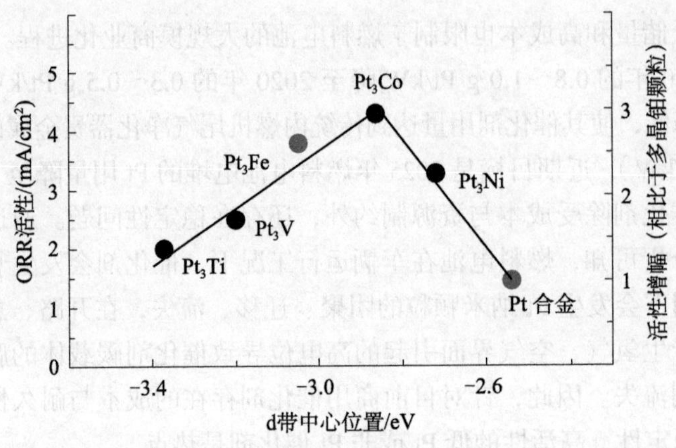

图 1-86 Pt_3M 表面 ORR 的实验测量面积比活性与 d 带中心位置之间的关系

位置能够带来最高的 ORR 活性，Fe、Co、Ni 是制备 Pt-M 合金电催化剂的良好过渡金属元素。

大量 Pt-M 合金电催化剂的制备与电化学性能研究表明，Pt-M 合金电催化剂的 ORR 活性与合金材料的组分和表面结构密切相关。例如，Yang 和斯塔门科维奇（Stamenkovic）等利用铂-镍合金纳米晶体的结构变化，制备了高活性与高稳定性的电催化剂（Stamenkovic et al.，2007；Carpenter et al.，2012；Zhang et al.，2010）。在溶液中，初始的 $PtNi_3$ 多面体经过内部刻蚀生成的 Pt_3Ni 纳米笼结构，使反应物分子可以从三个维度上接触催化剂。这种开放结构的内外催化表面包含纳米尺度上偏析的 Pt 表层，从而表现出较高的氧还原催化活性。与商业 Pt/C 相比，Pt_3Ni 纳米笼催化剂的质量比活性与面积比活性分别提高 36 倍与 22 倍。Chen 等利用酸处理和热处理两种方法制备了碳载 Pt_3Co 合金纳米颗粒的电催化剂（Chen et al.，2014），Co 的存在调节了 Pt-Pt 的原子间距离，从而使电催化剂显示出更高的 ORR 活性。得到的两种合金电催化剂在酸性电解质溶液中显示出优异的 ORR 活性，其面积比活性分别是 Pt 纳米颗粒的 2 倍和 4 倍。

Pt_3M 合金中，会出现过渡金属溶解过程（去合金化），造成催化剂表面粗糙，催化剂比表面积增大。美国通用汽车公司等研究了去合金催化剂（D-PtCo、D-$PtCo_3$、D-$PtNi_3$）微观形貌随合金前驱体尺寸以及去合金条件等的影响，发现空气条件下去合金程度大于氮气条件下去合金程度，颗粒尺寸大于 13 nm 的合金前驱体颗粒在空气下易形成多孔结构。另外，美国通用汽车公司等还研究了 D-$PtNi_3$/C 的 ORR 活性以及稳定性随去合金酸性溶液种类、去合金时间、去合金温度以及后续是否热处理等的变化情况。实验结果表明，D-$PtNi_3$/C 的 ORR 活性、H_2/ 空气燃料电池大电流功率输出性能以及耐久性受去合金条件的影响很大，目前在 ORR 活性与其耐久性方面，D-$PtNi_3$/C 能够达到美国能源部的性能要求，但是在大电流功率输出性能方面，其耐久性还欠缺，需要进一步研究其耐久性差的原因并寻找缓解措施。美国通用汽车公司等初步认为，Ni^{2+} 进入质子交换膜不是造成 H_2/ 空气大电流区域输出功率严重衰减的原因，铂合金比表面积减小（$<50\ m^2/g$ Pt）造成局域 O_2 传质阻力增大与严重衰减关系性更大，从而提出提高初始催化剂比表面积或者改善电极结构来缓解衰减的建议。

除了二元合金电催化剂的电化学活性比商业 Pt/C 电催化剂的性能有所提高，第三种或更多的过渡金属的加入，可能令二元合金催化剂的结构得到进一步的调整，使得多元合金电催化剂的活性和耐久性都有所提高。基于对双金属催化剂组分和结构的理解，针对高催化活性但是有溶解问题的Pt-Ni双金属体系，已有一系列的多组分研究。例如，米勒等将过渡金属掺杂的Pt_3Ni正八面体担载到碳材料上，得到了高活性的ORR电催化剂（Huang et al.，2015），并考察了过渡金属M（M=V、Cr、Mn、Fe、Co、Mo、W、Rh）对电催化剂性能的影响。结果表明，Mo-Pt_3Ni/C电催化剂具有较高的ORR性能，面积比活性和质量比活性分别为10.3 mA/cm^2和6.98 A/mg Pt，是商业Pt/C电催化剂的81倍和73倍，显著提高了三元电催化剂的ORR活性。根据理论计算，在氧化环境中Mo原子倾向于占据超级顶点/边缘的位点，Mo的掺杂可以降低表面Ni的平衡浓度，增加表面Pt/Ni-空位生成能，限制Ni原子在纳米粒子中的溶解速率。斯特拉瑟（Strasser）等则证实了当Rh的添加量超过3%时，Pt-Ni纳米颗粒即使经过30 000次电位循环，仍可以保持其八面体形状，而未添加Rh的Pt纳米颗粒经过8000次循环就已经破裂（Beermann et al.，2016；Lu et al.，2017）。在该研究中值得注意的是，是Pt原子的运动而不是Ni的溶解导致Pt-Ni纳米粒子八面体的分解。通过掺杂少量的Rh，可以降低Pt原子的迁移速率，抑制Pt-Ni八面体的分解。

2. Pt基核-壳结构电催化剂

核-壳结构电催化剂的制备是降低Pt基电催化剂中Pt用量的又一个有效方法（Oezaslan et al.，2013）。Pt基核-壳结构电催化剂是指以非Pt材料为核、以Pt或Pt-M为壳的电催化剂，该结构可以使Pt的活性位点充分暴露在电催化剂表面，提高贵金属Pt的利用率，并且内核原子与壳层Pt的协同作用（几何效应和电子效应）还可以进一步提高电催化剂的活性。几何效应和电子效应都可以改变Pt表面含氧物种的结合能，从而调变电催化剂的氧还原活性。核-壳结构电催化剂的制备方法如图1-87所示，主要包括电化学脱合金、Pt偏析法和沉积法。

去合金化法［图1-87（a）、图1-87（b）］是指首先制备Pt-M合金电催化剂，后用酸腐蚀或电化学腐蚀的方法将合金电催化剂的表面去合金化，即将表层合金中的过渡金属M腐蚀溶出，使电催化剂表层成为Pt壳层，形成以Pt-M合金为核、Pt为壳的核-壳结构电催化剂。例如，斯特拉瑟等首先制备了Pt-Cu合金纳米颗粒，后采用电化学腐蚀的方法将合金表层的Cu腐蚀溶解，得到核-壳结构电催化剂，该电催化剂具有高于商业Pt/C 4倍的ORR质量比活性（Mani et al.，2008）。

Pt偏析法［图1-87（c）、图1-87（d）］是指对Pt-M合金电催化剂进行

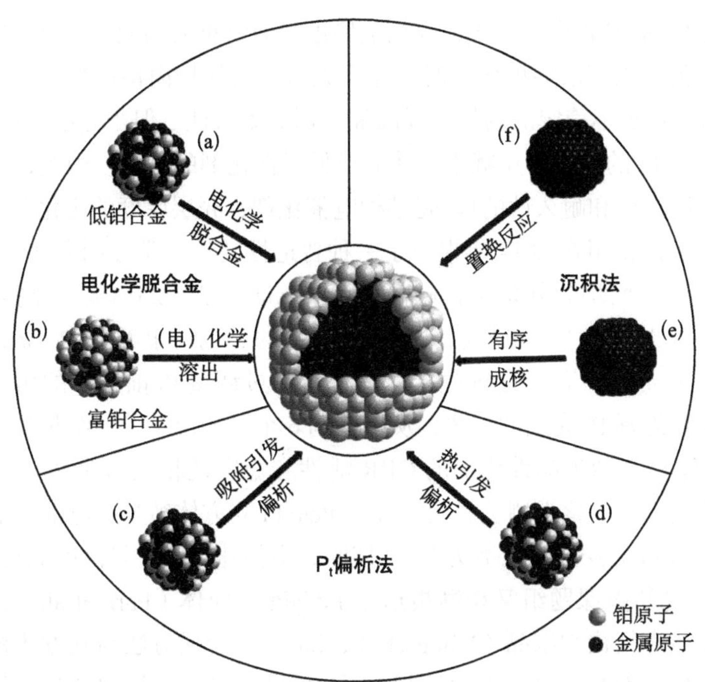

图 1-87 核－壳结构电催化剂基本合成方法的示意图

适当的处理，从而诱导表层合金发生偏析，使合金表面形成 Pt 原子层，获得核-壳结构电催化剂。例如，阿布鲁纳（Abruña）等制备了 Pt_3Co 合金电催化剂，然后对合金电催化剂进行热处理，使表层的合金结构中富集更多的 Pt，形成具有 2～3 个原子层的 Pt 壳。得到的电催化剂的质量比活性是未热处理的电催化剂的两倍，耐久性也得到显著提高，在 5000 圈的耐久性测试后催化剂的活性几乎没有降低（Wang et al.，2013）。

沉积法［图 1-87（e）、图 1-87（f）］是指分步制备电催化剂的内核和外壳：先制备非 Pt 纳米颗粒作为内核，再用化学还原法、原子层沉积法或欠电位沉积（under potential deposition，UPD）-置换（galvanic replacement）法在内核外沉积 Pt 外壳。其中原子层沉积法是指将 Pt 的气相前驱体通入反应器，其在非 Pt 纳米颗粒上发生化学吸附并反应而形成 Pt 纳米壳层的方法（Liu et al.，2016）。欠电位沉积-置换法是指利用电化学的方法在略正于某种金属的平衡电位，将此金属沉积于内核表面，然后利用 Pt 置换沉积在内核表面的金属，在内核表面形成 Pt 外壳。例如，采用欠电位沉积方法制备的 Pt-Pd-Co/C 单层核壳催化剂总质量比活性是商业催化剂 Pt/C 的 3 倍。

对于核-壳结构电催化剂，壳层的厚度是非常重要的因素，直接影响着催化剂的性能。当壳层厚度较小时，壳层无法完全保护内核；当壳层厚度较大时，可有效保护不稳定的内核，提高催化剂的耐久性，但当壳层厚度太大时，内核对壳层 Pt 的调节作用减小，从而降低了催化剂的活性。因此，为了得到同时具备高活性和耐久性的核-壳结构电催化剂，壳层厚度的优化尤为重要。

夏幼南课题组在 2014 年报道了一种普适性的动力学控制方法，在 Pd 立方体纳米晶表面均匀外延生长超薄 Pt 层（Xie et al., 2014）。在高温和低的 Pt 前驱体注射速率下，可控制已被还原沉积的 Pt 原子扩散至 Pd 立方体的整个表面，从而形成均匀的 Pt 壳层。进一步通过调整 Pt 前驱体的用量，可以精确地调控外延 Pt 壳层的层数，从 1 个到 6 个原子层不等。结果显示，Pd@Pt_{nL} 立方体核壳纳米晶催化剂的 ORR 活性和稳定性相比商业 Pt/C 催化剂有了显著的提高。电化学测试表明，综合 Pd@Pt 立方体纳米晶电催化剂的活性和寿命，当 Pt 壳层原子层数为 2～3 层时，电催化剂具有最高的氧还原性能。2015 年，夏幼南课题组又相继报道了 Pd@Pt 八面体（Park et al., 2015）和二十面体核壳结构纳米晶（Wang et al., 2015），合成方法与立方体核壳纳米晶完全一致，都是通过动力学控制 Pt 在 Pd 纳米晶表面均匀外延生长，只是 Pd 纳米晶种子的形貌不同而已。Pd@$Pt_{2.7L}$ 二十面体纳米晶催化剂展现出了最优的 ORR 性能，进一步证明 Pt 壳层为 2～3 层时最有益于氧还原催化反应（Wei et al., 2018）。其面积活性和质量活性分别达到了商用 Pt/C 催化剂的 8 倍和 7 倍，也明显强于同为 Pd@Pt 的立方体（Zhang et al., 2005）和八面体核壳纳米晶（Shao et al., 2013；Liu et al., 2017），说明催化剂的形貌对调控 ORR 性能有着重要的作用。10 000 个循环稳定性测试后，Pd@$Pt_{2.7L}$ 二十面体纳米晶催化剂的质量活性仍然为商用 Pt/C 的 4 倍，可见其良好的稳定性。

3. 非贵金属电催化剂

虽然关于 Pt 基电催化剂活性和耐久性的研究取得了显著的进展，但尚不能彻底解决 Pt 储量低、成本高的问题。因此，近年来非贵金属电催化剂受到了研究者的广泛关注，有望彻底解决燃料电池阴极用电催化剂大规模商业化的问题（Yang et al., 2013）。

1964 年，雅辛斯基（Jasinski）首次报道了 N_4-螯合物酞菁钴在碱性电解液中具有氧还原活性，从此开启了非贵金属电催化剂研究的新领域。最初，非贵金属电催化剂的研究主要集中在与酞菁钴具有相似结构的过渡金属大环化合物上，其结构特点是中心的过渡金属 M（M=Co、Fe、Ni、Mn）原子与周围的 4 个 N 原子配位，如卟啉、酞菁、四氮杂轮烯及其衍生物。该 $M-N_4$ 金属大环化合物具有 ORR 活性的原因是富集电子的过渡金属可以将电子转移至 O_2 的 π^* 轨道，从而减弱 O—O 键，促进了 ORR 的进行。然而，$M-N_4$ 金属大环化合物只能在碱性电解质溶液中稳定存在，在酸性电解质溶液中无法稳定存在，且 ORR 活性低。巴戈茨基（Bagotsky）等发现，简单的热处理可以有效提高电催化剂在酸性电解质溶液中的活性和稳定性。为了降低电催化剂的成本，研究人员陆续发展了以聚合物和过渡金属配合物为前驱体的非贵金属电催化剂，ORR 活性也得到了显著提高。

目前，非贵金属电催化剂的研究进入了快速发展时期，其中主要以过渡金属氧化物、含过渡金属的 N 掺杂碳材料和完全非金属的杂原子掺碳材料为代表。它们往往具有原料来源丰富、价格低廉、抗甲醇渗透等特点，在酸性或碱性条件下能达到与 Pt 相当的活性，被认为具有完全替代贵金属 Pt 作为氧还原催化剂的可能（Zeng et al.，2018）。

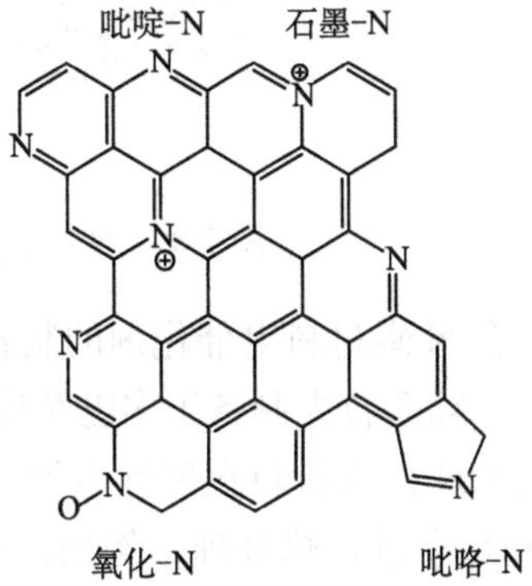

图 1-88　掺 N 石墨烯中 N 的四种形式

非贵金属电催化剂的前驱体在高温热处理过程中发生碳化，形成了稳定的石墨碳层结构，该石墨碳层中掺杂 N。掺杂 N 的形式包括吡啶-N（pyridinic-N）、吡咯-N（pyrrolic-N）、石墨-N（graphitic-N）和氧化-N（oxidized-N）（图 1-88）。不同类型的掺杂 N 位于石墨碳层的面内或边缘，增加了石墨碳层的缺陷。其中，吡啶-N 位于石墨碳层的边缘，与两个 sp^2C 原子相连，N 原子为石墨碳层的 π 体系提供一个 p_π 电子；石墨-N 位于石墨碳层的面内，即取代了石墨碳层中的一个 C 原子，使得三个相邻的六元环共用一个 N 原子，石墨-N 提供两个 p_π 电子。吡啶-N 和石墨-N 为 n 型掺杂，为非贵金属电催化剂提供了 ORR 活性位。拉·海（La Haye）等证明，掺杂了 N 的石墨碳层具有更高的表面极化作用，可以加快 ORR 过程的电子和质子的传递速度，从而提高电催化剂的 ORR 活性。2016 年，近藤（Kondo）等提出了活性位为与吡啶-N 相邻的 C 原子，ORR 机理如图 1-89 所示。N 的掺杂使得相邻的 C 原子成为路易斯碱活性位，O_2 可以吸附到该 C 原子上，并发生质子化，随后可以经过两种途径发生 ORR，即在同一活性位上发生直接四电子还原反应或在不同活性位上发生 2+2 的间接四电子还原反应。

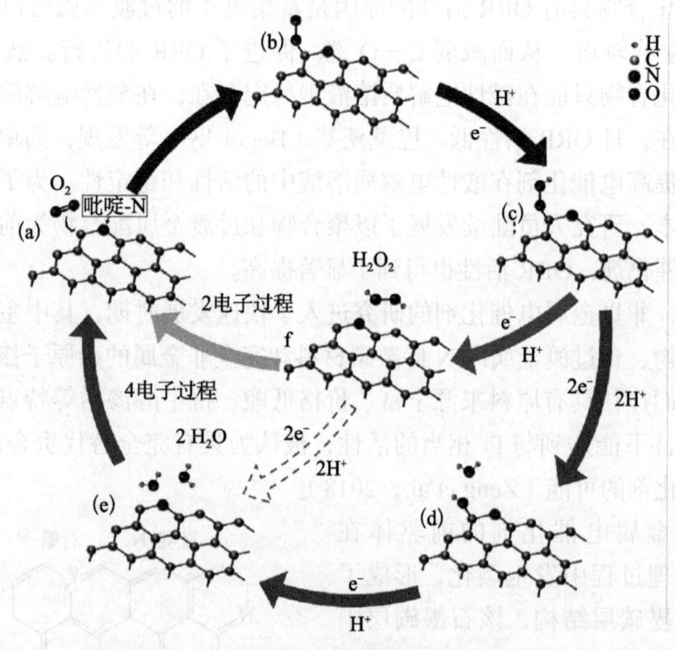

图 1-89 N 掺杂 C 材料上的氧化还原反应示意图

含 N 碳材料电催化剂的制备方法有：①原位制备含 N 碳材料电催化剂。例如，陆等首先制备了多巴胺聚合物球，后对其进行热处理，得到 N 掺杂的 C 纳米球，该材料在碱性电解液中具有较好的 ORR 活性。②将含 C 前驱体在 NH_3 中进行热处理。例如，余等将 C 纳米管和石墨烯在 NH_3 中进行热处理，得到的含 N 碳材料电催化剂具有高 ORR 活性。③将含 N 和含 C 的前驱体与硬模板结合，制备电催化剂。例如，Wei 等以蒙脱土（montmorillonite，MMT）片层为硬模板、聚苯胺为前驱体，热处理后除去硬模板，得到 N 掺杂的碳材料。由于 MMT 具有二维片层结构，使聚苯胺在热处理过程中碳化形成片层的类石墨烯结构，并且 N 的掺杂类型也以平面 N（吡啶-N 和吡咯-N）为主，该电催化剂在酸性电解质溶液中的 ORR 活性仅比商业 Pt/C 的半波电位低 60 mV。除了 N 元素，其他杂原子（如 F、P、S、B 等）掺杂的碳材料电催化剂也已经得到了广泛的研究，其中 S 掺杂的碳材料显示出较高的 ORR 活性和耐久性。两种或两种以上的杂原子共掺杂的碳材料同样具有较高的 ORR 活性。例如，S 和 N 共掺杂的碳材料可以显著提高电催化剂的 ORR 性能，N、O、S 三元素共掺杂的碳材料的 ORR 活性则高于 N、O 两种元素掺杂的碳材料。

除了 N 掺杂对活性位的影响，过渡金属对 ORR 活性也起到了至关重要的作用。M-N-C 非贵金属电催化剂的研究主要集中在前驱体的选择及其制备方法两个方面。制备电催化剂的前驱体主要包括金属大环化合物（金属卟啉、金属酞菁等）、含 N 聚合物（聚苯胺、聚苯二胺、聚吡啶、聚吡咯等）、配位化合物（金属盐与吡啶、联吡啶、乙二胺、邻菲罗啉等配体形成的配合物）。选择不同的前驱体制备的 M-N-C 非贵金属电催化剂具有不同的特点。金属大环化合物具有稳定的 M-N-C 的 ORR 活性位，热处理过程中大量活性位保留，使得非贵金属电催化剂具有高 ORR 活性；配位化合物则由于金属盐和配体的价格低廉，可以降低非贵金属电催化剂的成本；含 N 聚合物则由于苯环具有与石墨碳六元环相似的结构，碳化过程中更易形成石墨碳层，使非贵金属电催化剂具有更高的稳定性。前驱体的制备方法也是影响 M-N-C 非贵金属电催化剂 ORR 性能的重要方面，如碳载、非碳载及硬模板法。当以碳材料为非贵金属电催化剂的载体时，不同碳载体（形貌、粒径、比表面积不同）的选择对电催化剂的性能有着较大影响，如宋等在以金属配位化合物为前驱体制备电催化剂时，考察了石墨烯、炭黑、碳纳米管不同载体对电催化剂性能的影响，结果表明，对于该非贵金属电催化剂，以还原的氧化石墨烯为载体时，电催化剂显示出更高的 ORR 活性。

为了进一步提高非贵金属电催化剂活性位的密度，提高 ORR 性能，越来越多的研究者采用非担载的方法制备非贵金属电催化剂。例如，戴和冯等将金属卟啉聚合得到共价有机聚合物，直接碳化得到多孔状 M-N-C 非贵金属电催化剂，该催化剂显示出高 ORR 活性。王及刘等首先制备了金属有机框架材料，热处理后得到了高活性的 M-N-C 非贵金属电催化剂。共价有机聚合物和金属有机框架材料的制备都可以从分子角度上控制非贵金属电催化剂的结构和活性位排布。提高 M-N-C 非贵金属电催化剂活性位密度的另一有效方法是制备过程中加入硬模板（硅胶颗粒、介孔硅、分子筛、蒙脱土片层等）。高比表面积的硬模板的加入，可以有效控制非贵金属电催化剂的孔结构和比表面积，高比表面积的电催化剂暴露出更多的活性位，可以有效地提高 ORR 活性。例如，上海交通大学的冯新亮等以维生素 B_{12} 为前驱体，以硅胶颗粒、分子筛、蒙脱土片层为硬模板，制备了 Co-N-C 非贵金属电催化剂，其 ORR 活性均高于以炭黑为载体的电催化剂。对于 M-N-C 非贵金属电催化剂的性能表征，当前报道的电催化剂在酸性或碱性溶液中具有一定的 ORR 活性，并且在碱性电解质溶液中的相关研究取得了较大的进展，而酸性电解质溶液中的电催化剂活性与商业 Pt/C 电催化剂还有较大差距。

（二）质子交换膜材料

质子交换膜燃料电池由双极板、扩散层、催化层、质子交换膜以及相应的配套组件构成，其中质子交换膜作为核心部件之一，对质子交换膜燃料电池的各项参数都有着较大的影响。质子交换膜材料按照其膜材料的化学结构，可以分为全氟磺酸型质子交换膜、部分氟化质子交换膜、非氟质子交换膜以及复合型质子交换膜。尽管各类膜材料有着各不相同的化学结构，但是其传导质子的原理是基本一致的。除复合型质子交换膜外，其他膜材料均由均相的有机高分子材料组成，其高分子链由惰性的主链以及带有质子传输位点的侧链组成。侧链的质子传输位点往往是磺酸基团，在水溶胀的条件下可以发生质子的电离，质子在相邻质子传输位点之间可以转移，当这样的过程发生在整个膜内部时，即可实现长程的质子传输。复合型质子交换膜材料往往是在有机高分子材料的基础上引入各类添加剂，从而实现质子交换膜各项性能的提升的，但质子传输仍然是通过质子在质子传输位点之间的跃迁来实现的。

质子交换膜领域所开展的相关研究对降低燃料电池整体成本、拓展燃料电池在极寒与高温干旱地区的使用、提高燃料电池的能量转化效率，以及提升燃料电池的可靠性有着重大的意义。

世界上第一个燃料电池由英国律师出身的科学家威廉·罗伯特·格罗夫（William Robert Grove）爵士于 1839 年发明。在接下来的一个世纪，燃料电池没有任何的发展。通用电气公司从 20 世纪 50 年代开始发展燃料电池，而且获得了为 1962 年“双子座”太空计划完成能源任务的合同。1 kW 的“双子座”燃料电池系统的铂载量为 35 mg/cm^2，其在 0.78 V 下，电池的电流密度能达到 37 mA/cm^2。20 世纪 60 年代，改进的催化剂层由直接将聚四氟乙烯混合到催化剂里和电解质复合制成。当时通用电气公司就是采用这种方法制备燃料电池的。从 20 世纪 70 年代初开始，燃料电池工艺有了相当大的改善，尤其是全氟磺酸膜被大量应用。然而，燃料电池的最初研究和发展没有从美国联邦政府得到关注与资金，特别是美国能源部和相关工业企业都没有给予这方面的经费。直到几十年前，洛斯·阿拉莫斯国家实验室（Los Alamos National Laboratory，LANL）研究人员改进了方法使得铂所需的量降低，才使得质子交换膜燃料电池研究有了大的突破。值得注意的是，LANL 的雷斯特里克（Raistrick）想出了一个催化剂喷涂技术来制备电极。这种突破性的方法不仅可以大大提高活性催化剂的利用率，而且可以减少所需的铂贵金属的使用量。

燃料电池的重大发展离不开全氟磺酸材料在燃料电池中的应用。早期全氟磺酸的发展主要得益于氯碱工业的需要，用于分离阴阳极电解液以及氯气，同时传导钠离子。自 20 世纪 60 年代起，美国杜邦公司开始将全氟磺酸材料用于燃料电池，使得燃料电池有了长足的发展。发展至今，质子交换膜材料已经不局限于全氟磺酸型一种，各种新型膜材料开始应用于质子交换膜燃料电池。

1. 全氟磺酸型质子交换膜

全氟磺酸型质子交换膜由碳氟主链和带有磺酸基团的醚支链组成，是世界上最先进的质子交换膜，也是目前在 PEMFC 中唯一得到广泛应用的一类质子交换膜。以 Nafion® 系列产品为例，其结构式如图 1-90 所示，其中聚四氟乙烯骨架为疏水部分，带有磺酸基团的支链为亲水部分（Klaus & Chen，2008）。在全氟磺酸结构中，磺酸根通过共价键固定在聚合物分子链上，与质子结合形成的磺酸基团在含水的情况下可以解离出可以自由移动的质子。每个磺酸根周围大概可以聚集 20 个水分子，形成微观的含水区域，当这些含水区域互相连通时，可以形成贯穿整个质子交换膜的质子传输通道，从而实现质子的传输（Zawodzinski et al.，1993；Mauritz & Moore，2004）。

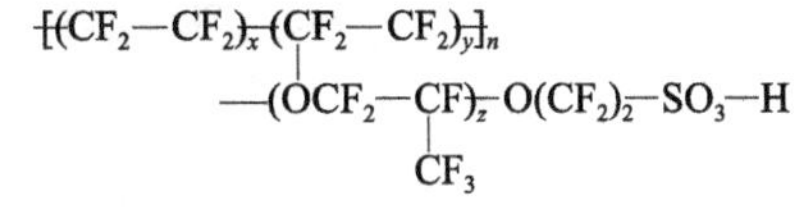

$$\mathrm{-[(CF_2-CF_2)_x-(CF_2-\underset{|}{C}F)_y]_n-} \quad \mathrm{-(OCF_2-\underset{\underset{CF_3}{|}}{C}F)_z-O(CF_2)_2-SO_3-H}$$

图 1-90　全氟磺酸树脂的结构式

目前关于质子交换膜的微观模型中，普遍认同的是离子团簇模型，如图 1-91 所示。疏水的聚四氟乙烯主链构成晶相疏水区，亲水的磺酸基团支链与水形成离子团簇，离子团簇的直径约为 4 nm，相邻团簇之间距离 5 nm，其间以直径 1 nm 的通道连接。基于这种模型，全氟磺酸膜在吸水后，水以球形区域在膜内分布，在球的表面，磺酸根形成固定的质子传输位点，游离的水合质子可以在这些位点之间进行传输，并通过离子团簇之间的通道形成贯穿的离子传输体系（Parthasarathy et al.，1992）。当量质量（EW，单位为 g/mmol）常用于表征全氟磺酸树脂的酸浓度，其数值等于含 1 mol 质子的干态膜的质量。此外，还常用离子交换容量（IEC，单位为 mmol/g）来表示全氟磺酸树脂的酸浓度，其数值等于 1 g 干态膜内质子的物质的量，IEC 与 EW 互为倒数。随着 EW 值的升高，单位质量内的质子传输位点数下降，膜的结晶度和刚性都会增加，而膜的吸水能力会下降，导致离子团簇的间距增加，最终导致质子传输能力的下降。EW 值过低则会导致质子交换膜的溶胀度提高，吸水量增加，尺寸稳定性与机械性能下降，甚至导致膜的溶解。因此，EW 需要控制在一定范围内，一般为 800～1500 g/mol。

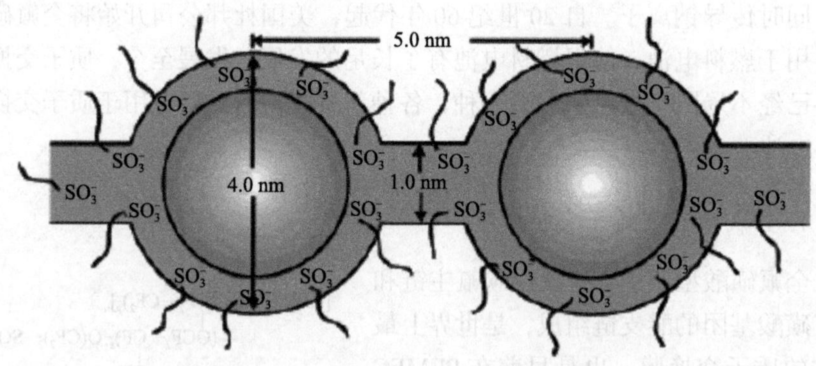

图 1-91　全氟磺酸膜的离子团簇模型

2. 部分氟化质子交换膜

部分氟化质子交换膜是指分子链同时含有碳氟链段与碳氢链段的质子交换膜（Kerres et al.，2008）。从化学键键能的角度来看，碳氟键的键能为 485 kJ/mol，而碳氢键键能常常在 350～435 kJ/mol，因此含有碳氟键的聚合物往往有更好的热稳定性与化学稳定性。部分氟化质子交换膜使用部分取代的氟化物代替全氟磺酸树脂，或者将氟化物与无机或其他非氟化物进行共混制膜。在过去的几十年，部分磺化的芳主链聚合物（HET）已被大量研究（Kerres et al.，2006）。得益于其侧链的系列氟化聚合物，这些聚合物显示出较好的化学和机械稳定性，被大量磺化以获得能够质子传导的膜。已被磺化的亚芳基聚合物有较高的质子传导能力。在已经磺化的亚芳基聚合物中有以下聚合物族：聚（亚苯基醚）、聚（2,6-二甲基-1,4-亚苯基）、聚（2,6-二苯基-1,4-亚苯基）、聚苯基喹喔啉、聚苯并咪唑（PBI）、聚苯醚、聚醚醚酮（PEEK）、聚酰亚胺（PI）和聚醚酰亚胺（Schöenberger et al.，2007；Arnett et al.，2007；Katzfuss et al.，2011）。这些燃料电池离子膜能达到很长的膜寿命，以磺化萘型聚酰亚胺为例，其在 60 ℃下寿命时间超过 3000 h。但是磺化芳主链聚合物也有很多待解决的问题，其中一个普遍问题是，这些离聚物膨胀性太强，从而在一定的磺化度（离子交换容量在 1.4～1.6 meq/g）或某一操作温度（60～80 ℃）下，它们会失去机械稳定性。因此，降低膜的溶胀度而不降低其质子传导性是部分氟化磺酸质子交换膜的发展需求。

3. 非氟质子交换膜

非氟质子交换膜实质上是碳氢聚合物膜。作为燃料电池隔膜材料，其价格便宜、加工容易、化学稳定性好、具有高的吸水率（Kim et al.，2016）。因为C—H键离解焓低，易被过氧化氢降解，严重危害其稳定性，因此，碳氢聚合物质子交换膜用于燃料电池的主要问题集中在化学稳定性的改良上。芳香聚酯、聚苯并咪唑、聚酰亚胺、聚砜、聚酮等由于具有良好的化学稳定性、耐高温性、环境友好以及成本低等优势，被大量研究通过质子化处理用于聚合物电解质膜燃料电池。目前研究较多的非氟质子交换膜主要是三类膜材料，即聚苯并咪唑、聚酰亚胺以及聚芳醚类聚合物。

聚苯并咪唑是一种常见的工程树脂，其主链上的咪唑环含有两个氮原子，能够与氢原子形成分子间氢键，如图1-92所示。通过这些氢键的“形成—断裂—重新形成”循环作用，咪唑环中的氮原子便可以作为固定的质子传输位点完成质子的传导（Li et al.，2009）。因此，聚苯并咪唑有作为燃料电池质子交换膜的可能性。聚苯并咪唑作为常见的工程塑料，具有较高的热稳定性和机械强度，制成薄膜后，其气体与甲醇的渗透率低，然而聚苯并咪唑中质子的离解度很低，聚苯并咪唑膜的质子传导率仅为10^{-7}～10^{-6} S/cm（Shannon & Prewitt，1969），因此聚苯并咪唑常常需要引入无机酸来实现质子的长程传输。在众多无机酸掺杂的复合膜中，磷酸掺杂的复合膜是聚苯并咪唑聚合物用于PEMFC的一种代表性技术，其质子传导率可以达到0.1 S/cm以上。由于聚苯并咪唑自身的玻璃化温度高达210 ℃，磷酸掺杂的聚苯并咪唑膜最高可以在210 ℃的温度条件下工作。但是，磷酸与聚苯并咪唑链之间没有共价键的结合且磷酸易溶于水，因此，随着燃料电池反应生产水，磷酸也会随之流失，导致电池性能下降。

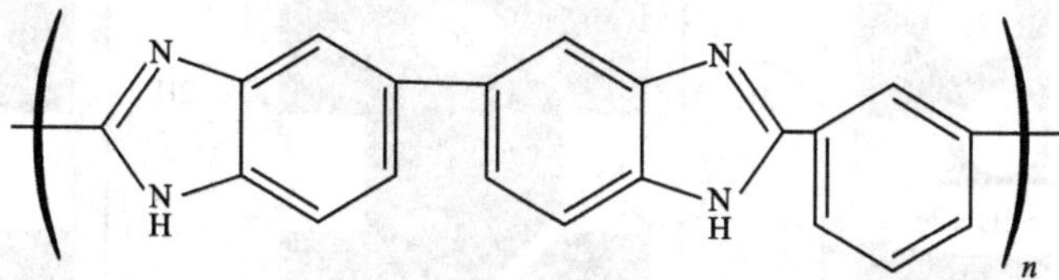

图1-92 聚苯并咪唑的化学结构

聚酰亚胺是目前综合性能最好的聚合物材料之一，能够耐400 ℃的高温，在航空、航天、微电子、隔膜等领域有着广泛的应用。在燃料电池领域，聚酰亚胺凭借良好的机械性能以及极佳的燃料阻隔性能，一直以来都是燃料电池质子交换膜领域的一个研究热点。

聚芳醚类聚合物包括聚苯硫醚（PPS）、聚醚砜（PES）、聚醚酮（PEK）、聚醚醚酮（PEEK）、聚醚酮酮（PEKK）、聚醚醚酮酮（PEEKK）、聚硫醚酮（PKS）、聚芳醚腈（PEN）等。聚芳醚类聚合物具有良好的热稳定性、力学性能以及化学稳定性，并且抗水解能力强，价格低廉，被认为是一类有潜在应用价值的质子交换膜材料。

4. 复合型质子交换膜

复合型质子交换膜是将质子交换膜聚合物与添加材料进行复合成膜，并且获取某一项或者几项性能的提升的一种质子交换膜。通常，复合型质子交换膜膜中添加的材料可以改善复合材料的力学性能、热稳定性以及电磁学性能，聚合物基体材料则提供一定的柔韧性、活性官能团等。

（三）双极板材料

燃料电池双极板又叫流场板，是电堆中的“骨架”，与膜电极层叠装配成电堆，在燃料电池中起到支撑膜电极，收集电流，为燃料、氧化剂、冷却液提供传质通道，分隔燃料和氧化剂等作用，是电堆中必不可少的关键功能和结构部件之一（图 1-93）。

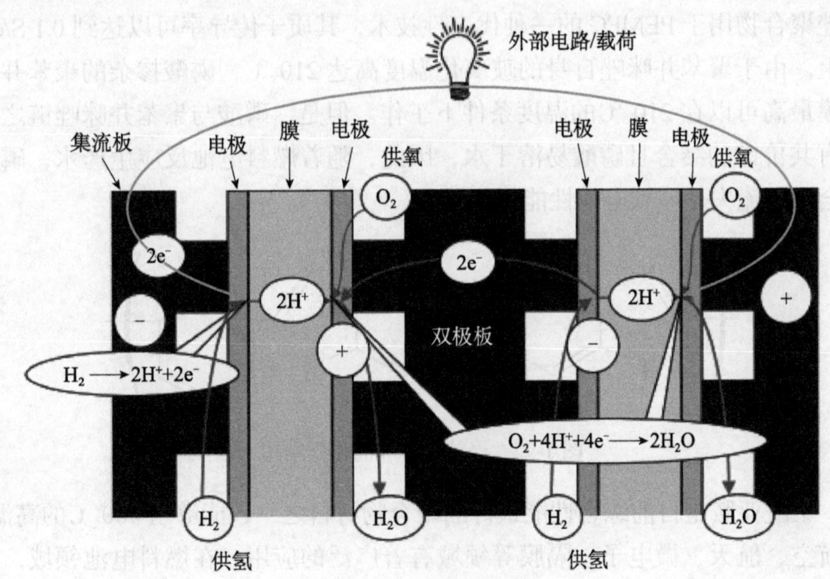

图 1-93　双极板结构示意图

在功能上，要求双极板材料是电与热的良导体，具有一定的强度以及气体致密性等；在稳定性上要求双极板在燃料电池酸性（pH = 2～3）、电位（约 1.1 V）、湿热（气液两相流，约 80 ℃）环境下具有耐腐蚀性且对燃料电池其他部件与材料的相容无污染性；在产品化上，要求双极板材料要易于加工、成本低廉。详细参数指标要求如表 1-7 所示。

表 1-7　双极板特性及指标要求

特性	指标要求
导电性	电导率 >100 S/cm；表面接触电阻 <10 mΩ/cm^2
阻气性	氢渗透率 <1.3×10^{-14} cm^3/（cm^2·s·Pa）
强度	弯曲强度 >25 MPa
导热性	热导率 >20 W/（m·K）（冷却遍布整板）； >100 W/（m·K）（仅边缘冷却）
加工精度	表面精度 <50 μm；公差 <0.05 mm
耐蚀性	腐蚀率 <1 μA/cm^2
重量	重量 <0.4 kg/kW；密度 <5 g/cm^3
成本	材料成本 + 制造成本 <3 美元 /kW

石墨材料在燃料电池的环境中具有良好的化学稳定性，同时具有很高的电导率，是目前质子交换膜燃料电池研究和应用中最广泛的材料。但由于石墨本身是多孔结构，必须对这些板进行浸渍处理，从而满足阻水隔气的要求。同时，传统石墨板的加工成本较高，强度较低，不利于电堆整体功率密度的提高和成本控制，严重制约了石墨板在车用领域燃料电池电堆的商业化推广。

除材料本身的特性要求之外，双极板表面的流场结构对实现双极板的功能也有重要的作用，其目的是确保反应气体均匀分布并去除反应生成的水。一般常见的设计包括直流道、蛇形流道、交指流道以及基于这些流道的改进型流道等。直通道的特点是结构简单易加工，流阻相对较小，适合用于较大活性面积的低压电池流场，但气体利用率较低，排水不佳。蛇形通道的特点是流速大、排水较好，多用于单电池测试，但进出口气体压差较大，不利于电流密度的均匀性分布和催化剂的充分利用。交指流道的特点是流道不连续，气体被强制通过扩散层，有利于提高气体利用率，提高功率密度，但这种结构对气体进气压力要求更高。每种流场结构都有各自的优缺点，难以兼顾，需要根据电池的实际应用场景和工况进行反向设计。双极板流场结构的合理设计直接关系到反应气体的运输、分配、液态水的排除、热量和电流的传递，进而影响电池的输出性能。

目前双极板材料的开发重点偏向于金属材料和复合材料两大类，双极板的种类、一般制法、特点与主要供应商如表1-8所示。

表1-8 双极板的种类、一般制法、特点与主要供应商

种类	一般制法	特点	主要供应商
石墨双极板	利用碳粉或石墨粉混合可石墨化树脂制备	优点： ①质轻； ②耐蚀性好； ③导电性。 缺点： ①脆性物质，易造成组装难度，厚度不易做薄； ②一般烧结成多孔性板，需添加添加物； ③石墨化时间长，机械加工难，价格贵	美国步高石墨公司、美国GrafTech国际有限公司，日本藤仓橡胶有限公司、日本Kyushu Refractories有限公司，英国Bac2有限公司，中国上海弘枫实业有限公司等
金属双极板	不锈钢、钛合金、铝合金等直接加工而成	优点： ①良好的电、热导体； ②机械性能佳，强度高； ③无孔，阻气性好。 缺点： ①密度高，较重； ②易腐蚀，需表面改性	瑞典Cell Impact公司，德国德纳股份有限公司、格雷伯机械技术有限公司，美国特来德斯通技术公司，以及中国上海治臻新能源股份有限公司、上海佑戈金属科技有限公司、爱德曼氢能源装备有限公司等
复合双极板	多层复合型：以薄金属为分隔板，以有孔薄碳板为流场板，用极薄导电胶黏合	优点：结合石墨板和金属板的优点。 缺点：制作烦琐。	加拿大巴拉德动力系统公司、中国新源动力股份有限公司
	复合材料型：热塑或热固性树脂料混合石墨粉、增强纤维等形成预制料，并固化、石墨化后成型	优点： ①注射成型，制造速度快； ②重量轻； ③耐蚀性佳。 缺点： ①导电性较差； ②机械强度低	

其中，金属双极板具有电导率高、价格低廉、工艺制法多样、高机械强度等优点。但在燃料电池内部工作环境下，常用金属如铝、钢、钛或镍容易被腐蚀，且溶解的金属离子会扩散到催化层和质子交换膜，造成导电性的降低和催化剂活性的损失，从而减少电池寿命。此外，双极板表面的腐蚀层还会增大接触电阻，造成电堆性能降低，因此金属板必须充分覆盖导电耐蚀

涂层。涂层本身的耐蚀性、致密性，以及与基材的热膨胀系数之差决定了涂层的有效性，其技术难点就在于成型技术和表面处理技术。其中以非贵金属（如不锈钢、钛）为基材，辅以表面处理技术是研究的热点，主要内容是要筛选易导电、耐腐蚀兼容的涂层材料与保证涂层致密、稳定的制备技术。

不锈钢双极板是目前主流开发的金属双极板，通常采用304/316 L不锈钢作为基底材料，成本低、易于加工。主要的制备工艺包括两种，即冲压焊接和电化学刻蚀。其中，根据涂层制备的先后顺序，可以将冲压焊接工艺再细分为三种。第一种是先对不锈钢板进行批量化涂层处理，然后进行冲压，完成后再通过激光焊接。优点是成本低，适合大批量生产；缺点是要求涂层具有非常好的塑性，在冲压过程中不被损坏，对涂层的要求较高，同时对焊接工艺也有较高要求。第二种是冲压焊接后进行涂层制备，这是最常用的工艺。优点是对涂层塑性要求不高，可以保证涂层的完整性；缺点是批量化生产效率低。第三种是冲压后进行涂层制备，最后进行焊接。该工艺导电性能较好，多用于对导电性能要求较高的电堆上；缺点是成本较高，焊接工艺要求与第一种方法类似。电化学刻蚀法则不需要进行焊接，镀膜后可直接使用，且流场成型不通过冷加工，无残余应力；缺点是相对于冲压双极板，其厚度/质量均远高于冲压板，因此制备成的燃料电池电堆比功率较低，一般应用在大巴、货车、运输车、固定燃料电池上。

目前国内从事涂层开发并获得成功应用的主要为大连理工大学材料科学与工程学院的林国强教授及其团队。该团队自“十五”计划期间起就联合中国科学院大连化学物理研究所和新源动力股份有限公司，在衣宝廉院士的引导下，共同承担我国车用燃料电池金属双极板的开发任务，对完成同期国家燃料电池方面的专项课题起到了技术支撑作用。“十一五”期间，在国家高技术研究发展计划（863计划）项目“低成本双极板及其制备技术”的资助下，该团队在表面改性工艺及涂层材料上取得突破，用脉冲偏压电弧离子镀技术涂镀碳铬纳米复合薄膜进行表面改性处理的不锈钢双极板，在导电、耐蚀及疏水等性能指标上均达到国际领先水平。“十二五”期间，在863计划重大项目“面向示范和产品验证车用燃料电池系统开发”子课题“燃料电池双极板表面改性与生产工艺研究”，以及江苏省重大创新载体、常州市重大公共技术服务平台建设项目的资助下，该团队在表面改性产业化装备及量产工艺上取得突破，经批量处理的不锈钢双极板在千瓦级车用燃料电池电堆运行环境下的性能表现优异，并能满足美国能源部提出的双极板性能要求，已成功应用于我国某汽车集团公司的燃料电池汽车项目。进入“十三五”时期后，团队又承担新一轮的国家重点研发课题“金属双极板涂层改性材料及制备技术”和“高稳定性高品质金属双极板表面改性量产工艺开发”，目前在第二代改性涂层材料和制备工艺，尤其是第二代大型量产装备开发上取得了重要进展，预期在项目完成时将使我国拥有完全自主知识产权的双极板改性处理连续生产万片级每年的专用设备及工艺技术，使我国的双极板改性量产装备及生产水平达到国际水平，以满足我国新能源汽车产业化发展的爆发性需求。

钛双极板采用冲压焊接的方式制备。钛合金韧性好，要达到燃料电池对双极板流场平整度的要求，对冲压工艺要求非常高。目前丰田公司已掌握该冲压工艺，国内钛合金冲压工艺大多不成熟，其生产的双极板平整度较差，缺陷较多，通常只能应用于无人机领域。

铝合金双极板同样采用冲压焊接的方法进行制备。铝合金易于成型，冲压工艺相对比较简单，但由于铝合金硬度低、熔点低，因此其表面改性难度较大。目前，国内采用铝合金作为双极板基体材料的企业主要为大连新源动力股份有限公司。

由于车辆空间限制（尤其是轿车），乘用车燃料电池具有高能量密度需求，金属双极板相较于石墨及复合双极板具有明显优势。例如，日本丰田公司的 Mirai 燃料电池汽车用金属双极板 PEMFC 模块的功率密度达到 3 kW/L，英国智能能源（Intelligent Energy）公司的新一代 EC200-192 金属双极板燃料电池模块的功率密度达到 5 kW/L。金属双极板使 PEMFC 模块的功率密度大幅提升，金属双极板已成为乘用车燃料电池的主流双极板。几乎各大汽车公司都采用金属双极板技术，如丰田公司、通用汽车公司、本田公司等。

复合材料双极板能较好地结合石墨板与金属板的优点，密度低、抗腐蚀、易成型，使电堆装配后达到更好的效果，但是目前其加工周期长、长期工作可靠性较差，因此没有被大范围推广。复合材料双极板近年来也开始有所应用，如石墨 / 树脂复合材料、碳 / 碳复合材料等，国内具备研制能力。

四、研究总结与展望

开发高活性和高稳定性的低铂、非铂催化剂对燃料电池的商业化与推广应用具有重大的现实意义。本节综述了一些最新的研究成果，介绍了铂黑催化剂、铂碳催化剂、铂合金催化剂、核-壳结构催化剂、铂单原子催化剂和非贵金属催化剂，并详细介绍了各类催化剂的制备工艺，对其高活性和高稳定性的机理做了简要描述。

低铂催化剂已成为降低燃料电池成本最有效的方法之一。尽管低铂催化剂的研究和商业化取得了较大的进步，但仍面临以下问题：①在保证催化剂活性和稳定性的前提下，如何进一步优化低铂催化剂的制备条件，探求简便、低成本和规模化的制备方法仍是研究者需要解决的关键问题；②在制备具有核-壳结构的低铂催化剂时，如何更有效地控制铂层的单分散性和实现铂层厚度的可控控制，从而能够精确地揭示催化剂的结构和活性之间的关系；③低铂催化剂的电池测试仍研究较少，而电极测试和实际的燃料电池测试具有较大的差异，研究者并不能从电极测试的结果来直接评判将催化剂应用于电池测试性能的优劣。因此，将低铂催化剂应用于电池测试中的研究也势在必行。如果以上问题得以有效解决，低铂催化剂实际应用于燃料电池将具有极为广阔的发展前景，燃料电池的大规模商业化将会早日实现。

非贵金属电催化剂替代铂基催化剂是燃料电池研究的最理想方向。经过研究者几十年的努力探索，非贵金属电催化剂无论在活性还是稳定性上都取得了较大的突破，并展现出了取代铂基催化剂的潜力。然而，与铂基催化剂相比，非贵金属催化剂在酸性条件下的活性，尤其是稳定性，还需进一步提高。为了让非贵金属电催化剂走向实用化，必须要对活性位结构有清晰认识，从而指导催化剂的理性设计和制备，进一步提高活性位的密度。当前虽然已有大量关于活性位结构的报道，但并未能达成一致性的结论，且争议很大。热解 M-N-C 非贵金属催化剂结构复杂，很可能存在多种类型的氧还原活性位，包括单原子分散金属中心与氮配位结构、非金属氮掺杂碳，以及碳（氮掺杂碳）包裹型金属纳米颗粒等。未来研究需要借鉴表面科学已取得巨大成功的模型催化剂研究思路，力求通过设计和制备结构明确可控的模型催化剂，结合原位谱学技术（如 X 射线近边吸收谱、X 射线精细结构谱、穆斯堡尔谱等）深入研究，观察和推断催化剂活性位点以及催化剂失活过程。此外，催化剂的性能不仅与催化活性位点密度相关，还受催化剂的导电性、孔结构、比表面积、亲疏水性、膜电极的制备方法和运行条件等特性影响。因此，在实践中需要把这些因素有机地综合起来考虑，才能共同提升非贵金属电催化剂的活性和稳定性。

质子交换膜是燃料电池的核心材料，质子交换膜的性能将直接影响燃料电池产业化进程和大规模应用。为了实现燃料电池的实用化与产业化，科研工作者在质子交换膜材料的制造工艺和材料改性方面已经进行了大量的研究。目前，进一步提高质子交换膜的使用耐久性、寿命和工作性能仍然是质子交换膜燃料电池产业化面临的主要任务；燃料电池质子交换膜市场还是一个新兴市场，国内外均未形成较大的规模。在燃料电池巨大的市场需求推动下，质子交换膜必将获得进一步发展。相信不久将会有更高性能、更低成本的质子交换膜产品问世，以大力推动燃料电池技术的发展及其产业化应用。

在质子交换膜燃料电池双极板所采用的三类材料中，石墨材料具有质量轻、耐腐蚀性好、导电性好等优点，但其脆性大、加工成本高；复合材料良好的耐腐蚀性、更轻的质量，是双极板材料发展的趋势之一，但目前生产的复合双极板的成本高、耐腐蚀性差，不能满足双极板的要求。目前双极板中使用最多的是金属材料，该材料具有强度高、加工性能好、导热导电性强等优点，但在燃料电池环境下耐腐蚀性较差，金属表面容易形成一层钝化膜。这层钝化膜虽然增强了金属材料的耐腐蚀能力，但是由于钝化膜的电导率低，使得金属双极板材料的电阻率升高，导致燃料电池的输出功率降低，且大量数据表明，金属表面钝化膜的导电性和耐腐蚀性成反比。实现金属材料的导电性和耐腐蚀性的合理匹配，在保证合理导电性的前提下实现双极板的高耐腐蚀性，保障整个体系的服役寿命提升，是金属双极板广泛应用的关键。

综上可知，通过在金属表面镀涂层可以使金属材料在燃料电池环境中的耐蚀性和电导率明显提高，且很多镀涂层后的金属材料可以满足双极板的性能要求，但是有些涂层材料与基体结合，表现出较差的耐腐蚀性或导电性。因此，金属材料要满足双极板的性能要求，必须选择性能优良的基体材料和与之相匹配的涂层。目前，金属基体材料中研究最多的有不锈钢、钛合金以及铝合金三种。不锈钢具有价格低、力学性能优异等优点，是基体材料中的首选。钛合金和铝合金的比强高、耐腐蚀性好，可以作为特殊用途的质子交换膜燃料电池的双极板材料。涂层材料种类很多，不同涂层材料与金属基体的匹配性和结合力各有差异，因此，寻找出一种适合基体的涂层材料是解决金属双极板耐腐蚀性和导电性问题的关键。在本节叙述的涂层材料中，有一种通过在 C 膜中掺杂 Cr 元素形成的涂层材料表现出十分优异的耐腐蚀性和导电性。较其他单层涂层，这种通过掺杂的方式形成的新型涂层材料的性能优异，并且摒弃了复合涂层工艺的复杂性，这种涂层可以作为金属双极板优良涂层的备选材料之一。在金属材料表面镀涂层虽然提高了双极板在燃料电池环境下的耐腐蚀性和电导率，但这种方法增加了双极板的制造成本和工艺的复杂性。如何在保证良好导电性和耐腐蚀性的前提下降低成本与工艺的复杂性，保障整个电池体系的服役寿命提升，是质子交换膜燃料电池下一步需要解决的问题。

五、学科发展政策建议与措施

我国政府一直关注和重视燃料电池汽车的发展，近些年更是密集出台了一系列氢能燃料电池汽车的支持政策。2015年工业和信息化部出台的《中国制造2025》中，“节能与新能源汽车”被列为重点发展领域，并且该文件也明确了“继续支持燃料电池汽车发展”“推动自主品牌节能与新能源汽车与国际先进水平接轨”的发展策略。2016年4月，国家发展和改革委员会、国家能源局颁布《能源技术革命创新行动计划（2016—2030年）》《能源技术革命重点创新行动路线图》，明确表示支持“氢能与燃料电池技术创新”。2016年7月28日，国务院发布《“十三五”国家科技创新规划》，提出发展“可再生能源和氢能技术”“突破……燃料电池动力系统……形成完善的电动汽车动力系统技术体系和产业链，实现各类电动汽车产业化”。2016年8月底，工业和信息化部公布《燃料电池汽车技术发展路线图》，明确指出：到2030年，我国燃料电池汽车规模预计将超过100万辆。2017年4月25日，工业和信息化部、国家发展和改革委员会、科学技术部联合印发《汽车产业中长期发展规划》，确定了多项产业发展任务和重点工程，包括创新中心建设工程、新能源汽车研发和推广应用工程等。此规划同时为我国汽车产业发展指明了前进的道路和目标，并将推动我国汽车产业进入转型升级、由大变强的战略机遇期。其中，在《汽车产业中长期发展规划》中明确了要逐步扩大燃料电池汽车试点示范范围。

对于燃料电池催化剂的研究，低铂催化剂已成为降低燃料电池成本最有效的方法之一。因此，将低铂催化剂应用于电池测试中的研究也势在必行。非贵金属电催化剂替代铂基催化剂是燃料电池研究的最理想方向，但目前仍处于基础研究状态。

对于质子交换膜材料的发展，近些年来，全氟磺酸质子交换膜在质子传导、化学耐腐蚀性、机械强度等方面都能够满足大部分需求，但是以Nafion®系列产品为代表的全氟磺酸质子交换膜仍然存在尺寸稳定性差、价格昂贵、燃料渗透率高等因素的限制。因此，针对高性能、高稳定性且廉价的质子交换膜的研究一直在广泛展开：利用新型的表征手段，从分子尺度入手，研究化学结构对质子传输的影响，从而进行结构设计；利用新型的制备手段，提升质子交换膜的性能，降低其生产成本，对推动燃料电池的普及具有重要意义。未来几年将是燃料电池技术从示范运行转向商业化的重要时期，质子交换膜的发展也必将迎来新的高峰。

对于双极板材料，不同双极板材料的主要挑战是如何平衡总体性能要求，实现长耐久性、量产能力和低成本。随着从实验室水平到量产规模的放大，品质要求和一致性控制也将变得更具挑战性。更简单的非原位测试（包括加速寿命衰减测试和加速压力测试），与接近实际应用的严格条件下电堆水平的长期原位测试一起，在加速极板和极板材料反复优化改进通向最终商业化生产和应用中起着极其关键的作用。

总之，燃料电池电动汽车动力性能好、充电快、续驶里程长、接近零排放，是未来新能源汽车的有力竞争者。国际上特别是日本的车用燃料电池技术链已趋于成熟，我国需要加大产业链建设，鼓励企业进行投入，发展批量生产设备，在产业链的建立过程中促进技术链的逐步完善。同时，还要继续在成本、寿命方面进行研发投入，鼓励创新材料的研制，加大投入，强化电堆可靠性与耐久性考核，为燃料电池汽车商业化形成技术储备。

黄　林（南京大学）、吴聪萍（南京大学）

本节参考文献

Arnett N Y, Harrison W L, Badami A S, et al. 2007 . Hydrocarbon and partially fluorinated sulfonated copolymer blends as functional membranes for proton exchange membrane fuel cells. Journal of Power Sources, 172: 20-29.

Beermann V, Gocyla M, Willinger E, et al. 2016. Rh-doped Pt-Ni octahedral nanoparticles: understanding the correlation between elemental distribution, oxygen reduction reaction, and shape stability. Nano Letters, 16: 1719-1725.

Carpenter M K, Moylan T E, Kukreja R S, et al. 2012. Solvothermal synthesis of platinum alloy nanoparticles for oxygen reduction electrocatalysis. Journal of the American Chemical Society, 134: 8535-8542.

Chen C, Kang Y J, Huo Z Y, et al. 2014. Highly crystalline multimetallic nanoframes with three-dimensional electrocatalytic surfaces. Science, 343: 1339-1343.

Huang X, Zhao Z, Cao L, et al. 2015. High-performance transition metal-doped Pt_3Ni octahedra for oxygen reduction reaction. Science, 348: 1230-1234.

Katzfuss A, Krajinovic K, Chromik A, et al. 2011. Partially fluorinated sulfonated poly (arylene sulfone) s blended with polybenzimidazole. Journal of Polymer Science Part A: Polymer Chemistry, 49: 1919-1927.

Kerres J, Schoenberger F, Chromik A, et al. 2008. Partially fluorinated arylene polyethers and their ternary blend membranes with PBI and H_3PO_4. Part Ⅰ. Synthesis and characterisation of polymers and binary blend membranes, Fuel Cells, 8: 175-187.

Kerres J A, Xing D M, Schöenberger F. 2006. Comparative investigation of novel PBI blend ionomer membranes from nonfluorinated and partially fluorinated poly arylene ethers. Journal of Polymer Science Part B: Polymer Physics, 44: 2311-2326.

Kim B, Kannan R, Nahm K S, et al. 2016. Development and characterization of highly conducting nonfluorinated di and triblock copolymers for polymer electrolyte membranes. Journal of Dispersion Science & Technology, 37: 1315-1323.

Klaus S R, Chen Q. 2008. Parallel cylindrical water nanochannels in Nafion fuel-cell membranes. Nature Materials, 7: 75-83.

Li Q F, Jensen J O, Savinell R F, et al. 2009. High temperature proton exchange membranes based on polybenzimidazoles for fuel cells. Progress in Polymer Science, 34: 449-477.

Liu H Y, Song Y J, Li S S, et al. 2016. Synthesis of core/shell structured Pd_3Au@Pt/C with enhanced electrocatalytic activity by regioselective atomic layer deposition combined with a wet chemical method. RSC Advances, 6: 66712-66720.

Liu J, Jiao M G, Lu L L, et al. 2017. High performance platinum single atom electrocatalyst for oxygen reduction reaction. Nature Communications, 8: 15938.

Lu B A, Sheng T, Tian N, et al. 2017. Octahedral PtCu alloy nanocrystals with high performance for oxygen reduction reaction and their enhanced stability by trace Au. Nano Energy, 33: 65-71.

Mani P, Srivastava R, Strasser P. 2008. Dealloyed Pt-Cu core-shell nanoparticle electrocatalysts for use in PEM fuel cell cathodes. Journal of Physical Chemistry C, 112: 2770-2778.

Mauritz K A, Moore R B. 2004. State of understanding of Nafion. Chemical Reviews, 104: 4535-4585.

Oezaslan M, Hasché F, Strasser P. 2013. Pt-based core-shell catalyst architectures for oxygen fuel cell electrodes. Journal of Physical Chemistry Letters, 4: 3273-3291.

Park J, Zhang L, Choi S I, et al. 2015. Atomic layer-by-layer deposition of platinum on palladium octahedra for enhanced catalysts toward the oxygen reduction reaction. ACS Nano, 9: 2635-2647.

Parthasarathy A, Srinivasan S, Appleby A J, et al. 1992. Temperature-dependence of the electrode-kinetics of oxygen reduction at the platinum Nafion® interface: a microelectrode investigation. Journal of Electrochemical Society, 139: 2530-2537 .

Schöenberger F, Hein M, Kerres J. 2007. Preparation and characterisation of sulfonated partially fluorinated statistical poly (arylene ether sulfone)s and their blends with PBI. Solid State Ionics, 178: 547-554.

Shannon R D, Prewitt C T. 1969. Effective ionic radii in oxides & fluorides. Acta Crystallographica Section B, 25: 925-946.

Shao M H, He G N, Peles A, et al. 2013. Manipulating the oxygen reduction activity of platinum shells with shape-controlled palladium nanocrystal cores. Chemical Communications, 49: 9030-9032.

Stamenkovic V R, Fowler B, Mun B S, et al. 2007. Improved oxygen reduction activity on Pt_3Ni (111) via increased surface site availability. Science, 315: 493-497.

Stamenkovic V R, Mun B S, Arenz M, et al. 2007. Trends in electrocatalysis on extended and nanoscale Pt-bimetallic alloy surfaces. Nature Materials, 6: 241-247.

Wang D L, Xin H L, Hovden R, et al. 2013. Structurally ordered intermetallic platinum-cobalt core-shell nanoparticles with enhanced activity and stability as oxygen reduction electrocatalysts. Nature Materials, 12: 81-87.

Wang X, Choi S I, Roling L T, et al. 2015. Palladium-platinum core-shell icosahedra with substantially enhanced activity and durability towards oxygen reduction. Nature Communications, 6: 7594.

Wei S J, Li A, Liu J C, et al. 2018. Direct observation of noble metal nanoparticles transforming to thermally stable single atoms. Nature Nanotechnology, 13: 856-861.

Xie S F, Choi S I, Lu N, et al. 2014. Atomic layer-by-layer deposition of Pt on Pd nanocubes for catalysts with enhanced activity and durability toward oxygen reduction. Nano Letters, 14: 3570-3576.

Yang X F, Wang A Q, Qiao B T. 2013. Single-atom catalysts: a new frontier in heterogeneous catalysis. Accounts of Chemical Research, 46: 1740-1748.

Zawodzinski T A, Derouin C, Radzinski S, et al. 1993. Water-uptake by and transport through Nafion® 117 membranes. Journal of the Electrochemical Society, 140: 1041-1047.

Zeng X, Shui J, Liu X, et al. 2018. Single-atom to single-atom grafting of Pt_1 onto Fe—N_4 center: Pt_1@Fe—N—C multifunctional electrocatalyst with significantly enhanced properties. Advanced Energy Materials, 8: 1701345.

Zhang J, Yang H Z, Fang J Y, et al. 2010. Synthesis and oxygen reduction activity of shape-controlled Pt_3Ni nanopolyhedra. Nano Letters, 10: 638-644.

Zhang J L, Vukmirovic M B, Xu Y, et al. 2005. Controlling the catalytic activity of platinum-monolayer electrocatalysts for oxygen reduction with different substrates. Angewandte Chemie International Edition, 44: 2132-2135.

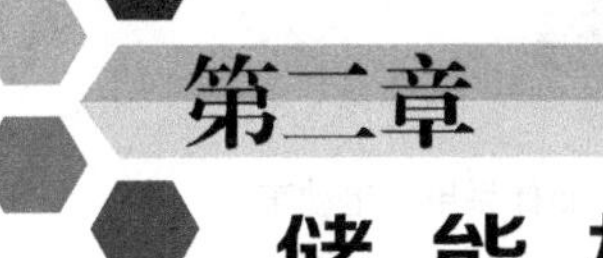

第二章 储能材料

第一节 锂离子电池关键材料

一、锂离子电池关键材料研究概述

自 20 世纪 90 年代日本索尼公司成功地将锂离子电池商业化以后，锂离子电池以其能量密度高、输出功率大、电压高、自放电少、工作温度范围宽、无记忆效应和环境友好等众多独特的优势成功占领了计算机、手机等电子产品的电源市场（Zhu et al., 2011）。近年来，随着电池技术的快速发展以及石油、天然气等不可再生资源的日益枯竭，锂离子电池已成为电动汽车、智能电网和大型储能基站等领域的重要组成部分（Goodenough & Park, 2013）（图 2-1）。但是，目前商业化锂离子电池远不能满足当前电子设备以及各种储能系统的需求，极大地限制了其进一步的规模化应用。高安全、长寿命、低成本、高比能是未来锂离子电池发展的必然趋势。目前，锂离子电池的正负极材料是限制其发展的首要因素。因此，不断突破正负极材料性能瓶颈是电池发展必须攻克的难关。在满足稳定、环保、安全和长寿命的基本条件下，提升电池能量密度的关键点主要集中在两方面：提高材料的比容量、提升电池的工作电压。

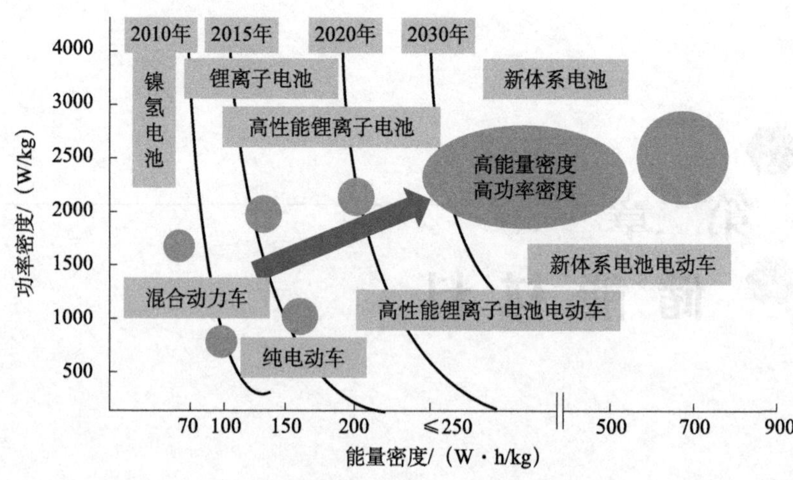

图 2-1　车用动力电池发展趋势

二、锂离子电池关键材料研究的历史及现状

（一）正极材料

锂离子电池正极材料钴酸锂（$LiCoO_2$）因其平稳的放电电压和良好的循环性能，已成功实现商业化应用（Yang et al.，2011）。然而，Co 资源紧缺、价格高昂、抗过充能力差以及污染环境等一系列问题严重限制了 $LiCoO_2$ 正极材料的进一步发展。因此，电池材料里面的 Co 元素逐步被替换成环境友好、价格低廉的 Mn 元素。尖晶石结构 $LiMn_2O_4$ 具有充放电电压高、安全性能优异等独特的优点，但是 $LiMn_2O_4$ 在充放电过程中会发生扬-泰勒效应（Jahn-Teller），导致尖晶石结构 $LiMn_2O_4$ 由立方相向四方相转变，最终导致容量急剧衰减。此外，当 $LiMn_2O_4$ 在 55 ℃以上的高温时，部分金属离子会溶解而导致循环性能进一步劣化。所以，$LiMn_2O_4$ 正极材料的大规模应用也面临诸多难题（Zhao et al.，2013）。锂离子电池正极材料 $LiFePO_4$ 由于具有安全性能高、环保、价格低廉、循环寿命长等特点，在国内作为锂离子动力电池正极材料已被成功地实现商业化应用，但是其橄榄石结构限制了材料的电运输能力，进而影响了锂离子电池的倍率性能（Bakenov & Taniguchi，2010）。对于材料的电子电导和离子扩散能力的提升，武汉理工大学麦立强课题组提出了多种有效的改善方法。该课题组创建了单根纳米线器件，并原位表征了材料电化学过程中的结构变化，发现了充放电过程中离子脱嵌导致材料结构劣化而使得材料的电导率下降的原因（Mai et al.，2010）。由此，该课题组从构筑

复杂微观结构着手，进行了结构稳定性提升的研究。通过不同的结构设计和复合，如新型分级异质结构、石墨烯半中空包覆结构和自缓冲纳米卷等，不同电极材料的导电性能有了大幅的提升（Cai et al.，2015；Mai et al.，2013；Yan et al.，2013）。

目前，我国电动汽车商用动力电池主要以 $LiFePO_4$ 为正极、石墨为负极，电池的能量密度仅有 120 W·h/kg 左右，远远低于国家规划 2020 年需要达到的 300 W·h/kg 的高能量密度目标。除 $LiFePO_4$ 以外，已商业化的低镍三元正极材料如 $LiNi_{1/3}Co_{1/3}Mn_{1/3}O_2$（NCM111）和 $LiNi_{0.5}Co_{0.2}Mn_{0.3}O_2$（NCM523）的比容量也只有 140～150 mA·h/g，也远远达不到国家规划的目标。目前，高镍类三元材料如 $LiNi_{0.8}Co_{0.1}Mn_{0.1}O_2$（NCM811）、$LiNi_{0.8}Co_{0.15}Al_{0.05}O_2$（NCA）的高理论容量（200～220 mA·h/g）和高的电压平台（约 3.7 V）使其成为下一代高能量密度锂离子电池正极的热门材料（图 2-2）。另外，富锂锰基正极材料如 $Li_2MnO_3·LiMO_2$（M=Ni、Co、Mn 等）的理论容量为 250～300 mA·h/g，电压平台为 4.2 V，有望成为下一代高性能锂离子正极材料。在富锂锰基正极材料中，锂离子数量的增加导致在高电压（4.5 V *vs.* Li^+/Li）下会发生氧的氧化还原，能够实现更高的容量，但也会导致其稳定性的降低。因此，如何提高富锂锰基材料的循环稳定性是下一代高性能锂离子电池正极材料的研究热点与难点。通过对该类正极材料的不断探索、提升和发展，如果能够选择合适的负极材料与这类正极材料匹配，将有望构建下一代高比能量锂离子电池并成功实现产业化应用。

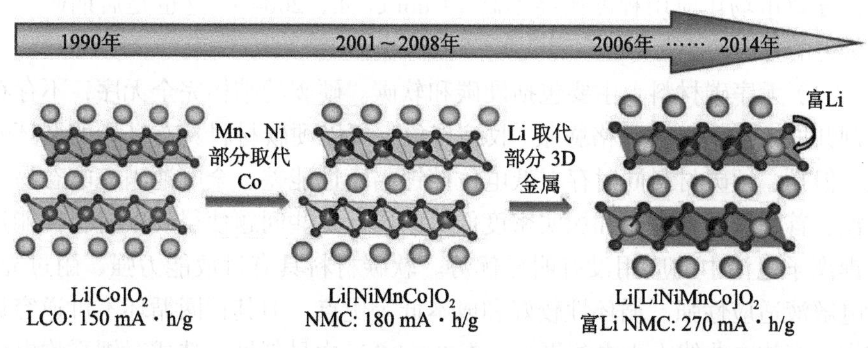

图 2-2　三元正极材料的发展趋势

钒系电极材料通常拥有比较高的理论容量，有望应用于锂离子动力电池和大规模储能等相关领域。纳米线钒系电极材料具有大的长径比、高的比表面积及各向异性等优良特性，作为电极材料时具有能量密度高、充电时间短的特点，适合制作大型锂离子动力电池，具有较高的市场化价值。武汉理工大学麦立强课题组面向我国动力电池、电动汽车技术发展、产业升级需求，结合我国在能源纳米材料制备技术方面的原始创新和在新型电池器件设计、电池封装技术方面的研究经验，以研发能量密度达 230 W·h/kg 的纯电动汽车动力电池为目标，重点突破纳米线钒系材料批量制备技术、动力电池器件集成技术、高性能封装技术，最终封装制备了高性能锂离子动力电池。相关成果也已成功应用于东风汽车集团有限公司新能源汽车动力电池的装车运行及中兴通讯股份有限公司高能动力电池的产品升级中。

（二）负极材料

锂离子电池负极材料是决定锂离子电池性能的另一个关键因素，目前商业化锂离子电池采用的负极材料主要包括以下几种。

（1）石墨类碳材料。主要分为天然石墨和人造石墨。天然石墨具有安全无毒、储量高以及成本低等优点，但是天然石墨粉表面活度不一致、晶粒较大、循环过程中晶体表面结构容易被破坏、表面固体电解质界面膜（SEI 膜）覆盖不均匀等问题会导致天然石墨类材料倍率性能较差以及首次库仑效率低。人造石墨由于晶粒较小、结晶取向度较小、石墨化程度较低，所以相较天然石墨在倍率性能和体积膨胀方面有所改善。尽管改性过后的人造石墨（如中间相碳微球）有着优异的电化学性能，但是其制备过程复杂且产率较低，导致市场化应用程度始终不高（Choi et al.，2004），这也是后期研究需要努力的方向。

（2）无序碳材料。主要包括硬碳和软碳。硬碳的结构完全无序，不存在溶剂共嵌入和明显的晶格膨胀 / 收缩现象，所以硬碳材料具有良好的循环性能。但是，硬碳材料同时存在低电位储锂倍率性能差、全电池满充电态易于析锂、首圈循环效率低和压实密度低等问题。这些问题使得硬碳材料在能量型锂离子电池中的应用没有明显优势。软碳材料具有过放能力强、耐过充、对电解液适应性强、循环性较好和成本低等优点，但其首圈循环不可逆容量较大，充放电曲线上无电位平台，在 0～1.2 V 内呈斜坡，造成对锂平均电位较高以至于锂离子电池端电压较低的问题（Lee et al.，2005）。

（3）钛酸锂材料，即 $Li_4Ti_5O_{12}$。其理论嵌锂容量为 175 mA·h/g，初次循环库仑效率可达 98.8%，且锂在嵌入脱出前后材料的体积变化不到 1%，是锂离子电池中罕见的零应变材料。经过修饰后，其导电性能、循环性能和倍率性能均得到了大幅提升。但是嵌锂态的 $Li_7Ti_5O_{12}$ 会与电解液发生化学反应导致胀气，使电池安全性能下降、容量衰减、寿命缩短。另外 $Li_4Ti_5O_{12}$ 的嵌锂电位过高，虽然避免了锂枝晶的产生，但是电池的能量密度由此会有一定程度的衰减。成熟的 $Li_4Ti_5O_{12}$ 制作工艺要求也苛刻，导致 $Li_4Ti_5O_{12}$ 的应用存在较高的技术门槛，主要市场为适合高功率锂离子电池应用的领域（Hao et al.，2006）。

（4）硅基材料。这也是目前锂离子电池负极材料的研究热点之一，主要包括碳包覆的氧化亚硅复合材料、碳包覆的纳米硅碳复合材料以及无定形的硅合金。硅负极材料因其较高的理论容量（形成 $Li_{4.4}Si$ 合金时约为 4200 mA·h/g）、环境友好、储量丰富等特点，很早就被研究人员考虑作为下一代高能量密度锂离子电池的负极材料（He et al.，2012）。目前硅基负极材料商业化应用必须解决材料储锂过程中体积变化巨大（大于300%）和首次库仑效率较低两大难题。硅在锂离子嵌入膨胀过程中，体积变化 100%～300%，导致材料内部产生较大的内应力。同时，体积膨胀又会导致新的 SEI 膜不断地生成，这将会不断消耗正极材料和电解液中有限的锂源，电池内阻将不断地增加，体积也会相应地膨胀，使得全电池鼓胀和循环性能衰减严重。另外，硅为半导体，导电性比石墨差，导致锂离子脱嵌过程中不可逆程度增大，进一步降低了其首次库仑效率。目前解决这些问题的方法主要集中在改进制备工艺、预锂化、缩小晶粒大小和降低结晶度等方面。另外，通过碳包覆、异质元素掺杂、设计微结构等方法也可以改善纳米硅碳负极材料的电化学性能。硅基负极材料具有无与伦比的高容量优势，目前一直是各大高校科研院所以及企业的研究热点。随着材料设计和制备工艺的不断提升和改进，硅基负极材料将会陆续地实现产业化应用（Zheng et al.，2014；Luo et al.，2015）。

三、锂离子电池关键材料研究目前面临的重大科学问题及前沿方向

在正极材料方面，高镍三元材料和富锂锰基材料是目前研究的热点，未来锂离子动力电池将由锰酸锂、磷酸铁锂为主的正极材料向三元材料、富锂锰基材料转移，通过研究者的共同努力，上述正极材料的电化学性能将会逐渐提高，并推动高镍三元正极材料广泛应用于锂离子电池。在负极材料方面，尽管如硅碳的新型负极材料拥有很大的应用潜力和实用价值，但是其产业化程度和技术成熟度与石墨类碳材料相比还有一定距离。无论是负极材料

还是正极材料，研究者应该不断地从其储锂机制、反应热力学/动力学机制、电化学反应过程、材料晶体结构的变化以及界面反应等基础科学问题入手，不断深入研究，揭示储能材料的相关理论反应机理机制，更好地解决储能领域面临的关键难题，进而突破锂离子电池领域的“卡脖子”技术。在产业化的过程中，材料的匹配性、服役与失效机制、综合性能指标改进、安全性、成本、批量化生产等问题，都是比较重要的环节。只有产、学、研深度地交流合作和努力后，高能量密度、高功率密度的新一代锂离子电池才能更好更快地被研发出来。高性能的新一代锂离子电池的关键在于新材料，研发出性能优异的新材料，是研究人员实现创新研究价值的真正舞台，也是企业提升核心竞争力、实现“弯道超车”的良好选择。

四、研究总结与展望

近年来，在全球面临能源危机、环境污染等生态问题限制的状况下，关键储能材料的研发与应用对建立高效、清洁、经济、安全的能源体系和实现新能源的可持续性发展尤为重要。尽管锂离子电池的发展道路依然艰难漫长和坎坷，但是由于二次能源以及新能源汽车、智能电网的蓬勃发展，锂离子电池产业必将在未来的20～30年内持续高速发展，这为我国锂离子电池关键材料的发展提供了很大的机遇。锂离子电池的重点研究领域是电极材料，中国、日本、韩国、美国、德国五国在锂离子动力电池的市场与技术方面有着绝对优势，但是锂离子电池材料的发展目前仍然没有结束。虽然日本、韩国和美国的企业在锂离子电池行业中始终拥有较大的话语权，但是中国仍有很多的发展机会来实现锂离子电池的“弯道超车”。

五、学科发展政策建议与措施

首先，继续加大对基础研究的支持力度，尤其是在颠覆性锂离子电池新材料的设计、创制和储能机制研究等方面，瞄准锂离子电池行业的关键科学问题，探索电极材料的储能本征机制，以推动锂离子电池行业的变革性发展。其次，不断加快锂离子电池行业的整体产业链布局，打造国家级矿产资源、原料、电极材料、电芯组装、模组装车运行的连续化、一体化产业园区建设，降低材料和单体电池的成本，推动锂离子动力领域工程技术方面的技术革新。此外，还需加强与日本、韩国等发达国家的新能源龙头企业合作，积极参与国际市场的竞争，这也是加速我国锂离子电池行业发展的良好途径。最后，加强技术攻关，寻求技术壁垒突破，争取更多的研究资金，培养自己的研发人才，只有培养出自己的研发人才、掌握自己的核心技术和关键技术，才可能实现真正的技术跨越。

麦立强（武汉理工大学）、张清杰（武汉理工大学）

本节参考文献

Bakenov Z, Taniguchi I. 2010. Electrochemical performance of nanocomposite $LiMnPO_4/C$ cathode materials for lithium batteries. Electrochemistry Communications, 12: 75-78.

Cai Z Y, Xu L, Yan M Y, et al. 2015. Manganese oxide/carbon yolk-shell nanorod anodes for high capacity lithium batteries. Nano Letters, 15: 738-744.

Choi W C, Byun D, Lee J K, et al. 2004. Electrochemical characteristics of silver- and nickel-coated synthetic graphite prepared by a gas suspension spray coating method for the anode of lithium secondary batteries. Electrochimica Acta, 50: 523-529.

Goodenough J B, Park K S. 2013. The Li-ion rechargeable battery: a perspective. Journal of the American Chemical Society, 135: 1167-1176.

Hao Y J, Lai Q Y, Lu J Z, et al. 2006. Synthesis and characterization of spinel $Li_4Ti_5O_{12}$ anode material by oxalic acid-assisted sol-gel method. Journal of Power Sources, 158: 1358-1364.

He Y, Yu X Q, Li G, et al. 2012. Shape evolution of patterned amorphous and polycrystalline silicon microarray thin film electrodes caused by lithium insertion and extraction. Journal of Power Sources, 216: 131-138.

Lee J, Lee S, Paik U, et al. 2005. Aqueous processing of natural graphite particulates for lithium-ion battery anodes and their electrochemical performance. Journal of Power Sources, 147: 249-255.

Luo F, Chu G, Xia X X, et al. 2015. Thick solid electrolyte interphases grown on silicon nanocone anodes during slow cycling and their negative effects on the performance of Li-ion batteries. Nanoscale, 7: 7651-7658.

Mai L Q, Dong Y J, Xu L, et al. 2010. Single nanowire electrochemical devices. Nano Letters, 10: 4273-4278.

Mai L Q, Wei Q L, An Q Y, et al. 2013. Nanoscroll buffered hybrid nanostructural VO_2 (B) cathodes for high-rate and long-life lithium storage. Advanced Materials, 25: 2969-2973.

Yan M Y, Wang F C, Han C H, et al. 2013. Nanowire templated semihollow bicontinuous graphene scrolls: designed construction, mechanism, and enhanced energy storage performance. Journal of the American Chemical Society, 135: 18176-18182.

Yang W S, Yang Z X, Qiao Q D. 2011. Improvement of structural and electrochemical properties of commercial $LiCoO_2$ by coating with LaF. Electrochimica Acta, 56: 4791-4796.

Zhao S, Chang Q J, Kai J, et al. 2013. Performance improvement of spinel $LiMn_2O_4$ cathode material by LaF_3 surface modification. Solid State Ionics, 253: 1-7.

Zheng J Y, Zheng H, Wang R, et al. 2014. 3D visualization of inhomogeneous multi-layered structure and Young's modulus of the solid electrolyte interphase (SEI) on silicon anodes for lithium ion batteries. Physical Chemistry Chemical Physics, 16: 13229-13238.

Zhu Y W, Murali S, Stoller M D, et al. 2011. Carbon-based supercapacitors produced by activation of graphene. Science, 332: 1537-1541.

第二节 锂硫电池材料

一、锂硫电池材料研究概述

近年来，全球面临的资源短缺和生态问题制约了人类的发展与社会的进步，锂硫电池作为一类新型的化学储能系统，因其比能量密度高和价格低廉而受到研究者的广泛关注。地球上的硫资源储量丰富，环境友好，以单质硫为正极、锂金属为负极的锂硫电池拥有 1675 mA·h/g 的高理论比容量，是目前以商业化钴酸锂为正极的锂离子电池（274 mA·h/g）比容量的 8 倍左右。因此，锂硫电池是目前锂电池领域内研究的热点和重点，有望成为最有发展潜力的下一代锂二次电池。锂硫电池在储能研究中的应用可以追溯到 20 世纪 70 年代，赫伯特（Herbert）和朱利叶斯（Juliusz）发现硫可以作为锂电池的正极材料（Herbert & Juliusz，1962）。随后饶等以硫作为正极，成功制备了一种金属-硫电池并计算其理论能量密度（Bhaskara & Lakshmanar，1968），其中电池中的电解液一般为碱性的高氯酸盐、碘化物、溴化物或氯酸盐，最优的电解液是以丙基、丁基和戊基胺作为溶剂的电解液。佩雷德（Peled）等采用有机醚作为锂硫电池电解液溶剂，使其性能得到了较大提升（Yamin & Peled，1983；Peled et al.，1989）。为了改善单质硫正极的导电性，佩雷德等提出将单质硫附着在多孔碳表面，成功制备了可充电的二次锂硫电池。

传统的锂硫电池主要由金属锂负极、有机电解液以及含硫的复合正极材料组成。锂硫电池具有十分有趣的充放电过程，因为硫处于充电状态，所以锂硫电池的激活是以放电开始的。在放电反应开始时，金属锂在负极被氧化生成锂离子和电子，其中锂离子通过内部电解液从负极自发地扩散到正极并发生反应，而电子则通过外部电路传输到正极，由此形成电流。硫在正极得到电子并与锂离子反应被还原成硫化锂。

然而，锂硫电池目前仍然存在硫导电性差、穿梭效应严重及体积膨胀等问题，导致其活性物质的损失和较低的库仑效率。其中，针对锂硫电池面临的穿梭效应严重问题，需要设计开发一种新的强导电极性材料来提升其综合性能（Bruce et al., 2011；Manthiram et al., 2015；Xu et al., 2015；Yang et al., 2013；Yin et al., 2013）。

二、锂硫电池材料研究的历史及现状

（一）正极材料

在过去的几十年里，人们探索了各种不同的技术来解决锂硫电池面临的问题，并取得了实质性的突破，主要思路为设计高导电性、高比表面积、高化学稳定性的硫正极的载体材料。

单质硫绝缘的特性使其不能单独用作电池正极材料。因此，必须通过提高电导率来提高正极活性物质的利用率及电池的倍率性能，同时还要保持正极材料的结构稳定性，抑制容量的不可逆损失，从而提高电池的循环稳定性。当前，对锂硫电池正极材料的研究主要集中在碳/硫复合材料、导电聚合物/硫、金属氧化物/硫三个方面。因而，提高正极活性物质硫的利用率和库仑效率、增强电池的循环稳定性成为锂硫电池的重点研究内容。

碳/硫复合材料是自从锂硫电池问世以来研究最为广泛的热点材料，主要有以下几个优点：①碳材料具有优异的导电性能，有利于提升整个电极的导电性能；②碳材料的多孔结构有利于负载更多的活性硫、提高电池的能量密度，有利于电解液的存储以及渗透到活性材料表面，还有利于放电反应快速彻底进行；③碳材料较高的比表面积对多硫化合物的吸附作用明显，减少了因多硫化合物的溶解引发的穿梭效应，提高了整个电池的循环性能和安全性能。华中科技大学黄云辉课题组（Li et al., 2014）通过制备有序微-介孔碳复合材料，用其作为装载硫材料合成了具有壳-核结构的碳/硫复合材料（图 2-3）。该材料在倍率电流为 0.5 C 时的可逆比容量高达 837 mA·h/g；200 次循环后的容量保持率为 80%，每圈循环仅衰减 0.1%。

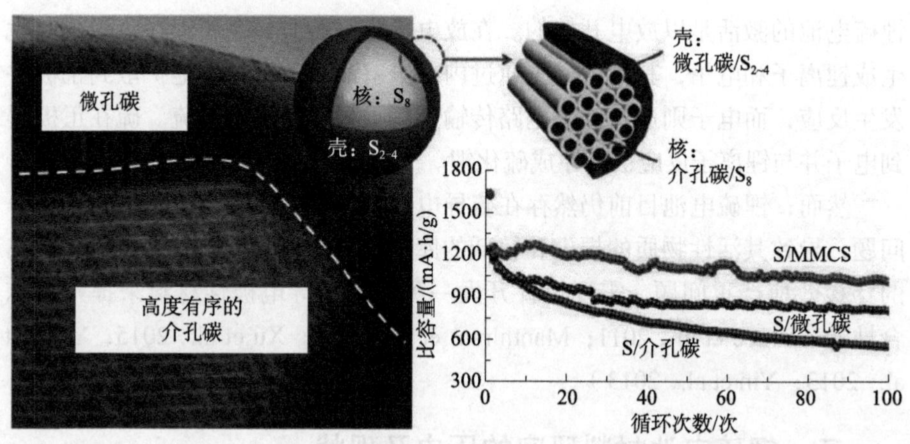

图 2-3　有序微－介孔碳 / 硫复合材料及循环特性曲线图

导电聚合物是一类很有前途的锂硫电池正极材料，主要原因表现为以下几个方面：①材料内部良好的电子传导性，能够有效提升在电化学反应过程中的电子传导能力，提升锂硫电池充放电倍率特性；②通过交联聚合制备的导电聚合物具有很强的弹性、韧性良好的延展性，可以缓冲循环过程中硫的体积膨胀，抑制长循环过程中活性材料的脱落；③导电聚合物丰富多样的官能团能够有效吸附多硫化合物，防止多硫化合物的溶解和扩散，提升电极材料的循环性能。目前的相关研究主要集中在将导电聚合物（如聚噻吩、聚苯胺、聚吡咯等）涂覆在硫碳电极表面，以改善电极材料的循环稳定性。除涂覆方法之外，也可以通过球磨复合、原位化学氧化聚合以及热处理复合等方法制备导电聚合物 / 硫复合材料。

金属氧化物中，金属离子与氧离子之间具有很强的离子键，使得其不溶于大部分的有机溶剂。氧化物表面也因为氧离子的存在具有很强的极性。在锂硫电池研究初期，对于非导电的氧化物纳米材料，通常以添加剂的形式直接添加到正极材料中，其质量分数一般小于 10%。这些金属氧化物的极性表面可以有效吸附多硫化物，从而抑制多硫化物的溶解扩散问题。另外，部分金属氧化物具有半导体特征，导电性较好，所以可以用作锂硫电池的导电基底。金属氧化物材料作为锂硫电池电极材料的包覆层能够有效防止多硫化合物的溶解和扩散（Jiang et al.，2012；Wang et al.，2012）。最经典的研究是美国斯坦福大学崔屹团队开发的硫–二氧化钛蛋黄–壳核结构（Seh et al.，2013）。首先，单分散的纳米硫颗粒通过硫代硫酸钠和盐反应合成；其次，利用水解溶胶–凝胶前驱体的方式，将制备的 TiO_2 包覆在纳米硫颗粒表面；

最后，通过硫的部分溶解，在 TiO_2 壳体和硫核之间形成一个半中空结构。第一，预留的空间可有效协调硫在锂化过程中的体积膨胀；第二，TiO_2 壳体为半导体，可以增强单质硫材料的电导性能。相比于未处理的材料，该活性材料表现出了极好的循环稳定性和倍率性能。

武汉理工大学麦立强课题组采用一种简单有效的方法成功制备出具有高导电性、极性的多孔三维石墨烯负载氮化钛纳米线（Li et al.，2018）（图 2-4）。石墨烯网络可提供高效的电子传导，优化离子扩散；氮化钛纳米线与多硫化物之间的强作用力有利于固硫；该材料应用于锂硫电池时，在 5 C 倍率下，比容量维持在 676 mA · h/g，1 C 倍率下 200 次循环后，比容量保持在 957 mA · h/g。这一研究将为同时实现锂硫电池高能量密度和高功率密度提供相关理论研究基础。

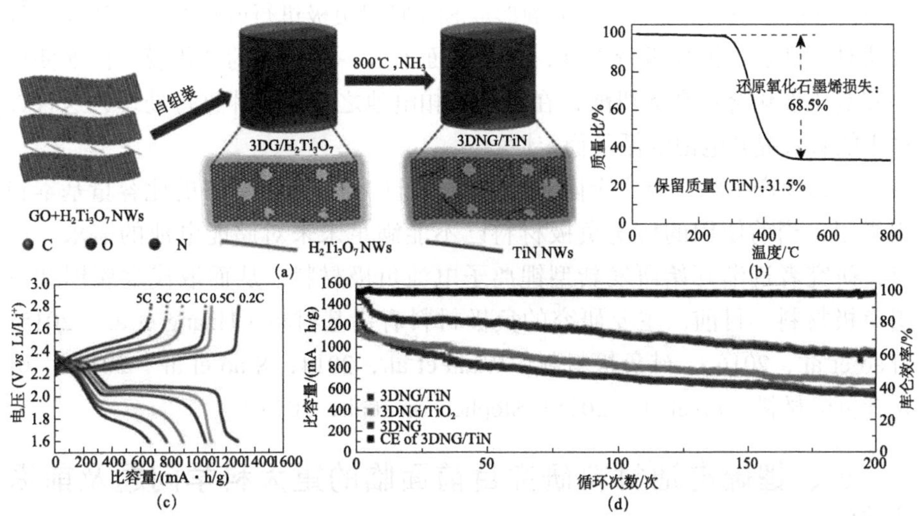

图 2-4 高导电性的多孔三维石墨烯负载氮化钛纳米线复合材料及其性能曲线图

（a）制备方法示意图；（b）热重分析；（c）电池容量分析；（d）电池稳定性分析

（二）负极材料

目前，研究最为成熟和商业化应用最广泛的负极材料主要是石墨负极。锂离子嵌入石墨材料的层间，形成锂和石墨的二元层间化合物（b-GIC），当锂离子完全嵌入时，最终形成 LiC_6，其理论比容量达到 372 mA · h/g。Xu 课题组（Xu et al.，2015）先用赫默斯法将普通石墨预氧化后突然经过高温还原，获得了膨胀的石墨，该材料为层状结构，其层间距达到 0.43 nm，在

20 mA/g 的电流密度下，放电可逆比容量接近 300 mA·h/g，在 200 mA/g 下循环 2000 次，比容量仍很稳定。部分研究者正极采用硫化锂材料、负极采用无定形碳材料来制作锂硫电池，无定形碳材料的层状结构能够储存锂离子。其结构的缺陷较多，具有更多的锂离子活性位点；相对电压更低，存储容量更大。

在锂硫电池中，负极材料一般选择金属锂，在负极中金属锂发生氧化反应，失去一个电子后变成锂离子，锂离子通过浓度差扩散到电解液后穿过隔膜到达正极，然后硫原子与锂离子生成硫化锂。在电极反应中，电子通过外电路做功，锂离子通过内部浓度差扩散。对金属锂进行表面修饰，如石墨烯包覆和在电解液中加入添加剂是目前行之有效的方法。张继光等在电解液中加入了少量的硝酸锂后可在锂负极表面形成致密的 SEI 膜（Ding et al.，2013）。除在电解液中加入添加剂形成 SEI 膜对负极进行改性之外，还有一种方法是对锂负极锂片进行处理，在其表面形成一层稳定的钝化膜。比如使用一些氧化性化合物和无机酸，在组装纽扣电池之前对锂片进行处理，这种方法也能有效提升电池的循环稳定性。

金属锂产生的严重安全隐患，以及传统的石墨负极材料的比容量基本趋于极限，当前使用的传统负极材料已不能满足未来对储能电池的需求。所以，研究者逐渐开始研发新型锂离子电池负极材料，从而取代金属锂和石墨负极材料。目前，主要研究的负极材料有锡基材料（Huang et al.，2015；Zhao et al.，2016）、硅负极材料（Wada et al.，2014；Xiao et al.，2015）、硫化锂负极材料（Jin et al.，2015；Stephenson et al.，2014）。

三、锂硫电池材料研究目前面临的重大科学问题及前沿方向

锂硫电池是下一代最有应用前景的高能量密度二次存储电池，但要实现进一步应用，还需要面临诸多挑战，要解决许多关键和制约性的问题，现存的主要问题有以下几个方面（Chung et al.，2017；Li et al.，2016；Rosenman et al.，2015）。

（1）锂硫电池的活性物质硫虽然廉价易得、成本比较低，但存在自身导电性差的问题，导致电池内阻过大，不利于电子的传输，影响库仑效率。放电产物硫化锂的导电性较差，使得锂离子在硫化锂中的扩散率很低，而且一旦硫化锂生成并覆盖在材料上，就会影响锂离子后续在材料中的嵌入和脱出。同时，由于硫化锂具有绝缘性，其从导电基底上脱离会导致在随后的电

化学反应中失去活性，造成活性物质的损失，从而影响电池容量和寿命。

（2）在放电过程中，硫转化为硫化锂后会发生严重的体积膨胀，对电极造成破裂、脱落及与集流体分离等严重且不可逆的破坏。此外，硫导电性较差，从导电基底脱落会降低有效的电子传输，造成容量的不可逆损失。

（3）电池循环过程的中间产物可溶，会溶解于电解液，并不断向锂片处扩散和往来。随着反应的进行，这种往来会造成穿梭效应，因此硫的流失不可避免。另外，随着多硫化锂在电解液中的溶解，整个电解液的黏度会增大，从而增加离子扩散的阻力，减缓离子传输速率，进而导致反应动力学严重滞后。

为解决上述问题，研究者通过提高活性物质的导电性（Jin et al.，2012）、控制中间产物多硫化物的迁移（Diao et al.，2013）、增强正极导电材料的韧性（Choi et al.，2014）来提高锂硫电池的性能，采用硫与碳材料（Su & Manthiram，2012；Oschatz et al.，2013；Wang et al.，2013；Yang et al.，2014）、导电聚合物材料（Li et al.，2013；Zhou et al.，2013）、金属氧化物材料（Yang et al.，2016；Ma et al.，2015）复合为解决手段来解决锂硫电池中的一系列问题。利用多硫化物与极性宿主之间的吸附作用及结构设计（物理固硫）等方法提升硫的负载量是锂硫电池的重要研究方向之一。

四、研究总结与展望

目前市场对电池的要求越来越高，如高安全性、长循环性能、低成本、高比能量等（Oschatz et al.，2013）。尽管锂硫电池在能量密度和材料成本方面拥有极大的优势，但是要使锂硫电池得以实际应用，仍存在一些问题需要解决。例如，在室温下，活性物质硫电导率低；锂硫电池在充放电过程中生成的中间产物聚硫化锂极易溶于电解液中，并在浓度梯度和电场力的作用下扩散到锂负极上，造成活性物质损失；放电终产物沉积在导电材料表面，造成电极体积的变化（Liang et al.，2015；Wang et al.，2015）。上述问题严重影响了锂硫电池的电化学性能，使得实验制备的锂硫电池的容量很难达到其理论比容量，循环性能也不如目前已经商业化应用的其他类型锂离子电池。因此，针对以上锂硫电池存在的主要问题，寻找能够有效改善并解决这些问题的方案，是推动锂硫电池商业化应用的关键。

硫材料本身导电性较差，多硫化合物在电解液中的溶解、扩散和充放电过程中较大的体积变化（80%），导致了锂硫电池较差的循环性能和安全隐患，影响其商业化进程。因此，锂硫电池正极材料的研究仍极具潜力，如锂硫电池充放电机理、锂负极的枝晶问题、电极材料的制备、正极活性材料硫的负载量以及电解液和隔膜等。锂离子电池在商业市场上实际应用的前提是必须解决上述制约锂硫电池发展的关键科学问题，进而改善锂硫电池的循环稳定性与能量密度，实现锂硫电池产业化的发展。在锂硫电池的相关研究中，不应局限于解决硫单质正极的某一个缺陷，而是应将问题统一整合在一起，提出具体有效的整体解决方案，在改善电极材料倍率性能和循环性能的同时，提高载硫量，实现电池性能的协同优化与提升。

五、学科发展政策建议与措施

我国在锂硫电池技术方面已经取得诸多先进的科研成果，伴随着材料、化学等相关学科的发展，锂硫电池建设已被纳入重要议程。应运用锂硫技术，将其应用到我国各行各业的储能领域中，实现清洁高效能源的转化与利用，更好地促进人类的发展和社会的进步。同时，还应该立足国情，紧紧围绕锂硫电池应用型储能的实际定位，以社会需求为导向，加强学科建设，培养综合素质高，面向生产、管理和服务一线的科研应用型人才。

将锂硫电池作为能源方向重点发展的学科与方向，完善机制，优化资源配置，集中国家和地方的有限财力，提高建设效益；适时调整优化国家储能领域重点学科建设方向和结构，逐步在全国范围内形成布局合理、各具特色和优势、全面支撑各行业和区域发展的国家储能重点学科体系，满足国家的重大需求。

麦立强（武汉理工大学）

本节参考文献

Bhaskara F, Lakshmanar R M, 1968. Organic electrolyte cells. U.S. Patent: 3413154.

Bruce P G, Freunberger S A, Hardwick L J, et al. 2011. Li-O_2 and Li-S batteries with high energy storage. Nature Materials, 11: 19-29.

Choi K M, Jeong H M, Park J H, et al. 2014. Supercapacitors of nanocrystalline metal-organic frameworks. ACS Nano, 8: 7451-7457.

Chung S, Han P, Chang C H, et al. 2017. A shell-shaped carbon architecture with high-loading capability for lithium sulfide cathodes. Advanced Energy Materials, 7: 1700537.

Diao Y, Xie K, Xiong S Z, et al. 2013. Shuttle phenomenon—the irreversible oxidation mechanism of sulfur active material in Li-S battery. Journal of Power Sources, 235: 181-186.

Ding F, Xu W, Graff G L, et al. 2013. Dendrite-free lithium deposition via self-healing electrostatic shield mechanism. Journal of the American Chemical Society, 135: 4450-4456.

Herbert D, Juliusz U, 1962. Electric dry cells and storage batteries. U.S. Patent: 3043896.

Huang X K, Cui S M, Chang J B, et al. 2015. A hierarchical tin/carbon composite as an anode for lithium-ion batteries with a long cycle life. Angewandte Chemie International Edition, 54: 1490-1493.

Jiang J, Li Y Y, Liu J P, et al. 2012. Recent advances in metal oxide-based electrode architecture design for electrochemical energy storage. Advanced Materials, 24: 5166-5180.

Jin J, Wen Z Y, Liang X, et al. 2012. Gel polymer electrolyte with ionic liquid for high performance lithium sulfur battery. Solid State Ionics, 225: 604-607.

Jin R C, Yang L X, Li G H, et al. 2015. Hierarchical worm-like CoS_2 composed of ultrathin nanosheets as an anode material for lithium-ion batteries. Journal of Materials Chemistry A, 3: 10677-10680.

Li W Y, Zhang Q F, Zheng G Y, et al. 2013. Understanding the role of different conductive polymers in improving the nanostructured sulfur cathode performance. Nano Letters, 13: 5534-5540.

Li Z, Jiang Y, Yuan L X, et al. 2014. A highly ordered meso@microporous carbon-supported sulfur@smaller sulfur core-shell structured cathode for Li-S batteries. ACS Nano, 8: 9295-9303.

Li Z, Zhang J T, Guan B Y, et al. 2016. A sulfur host based on titanium monoxide@carbon hollow spheres for advanced lithium-sulfur batteries. Nature Communications, 7: 13065.

Li Z H, He Q, Xu X, et al. 2018. A 3D nitrogen-doped graphene/TiN nanowires composite as a strong polysulfide anchor for lithium-sulfur batteries with enhanced rate performance and high areal capacity. Advanced Materials, 30: e1804089.

Liang X, Hart C, Pang Q, et al. 2015. A highly efficient polysulfide mediator for lithium-sulfur batteries. Nature Communications, 6: 5682.

Ma L, Wei S Y, Zhuang H L, et al. 2015. Hybrid cathode architectures for lithium batteries based on TiS_2 and sulfur. Journal of Materials Chemistry A, 3: 19857-19866.

Manthiram A, Chung S, Zu C X. 2015. Lithium-sulfur batteries: progress and prospects. Advanced Materials, 27: 1980-2006.

Oschatz M, Thieme S, Borchardt L, et al. 2013. A new route for the preparation of mesoporous carbon materials with high performance in lithium-sulphur battery cathodes. Chemical

Communications, 49: 5832-5834.

Peled E, Sternberg Y, Gorenshtein A, et al. 1989. Lithium-sulfur battery: evaluation of dioxolane-based electrolytes. Journal of the Electrochemical Society, 136: 1621-1625.

Rosenman A, Markevich E, Salitra G, et al. 2015. Review on Li-sulfur battery systems: an integral perspective. Advanced Energy Materials, 5: 1500212.

Seh Z W, Li W Y, Cha J J, et al. 2013. Sulphur-TiO_2 yolk-shell nanoarchitecture with internal void space for long-cycle lithium-sulphur batteries. Nature Communications, 4: 1331-1336.

Stephenson T, Li Z, Olsen B, et al. 2014. Lithium ion battery applications of molybdenum disulfide (MoS_2) nanocomposites. Energy & Environmental Science, 7: 209-231.

Su Y S, Manthiram A. 2012. Lithium-sulphur batteries with a microporous carbon paper as a bifunctional interlayer. Nature Communications, 3: 1166.

Wada T, Ichitsubo T, Yubuta K, et al. 2014. Bulk-nanoporous-silicon negative electrode with extremely high cyclability for lithium-ion batteries prepared using a top-down process. Nano Letters, 14: 4505-4510.

Wang C, Wan W, Chen J T. 2013. Dual core-shell structured sulfur cathode composite synthesized by a one-pot route for lithium sulfur batteries. Journal of Materials Chemistry A, 1: 1716-1723.

Wang J L, He Y S, Yang J. 2015. Sulfur-based composite cathode materials for high-energy rechargeable lithium batteries. Advanced Materials, 27: 569-575.

Wang Z Y, Zhou L, Lou X W. 2012. Metal oxide hollow nanostructures for lithium-ion batteries. Advanced Materials, 24: 1903-1911.

Xiao X C, Zhou W D, Kim Y N, et al. 2015. Regulated breathing effect of silicon negative electrode for dramatically enhanced performance of Li-ion battery. Advanced Functional Materials, 25: 1426-1433.

Xu J T, Wang M, Wickramaratne N P, et al. 2015. High-performance sodium ion batteries based on a 3D anode from nitrogen-doped graphene foams. Advanced Materials, 27: 2042-2048.

Xu R, Lu J, Amine K. 2015. Progress in mechanistic understanding and characterization techniques of Li-S batteries. Advanced Energy Materials, 5: 1500408.

Yamin H, Peled E. 1983. Electrochemistry of a nonaqueous lithium/sulfur cell. Journal of Power Sources, 9: 281-287.

Yang X, Zhang L, Zhang F, et al. 2014. Sulfur-infiltrated graphene-based layered porous carbon cathodes for high-performance lithium-sulfur batteries. ACS Nano, 8: 5208-5215.

Yang Y, Zheng G Y, Cui Y. 2013. Nanostructured sulfur cathodes. Chemical Society Reviews, 42: 3018-3032.

Yang Z Z, Wang H Y, Lu L, et al. 2016. Hierarchical TiO_2 spheres as highly efficient polysulfide host for lithium-sulfur batteries. Scientific Reports, 6: 22990.

Yin Y X, Xin S, Guo Y G, et al. 2013. Lithium-sulfur batteries: electrochemistry, materials, and prospects. Angewandte Chemie International Edition, 52: 13186-13200.

Zhao K N, Zhang L, Xia R, et al. 2016. SnO_2 quantum dots@graphene oxide as a high-rate and long-life anode material for lithium-ion batteries. Small, 12: 588-594.

Zhou W D, Yu Y C, Chen H, et al. 2013. Yolk-shell structure of polyaniline-coated sulfur for lithium-sulfur batteries. Journal of the American Chemical Society, 135: 16736-16743.

第三节　锂–空气电池材料

一、锂–空气电池材料研究概述

目前，研究者关注的新型二次电池主要包括锂硫电池、锂–空气电池、钠硫电池、钠离子电池、钠–氯化镍电池等。在这些体系中，锂–空气电池有着一项特别的优势：它的理论能量密度高达 11 400 W·h/kg，远超其他电池，几乎可与汽油相媲美（图 2-5）。如果能够实现锂–空气电池的实用化，不仅电动汽车的长距离行驶能力将大幅改善，更多需要使用高容量电池的行业也将受益匪浅。同时，锂–空气电池的结构使其具有成本低、安全性好、污染小等优点。

1976 年，利陶尔（Littauer）等（Littauer & Tsai，1976）提出了一种新的电池体系：使用金属锂作负极、空气中的氧气作正极活性物质，在水系电解液中进行电化学循环。由于早期的二次电池多使用含电解质的水溶液作为电解液，锂–空气电池起初被提出概念时也使用了水系电解液。由于实验条件的限制，该电池的金属锂负极会被扩散出的水系电解液腐蚀掉，导致电池无法稳定工作。其拙劣的储能表现未能激起研究者太多的兴趣，此后一段时间，有关锂–空气电池的研究鲜有报道。

锂–空气电池是一种以金属锂为负极、多孔导电材料为正极，使用空气中的氧气作为正极反应物质的金属–空气电池。在放电时，负极金属锂失去电子后变为锂离子。锂离子游离到电解液中，穿过隔膜后迁移至正极；同时电子经外电路也流动至正极。二者与此处溶解的氧气相结合，发生氧化还原反应后生成放电产物。在充电时，正极发生水氧化反应，放电产物被分解为锂离子、氧气和电子。氧气向外界逃逸，而锂离子和电子则回到负极并重新生成金属锂。如此循环往复，完成锂–空气电池中化学能和电能的转化。

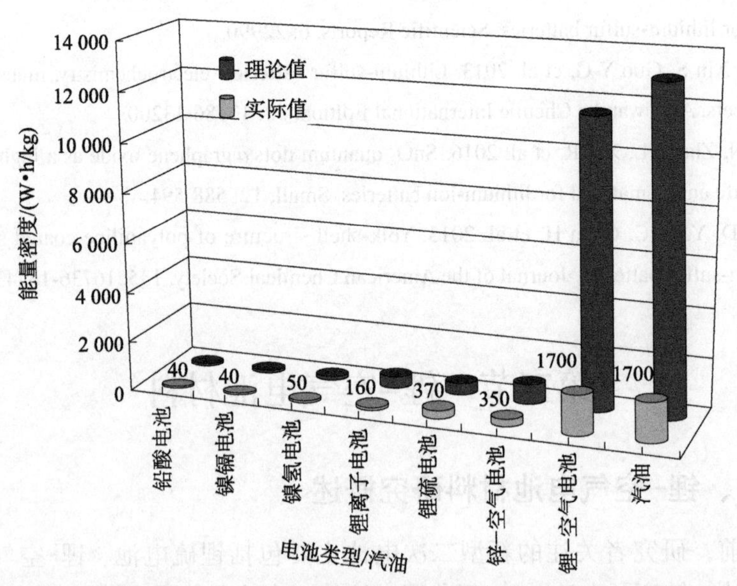

图 2-5 多种二次电池与汽油的能量密度对比（Girishkumar et al.，2010）

1996 年，亚伯拉罕（K. M. Abraham）等（Abraham & Jiang，1996）首次报道了一种使用有机电解液的锂-空气电池，他们以添加有机溶剂和锂盐的凝胶聚合物作为电解质，以酞菁钴作为空气电极的催化剂，在高纯氧的气氛下进行电池测试。该电池开路电压接近 3 V，工作电压为 2.0～2.8 V，比容量可达 1400 mA · h/g，远高于常规的锂离子电池体系。研究者认为，放电机理是锂离子和氧气在碳基空气电极上生成过氧化锂，而空气电极中的气孔被反应产物过氧化锂沉积阻塞并最终导致放电结束。2002 年，里德（J. Read）发表了关于锂-空气电池放电机理、电极材料以及电解液组成等方面的研究（Read，2002）。他详细研究了多种因素对电池放电容量、倍率性能以及循环性造成的影响，认为电解液的组成对电池的性能和放电产物的沉积行为有重要影响，并首次提出以醚类溶剂作为锂-空气电池的电解液，通过这一方式，可达到 2800 mA · h/g 的比容量。2006 年，布鲁斯（P. G. Bruce）等（Ogasawara et al.，2006）在空气电极中引入了 α-MnO_2 纳米线作为催化剂，达到了 3000 mA · h/g 的比容量。同时，在限制比容量为 600 mA · h/g 的情况下，实现了超过 50 次的循环，并保持比容量不衰减。此时，真正具备实际意义的可循环锂-空气电池已初步形成。也是从这一时期开始，全球各地大量的研究者加入锂-空气电池的研究热潮中。同时，各大商业公司也表现出了对锂-空气电池的兴趣。2009 年，IBM 公司宣布与美国阿贡国家实验室合作，共同开展“将锂-空气电池应用于电动汽车”研发项目，从此掀起了锂-空气电池基础研究热潮。2012 年，IBM 公司再次宣布与日本旭化成（Asahi Kasei）公司和中央玻璃（Central Glass）有限公司合作，共同开发“电池 500”（Battery-500）项目，进一步拓展锂-空气电池应用方面的研发。2013 年，IBM 公司与宝马汽车公司共同合作开发锂-空电池，在 2020 年底前实现锂-空气二次电池的商业化。从此，锂-空气电池研究开启了继二次锂离子电池之后的后锂离子二次电池充电技术的新时代。

二、锂-空气电池材料研究的历史及现状

（一）正极材料

锂-空气电池的正极材料是锂-空气电池的核心，正极材料本身的性质和结构对锂-空气电池的性能有很大的影响。正极材料的孔可以为氧气的传输提供通道，同时研究发现，当正极孔道完全被放电产物所填充时，放电过程将终止（王芳等，2015）。因此，碳多孔材料因其孔隙率高、导电性好、成本低、易加工等特点，被广泛用作空气电极，其中锂-空气电池所常用的碳材料多为具有良好导电性和高孔容的炭黑（罗仲宽等，2015）。此外，空气电极的材料还有多种，可分为多孔碳材料和多孔非碳材料。

目前，发展多孔碳材料来改善整个体系稳定性的研究已成为该领域的热点研究课题。一般来说，电极的孔容越大比容量越大，主要因为较大的孔容允许通过的放电产物更多，从而使电池能够提供更高的能量（Liu et al.，2015）。研究人员还发现，良好的孔隙结构对电池内的氧扩散有一定的帮助，而电池中的氧扩散能力决定了电池的容量。例如，2010 年埃斯瓦兰（Eswaran）课题组（Eswaran et al.，2010）设计出一款锂-空气电池，其空气电极由双层碳电极组成：活化层和具有高比表面积碳的扩散层。其中，活化层含有碳、α-二氧化锰和聚四氟乙烯三种材料，扩散层则由多孔碳和聚四氟乙烯两种物质构成，在特定的电流密度下放电时，该类型电池比容量达到 3100 mA · h/g。除此之外，多孔碳空气电极的性能还与孔径、比表面积、碳电极厚度等因素有关。

随着科学家对多孔碳材料的研究越来越成熟，研究发现，多孔碳材料并没有那么完美。2012 年，布鲁斯团队（Thotiyl et al.，2013）研究发现，当电池的工作电压超过 3.5 V 时，多孔碳易分解，发生副反应，严重影响电池整体性能，因此采用了非碳类多孔材料（纳米多孔金）作为空气电极正极

材料。2014年，孙克宁课题组（Zhao et al.，2014）指出，正极碳材料的分解可通过催化电解质降解来进一步损害电池的可循环性，因此他们研发了一种既不含碳又不含黏结剂的二氧化钛作为锂-空气电池空气电极基底。研究结果表明，在1 A/g或5 A/g的大电流密度下，该锂-空气电池均可循环140多次。

（二）负极材料

负极材料作为金属-空气电池的重要组成部分，其活性、组成及形貌对金属-空气电池的反应机理和性能有着直接的影响，同时负极材料的价格也直接影响着电池的成本。因此，开发价廉高效的负极材料，对改善电池性能和降低电池成本都起着至关重要的作用。目前改善负极材料性能使用最多的方法是用多元材料代替单一金属负极。

早在2000年，Shi课题组（Shi et al.，2001）首次报道了使用直接合金法或动力学控制的气相沉积（kinetically controlled vapor deposition，KCVD）方法制备Li-Mg合金，并将该合金作为锂电池负极。结果表明，在分别使用这两种方法制备的Li-Mg合金作负极时，电池在循环过程中其负极（合金）上形成的枝晶状物质明显减少了，特别是通过KCVD方法制备的Li-Mg合金，锂的扩散系数远大于其他夹层电极材料的扩散系数。他们还指出，KCVD方法是制备合金非常实用的方法，而且可以通过控制基片温度改变合金电极的微观结构和形态。该方法不仅可以在室温甚至更低温度下在基片上改变生成Li-Mg合金的形态，还提供了消除循环过程中枝晶形成的方法。另外，该合金作为电池负极，比单一锂金属作为电池负极时具有更好的循环性能和倍率性能。另外，对单一金属负极材料进行改性也是降低负极自腐蚀、改善负极活性和提高锂-空气电池性能的好方法。改性后的负极一般具有更负的电位、更高的利用效率，因此具有更理想的放电性能，解决了电池容量损失严重和负极利用率低等问题。

三、锂-空气电池材料研究目前面临的重大科学问题及前沿方向

锂-空气电池具有极高的理论比容量且环境友好，有望作为一种清洁可再生能源来替代传统化石燃料使用，具有非常广阔的应用前景。自从20世纪70年代被提出概念后，锂-空气电池已经历了40多年的发展历史。在此期间，

有大量的研究者对这一电池体系进行过多方面的优化。近年来，锂-空气电池研究取得了一些突破性进展，然而对于未来产业化发展道路来说仍处于初级阶段，制约其发展和应用的主要因素或问题可归结于以下几点。

首先，在锂-空气电池目前的研究中，大部分充放电和电化学性能测试都是在封闭的纯氧环境中完成的，甚少会在空气中直接进行测试，原因是空气中含有水和二氧化碳，会进入电池中对电池造成不良的影响，水会对金属锂负极造成腐蚀，二氧化碳则会与电池中的化学物质发生反应，如与过氧化锂反应生成难以溶解的碳酸锂等，从而影响电池的容量和寿命。因此，寻找合适的防水透氧膜对于锂-空气电池在空气中的实际应用来说十分重要。

其次，目前常用的非水性（有机）锂-空气电池电解液如碳酸酯类和醚类，在氧气的环境中易于发生分解和挥发现象，该问题会导致锂-空气电池的比容量降低和放电提前终止。作为最常见的固态电解质启动子，与电解液和锂负极之间表现出的不稳定性也影响着电池的性能。同时，锂盐和溶剂的选择直接影响着氧气的溶解度和扩散率，因此需要开发更加稳定高效的锂-空气电池电解质来保证电池的高容量和长久稳定性。

最后，锂-空气电池充放电过程中的氧化还原反应和水氧化反应极化效应严重，电位过高不仅会导致电解液的分解，发生不可逆转的副反应，还会使锂-空气电池的循环效率降低，过早地终止放电。在有机体系的锂-空气电池中，生成的放电产物过氧化锂不溶于电解质，会堵塞正极的孔隙，降低氧气的扩散率，因此寻找兼具高效的双相催化作用和高比表面积的催化剂来降低锂-空气电池的极化、提高电池的电化学性能是锂-空气电池研究中亟须解决的问题。

在锂-空气电池的各个部件中，空气正极上存在的问题最突出，对锂-空气电池性能的影响也最大。因此，需要使用合适的空气正极材料，以减缓电池充放电过程中的极化现象，从而提高电池的整体性能。对于锂-空气电池来说，其理想的正极材料应满足以下条件：①较高的本征氧化还原反应/水氧化反应催化活性，以缓和充放电过程中的极化效应；②合理的微结构设计，有利于放电产物的沉积以及氧气和锂离子在充放电过程中的扩散传输；③在长期的工作过程中保持足够稳定性，从而尽可能延长电池的循环寿命；④足够的电子电导率，确保电极表面的电化学反应能够有效地进行。

四、研究总结与展望

近年来，具有质轻、价廉、无毒、高效等优点的金属-空气电池面世，引发了各国研究者的研究热潮，其中锂-空气（或锂-氧气）电池因具有最大理论比容量而备受关注。

虽然锂-空气电池是绿色电池，且在能源领域具有巨大的发展潜力，但锂-空气电池的发展依旧命途多舛。由起初会被自身组分腐蚀至无法工作的“残疾”体系，到现在可以稳定循环充放电的高比能量电池，背后是众多研究者的共同努力。然而，目前锂-空气电池还面临许多关键问题：锂-空气电池充放电机理复杂且尚未被完全澄清；一些关键性能仍需提升；真正实用的电极材料和电解液体系有待开发。现在的锂-空气电池仍处于初期研究阶段，有待进一步深入研究。优化催化剂，控制放电产物，实现更低的电压极化和更长的循环寿命，解决金属锂负极表面生成氢氧化锂致密层阻隔或降低离子、电子、氧传输等问题仍然是研究锂-空气电池的热点内容。锂-空气电池充放电反应条件要求苛刻等因素也使大多数成果并不能有效地应用到正常的大气环境中。因此，锂-空气电池产业化仍有很长的路要走。

五、学科发展政策建议与措施

围绕锂-空气电池领域，应加深调查和广泛研讨，分析有关领域的发展态势和规律，提炼关键学科理论和技术问题，提出学科创新发展的新思想和新方法，共同促进锂-空气电池这一研究领域的发展与进步。促进锂-空气电池学科发展的相关建议如下：遵循材料规律，实行分类分层发展；注重性能提升，强化特色发展；突破学科壁垒，实现统筹发展；强化政策导向，引领特色发展。

麦立强（武汉理工大学）、张清杰（武汉理工大学）

本节参考文献

罗仲宽，尹春丽，吴其兴，等. 2015. 有机电解液型锂空气电池空气电极研究进展. 深圳大学学报理工版，32: 111-120.

王芳，李豪君，刘东，等 . 2015. 高性能非水性体系锂空气电池研究进展 . 稀有金属材料与工程，44: 2074-2080.

Abraham K M, Jiang Z. 1996. A polymer electrolyte-based rechargeable lithium/oxygen battery. Journal of the Electrochemical Society, 143: 1-5.

Eswaran M, Munichandraiah N, Scanlonb L G. 2010. High capacity Li-O_2 cell and electrochemical impedance spectroscopy study. Electrochemical & Solid-State Letters, 13: A121-A124.

Girishkumar G, McCloskey B, Luntz A C, et al. 2010. Lithium-air battery: promise and challenges. Journal of Physical Chemistry Letters, 1: 2193-2203.

Littauer E L, Tsai K C. 1976. Anodic behavior of lithium in aqueous electrolytes. Journal of the Electrochemical Society, 123: 771-776.

Liu T, Leskes M, Yu W J, et al. 2015. Cycling Li-O_2 batteries via LiOH formation and decomposition. Science, 350: 530-533.

Ogasawara T, Débart A, Holzapfel M, et al. 2006. Rechargeable Li_2O_2 electrode for lithium batteries. Journal of the American Chemical Society, 128: 1390-1393.

Read J. 2002. Characterization of the lithium/oxygen organic electrolyte battery. Journal of the Electrochemical Society, 149: A1190-A1195.

Shi Z, Liu M L, Naik D, et al. 2001. Electrochemical properties of Li-Mg alloy electrodes for lithium batteries. Journal of Power Sources, 92: 70-80.

Thotiyl M M O, Freunberger S A, Peng Z Q, et al. 2013. The carbon electrode in nonaqueous Li-O_2 cells. Journal of the American Chemical Society, 135: 494-500.

Zhao G Y, Mo R W, Wang B Y, et al. 2014. Enhanced cyclability of Li-O_2 batteries based on TiO_2 supported cathodes with no carbon or binder. Chemistry of Materials, 26: 2551-2556.

第四节　全固态锂离子电池材料

一、全固态锂离子电池材料研究概述

随着能源危机和环境污染的问题日益加剧，社会的进步和人类的发展越来越依赖于清洁、可再生能源。在实际的应用中，太阳能、风能、水能等可再生能源需要被转化为电能才能被人们广泛地利用和传输。目前，作为洁净能源代表之一的锂离子电池因为具有能量密度高、输出功率大、电压高、自放电小、工作温度范围宽、无记忆效应和环境友好等优点，已被广泛应用于各种微纳器件、便携式电子设备、电动汽车、轨道交通、航空航天和大规模储能等领域。目前，商业化的锂离子电池采用易燃的液态有机电解质，该电解质和电极材料在充放电过程中易发生副反应，导致电池容量的不可逆衰减，同时电池在使用过程中，其液态有机电解质会发生挥发、干涸、泄漏等现象，从而影响锂离子电池的使用寿命。另外，在锂金属作为负极材料的锂离子电池循环过程中，金属锂表面电流密度和锂离子分布不均匀等因素会造成金属锂表面孔洞或枝晶的产生，枝晶会刺穿隔膜，带来电池短路、热失控和着火爆炸等一系列安全隐患。考虑到安全问题是储能的关键，用固态电解质代替液态电解质是获得高安全性、高能量密度和优异循环寿命的全固态锂离子电池的根本途径（Manthiram et al.，2017；任耀宇，2017）。全固态锂离子电池作为新型锂离子电池的代表，具有安全性高、能量密度高、工作温度范围广等显著优点，可从本质上解决锂离子电池的安全性问题，同时能进一步提高锂离子电池的实用价值。

二、全固态锂离子电池材料研究的历史及现状

（一）固态电解质

1992 年，美国橡树岭国家实验室在高纯氮气气氛中通过射频磁控溅射装置溅射高纯 Li_3PO_4 靶制备得到锂磷氧氮（LiPON）电解质薄膜。该材料的室温离子电导率为 2.3×10^{-6} S/cm，电化学窗口为 5.5 V *vs.* Li^+/Li，热稳定性较好，并且与 $LiCoO_2$、$LiMn_2O_4$ 等正极以及金属 Li、Li 合金等负极相容性良好（Hamon et al.，2006）。锂镧钛氧是在钙钛矿型固体电解质 $CaTiO_3$ 的基础上通过 Li^+ 和 La^{3+} 取代 Ca^{2+} 得到的，结构通式为 $Li_{3x}La_{2/3-x}TiO_3$（$0.04 < x < 0.17$，简称为 LLTO），不同的组分和制备条件会影响其晶体结构，其室温离子电导率最高可达 1.4×10^{-3} S/cm（其中 $3x = 0.34$）（Inaguma & Itoh，1996）。石榴石型固体电解质最早由韦普纳（Weppner）等发现，其化学通式为 $Li_5La_3M_2O_{12}$（M = Nb、Ta），利用 Zr^{4+} 替换 M 位元素得到 $Li_7La_3Zr_2O_{12}$（LLZO），室温离子电导率被提高至 3.0×10^{-4} S/cm，异价 Zr 元素的引入不但提高了载流子 Li^+ 的浓度，而且扩大了传输通道，削弱了晶格骨架对 Li^+ 的束缚作用（Murugan et al.，2007）。伴随着固态电解质相关材料的研究与发展，越来越多高离子电导率的无机固态电解质逐步出现在大众的视野中，全固态锂离子电池的实际应用得到逐步实现（图 2-6）。

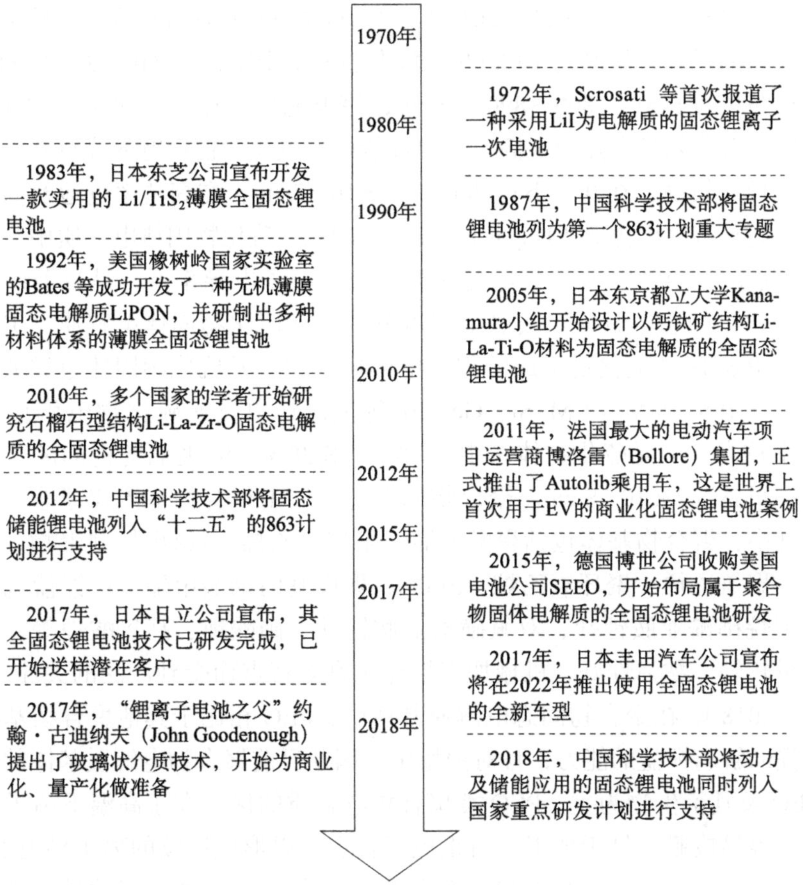

1970年

1972年，Scrosati 等首次报道了一种采用LiI为电解质的固态锂离子一次电池

1980年

1983年，日本东芝公司宣布开发一款实用的 Li/TiS_2 薄膜全固态锂电池

1987年，中国科学技术部将固态锂电池列为第一个863计划重大专题

1990年

1992年，美国橡树岭国家实验室的Bates 等成功开发了一种无机薄膜固态电解质LiPON，并研制出多种材料体系的薄膜全固态锂电池

2005年，日本东京都立大学Kanamura小组开始设计以钙钛矿结构Li-La-Ti-O材料为固态电解质的全固态锂电池

2010年

2010年，多个国家的学者开始研究石榴石型结构Li-La-Zr-O固态电解质的全固态锂电池

2011年，法国最大的电动汽车项目运营商博洛雷（Bollore）集团，正式推出了Autolib乘用车，这是世界上首次用于EV的商业化固态锂电池案例

2012年

2012年，中国科学技术部将固态储能锂电池列入“十二五”的863计划进行支持

2015年

2015年，德国博世公司收购美国电池公司SEEO，开始布局属于聚合物固体电解质的全固态锂电池研发

2017年

2017年，日本日立公司宣布，其全固态锂电池技术已研发完成，已开始送样潜在客户

2017年，日本丰田汽车公司宣布将在2022年推出使用全固态锂电池的全新车型

2018年

2017年，“锂离子电池之父”约翰·古迪纳夫（John Goodenough）提出了玻璃状介质技术，开始为商业化、量产化做准备

2018年，中国科学技术部将动力及储能应用的固态锂电池同时列入国家重点研发计划进行支持

图 2-6 全固态锂离子电池发展大事件图（许晓雄和李泓，2018）

目前已开发的主要固态电解质可分为两大类，即无机固态电解质和聚合物电解质（Fergus，2010）。对于全固态锂离子电池，特别是能适用于未来电动汽车、大规模储能应用的体型电池，采用的固态电解质应满足以下要求：①具有高的室温电导率（$>10^{-4}$ S/cm）；②电子绝缘；③电化学窗口宽；④与电极材料相容性好；⑤热稳定性好、耐潮湿环境、机械性能优良；⑥原料易得、成本较低、合成方法简单。无机固态电解质是无机全固态锂电子电池的核心，主要分为氧化物电解质和硫化物电解质。氧化物电解质主要分为超离子导体（NASICON）结构类型和石榴石型两种。近年来，研究者针对氧化物固态电解质开展了大量的研究工作，其制备的固态电解质锂离子电导率为 10^{-7}～10^{-3} S/cm，在空气中的稳定性较好（Aono et al.，1993）。南策文院士团队制备了新型 $Li_{6.75}La_3Zr_{1.75}Ta_{0.25}O_{12}$ (LLZTO) 无机固态电解质，离子电导率达

到 5×10^{-4} S/cm（Zhang et al.，2017）。中国科学院物理研究所提出并验证了原位固态化的设想，研制的 10 A·h 软包电芯能量密度达到 310～390 W·h/kg，体积能量密度达到了 800～890 W·h/L，该电池可以在室温和 90 ℃下循环。硫化物电解质除具有热稳定高、安全性能好、电化学窗口宽的优点外，其锂离子电导率较高，在 0 ℃下可达到 10^{-4}～10^{-2} S/cm，在高功率电池及高低温电池方面具有突出优势（Hayashi et al.，2003）。硫化物固体电解质中，Li_2S-P_2S_5 体系离子电导率较高、电化学窗口宽、电子电导率低，是目前研究最多的硫化物固态电解质。按照组成可分为二元硫化物固体电解质（主要由 Li_2S 和 P_2S_5 两种硫化物组成的固体电解质）和三元硫化物固体电解质［主要有 Li_2S-P_2S_5 和 MS_2（M=Si、Ge、Sn 等）］，其中三元硫化物电解质离子电导率较高。二元硫化物体系中，70% Li_2S-30% P_2S_5 玻璃陶瓷的离子电导率最高可达 3.2×10^{-3} S/cm。聚合物固态电解质由聚合物基体和锂盐络合而成。其中，聚合物基体包括聚酰亚胺、聚偏氟乙烯–六氟丙烯共聚物、聚丙烯腈、聚氯乙烯、聚环氧乙烷（PEO）、聚甲基丙烯酸甲酯等；锂盐包括双三氟甲烷磺酰亚胺锂盐、双氟磺酰亚胺锂盐、硝酸锂、高氯酸锂等。由于其质量较轻、黏弹性好、机械加工性能优良等特点而受到广泛的关注（Fan et al.，2018）。在聚合物固态电解质基体中，PEO 相比于其他聚合物基体具有更强的解离锂盐的能力，且对锂稳定，因此目前聚合物电解质基体的研究以 PEO 及其衍生物为主。PEO 类聚合物电解质的优点在于高温下离子电导率高，容易成膜，易于加工，与正极复合后可以形成连续的离子导电通道，正极面电阻较小；缺点在于 PEO 的氧化电位在 3.8 V，需经过改性处理，进而拓宽电压窗口。PEO 基电解质结晶度高，导致室温下电导率低，仅 10^{-7}～10^{-6} S/cm，因此工作温度通常较高，约为 60 ℃，并需要更加完备的电池管理系统。聚合物自身的柔性还能改善全固态锂离子电池的界面问题，古迪纳夫课题组设计的聚合物 / 无机 / 聚合物夹层电解质结构可以改变电极 / 聚合物界面处的双层电场，阻止阴离子运输，从而有效改善电池的库仑效率（Zhou et al.，2016）。

（二）正极材料

全固态锂离子电池正极一般采用复合电极，除电极活性物质外，还包括固态电解质和导电剂，在电极中起到传输离子和电子的作用。$LiCoO_2$、$LiFePO_4$、$LiMn_2O_4$ 等氧化物正极在全固态锂离子电池中的应用较为普遍。为了进一步提高全固态锂离子电池的能量密度及电化学性能，新型高能量正极材料也在被积极地研究和开发中，主要包括高容量的三元正极材料和 5 V 高电压材料等（Hovington et al., 2015）。三元材料的典型代表是 $LiNi_{1-x-y}Co_xMn_yO_2$（NCM）和 $LiNi_{1-x-y}Co_xAl_yO_2$（NCA），均具有层状结构，且理论比容量高（约 200 mA·h/g）。除了氧化物正极，硫化物正极也是全固态锂离子电池正极材料的一个重要组成部分，这类材料普遍具有高的理论比容量，比氧化物正极高出几倍甚至一个数量级，与导电性良好的硫化物固态电解质匹配时，由于化学势相近，不会造成严重的空间电荷层效应，因此制备出的全固态锂离子电池有望满足高容量和长寿命的使用要求。

（三）负极材料

金属锂因其高容量和低电位的优点成为全固态锂离子电池最主要的负极材料之一，然而金属锂在循环过程中会有锂枝晶的产生，不但会使可供嵌 / 脱的锂量减少，更严重的是会引发短路等安全问题（Kim et al., 2015）。加入其他金属与锂组成合金是解决上述问题的主要方法之一，这些合金材料一般都具有高的理论比容量，并且金属锂的活性因其他金属的加入而降低，可以有效控制锂枝晶的生成和电化学副反应的发生。碳族的碳基、硅基和锡基材料是全固态锂离子电池另一类重要的负极材料。碳基以石墨类材料为典型代表，石墨碳具有适合于锂离子嵌入和脱出的层状结构，具有良好的电压平台，充放电效率在 90% 以上，然而理论比容量较低（仅为 372 mA·h/g）是这类材料最大的不足。氧化物负极材料主要包括金属氧化物、金属基复合氧化物和其他氧化物。典型的氧化物负极材料有 TiO_2、MoO_2、In_2O_3、$A1_2O_3$、Cu_2O、VO_2、SnO_x、SiO_x 等，这些氧化物均具有较高理论比容量，然而在从氧化物中置换金属单质的过程中，大量的锂被消耗，造成巨大的容量损失，并且循环过程中伴随着巨大的体积变化，造成电池的失效，通过与碳基材料的复合，可以改善这一问题。武汉理工大学麦立强团队研制出了一种锂金属界面层（GZCNT）（Zhang et al., 2018），利用亲锂–憎锂的梯度策略，首次构建了锂金属界面层，有效地抑制了锂枝晶的生长。GZCNT 梯度膜由亲锂的氧化锌 / 碳纳米管底层、憎锂的碳纳米管顶层，以及中间过渡层有序构成。亲锂的底层与金属锂紧密结合，可促进稳定的固态电解质膜的形成，抑制金属锂和亲锂层层间形成锂枝晶，顶层的憎锂层因具有较大的模量可以抑制锂枝晶的进一步生长，而中间的过渡层又可以防止因亲锂、憎锂的突然转变而产生明显的锂枝晶分级层，从而确保金属锂负极的超长循环。

三、全固态锂离子电池材料研究目前面临的重大科学问题及前沿方向

固态电解质的离子电导率较低，导致全固态锂离子电池内部存在较大的阻抗，大大影响着全固态锂离子电池的实际应用。为此，复合固态电解质成为一个新的研究方向，新型复合固态电解质具备无机固态电解质的高离子电导率和聚合物电解质的柔性，目前已经取得了显著的成绩。2015 年，美国斯坦福大学的崔屹课题组首先报道了 $Li_{0.33}La_{0.557}TiO_3$ 纳米线 / 聚丙烯腈复合固态电解质在室温下具有 2.4×10^{-4} S/cm 的离子电导率（Liu et al.，2015）。中国科学院上海硅酸盐研究所的郭向欣团队制备了基于聚环氧乙烷 / 锂镧锆氧复合固态电解质的 2 A · h 级的固态锂离子电池。中国科学院青岛生物能源与过程研究所的崔光磊团队开发了聚丙烯碳酸酯、纤维素、锂镧氧复合的固体电解质，研发的电池能量密度达到了 300 W · h/kg，并首次在马里亚纳海沟完成了深海测试。但他们对此种电解质中的锂离子传输机制尚不明确，需要后续进一步深入探讨。

除现有的固态电解质离子电导率没有完全满足实用需求外，全固态锂离子电池中，界面对电池性能的影响也至关重要（Xu et al.，2018）。第一，大多数固态电解质的离子电导率远远低于液态电解质。虽然一些无机硫化物固态电解质的离子电导率已经可以与有机液态电解质相媲美，但不稳定的界面问题尤为严重。第二，固态电解质和电极之间的界面相容性通常较差，是全固态锂离子电池发展的又一大问题，并且严重影响全固态锂离子电池的实际应用，复杂的界面层对界面电阻和全固态锂离子电池整体性能影响显著。第三，锂枝晶仍然是全固态锂离子电池的一大挑战，即使是机械强度高的无机固态电解质，锂枝晶也可以从缺陷处生长。第四，尽管固体电解质和电极之间 SEI 膜的形成已被广泛研究，但关于 SEI 膜在循环过程中的形成机理以及微观结构仍然没有得到完整的诠释。固体电解质，尤其是在陶瓷电解质中有大量的晶界存在，因为较高的晶界电阻不利于锂离子在正负极之间传输，而且通常晶界电阻高于材料本体电阻，因此晶界电导率对固体电解质总电导率有显著影响。针对这些问题的研究思路包括：对电极材料进行表面修饰处理、对电解质进行掺杂改性制备复合电解质、在界面增加柔性缓冲层、将电极材料纳米化、开发新型或优化现有电极材料以减小体积效应等。

此外，电极材料容量衰减快、机理尚不明确是全固态锂离子电池等电化学储能领域的国际性难题，传统的非原位表征方法难以揭示电化学储能器件

容量衰减的本质原因，同时也难以实现全固态锂离子电池中界面的精确研究。为了解决上述问题，武汉理工大学麦立强团队构建了世界上第一个单根纳米线固态电化学器件（Mai et al.，2010），原位表征并揭示了固态电池界面结构及演化过程；针对电化学能源领域容量衰减快且机理尚不明确的关键问题，率先将纳米器件引入储能研究，提出并组装了可同时用于微纳系统支撑电源及原位检测微纳电池性能的单根纳米线全固态锂离子电池，首次通过原位表征建立了纳米线的电输运、结构与电极充放电状态的直接联系，揭示了纳米线的本征电化学行为及容量衰减、性能劣化的本质。该研究对解决电化学储能材料容量衰减快、电导率低等关键问题具有突破性意义（图 2-7）。

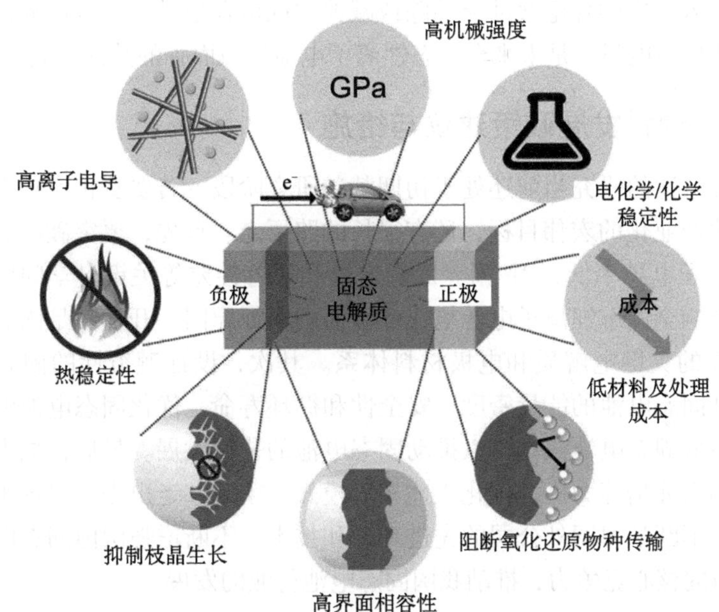

图 2-7　全固态锂离子电池的要求及挑战（Fan et al.，2018）

四、研究总结与展望

全固态锂离子电池安全性高、能量密度高，是新能源汽车电池极有希望的发展方向，发展前景广阔。目前，具有潜力的固态电解质材料可分为聚合物、硫化物和氧化物。其中，基于前两种材料的体型电池以及基于氧化物的薄膜电池已经率先进入商业化应用阶段。然而，全固态锂离子电池要最终实现锂电池高能量密度、高功率、宽工作温度范围和高安全性的目标，仍有一些问题有待解决。

这些挑战中的大多数都与固-固界面有着根本的关系，这限制了全固态锂离子电池的实际应用。目前的研究逐渐从寻求更好的固态电解质转向解决固态电池的界面电阻和界面稳定性问题。对于正极-电解质界面，在硫化物电解质和高压阴极之间的界面处形成高电阻界面与贫锂层是一个关键问题。对于负极-电解质界面，主要问题是锂枝晶的生长并通过固体电解质导致不必要的渗透，以及硫化物电解质和锂负极之间的副反应。由于难以同时实现复合电极内的连续且有效的离子和电子传输，因此也不应忽略颗粒间的界面。此外，计算模拟能帮助我们更好地了解固态电池的界面，更合理地进行材料和结构的设计，在协助传统实验中起着越来越重要的作用。因此，如何通过引入稳定的导电缓冲层消除或减弱空间电荷层效应，抑制界面层的生成，降低界面电阻，是未来全固态锂离子电池领域面临的共同挑战。

五、学科发展政策建议与措施

固态电池的研究当前还处于初期基础研究阶段，若要真正实现全固态锂离子电池产业化的宏伟目标，还有很长的路要走。首先，要突破固态电池电解质的离子电导率低、固-固界面稳定性和兼容性差等关键科学问题，探索并揭示影响全固态锂离子电池电化学性能的本征原因，开发满足固态电池动力学特性的关键电解质和电极材料体系。其次，设计颠覆性的固态电池结构，提升固态电池的能量密度、安全性和循环寿命，优化固态电池电芯制备工艺，降低固态电池成本，以推动固态电池的快速发展。最后，加大固态电池领域的人才培养力度，优化人才培养模式，秉承“走出去、引进来”的战略思想，不断学习国外公司的先进理念和技术，不断增强国内固态电池领域科研队伍的核心竞争力，推动我国固态电池行业的发展。

麦立强（武汉理工大学）

本节参考文献

任耀宇. 2017. 全固态锂电池研究进展. 科技导报，35: 26-36.

许晓雄，李泓. 2018. 为全固态锂电池“正名”. 储能科学与技术，7: 1-7.

Aono H, Sugimoto E, Sadaoka Y, et al. 1993. The electrical properties of ceramic electrolytes for $LiM_xTi_{2-x}(PO_4)_{3+y}$-$Li_2O$, M=Ge, Sn, Hf, and Zr systems. Journal of the Electrochemical Society, 140: 1827-1832.

Fan L, Wei S Y, Li S Y, et al. 2018. Recent progress of the solid-state electrolytes for high-energy metal-based batteries. Advanced Energy Materials, 8: 1702657.

Fergus J W. 2010. Ceramic and polymeric solid electrolytes for lithium-ion batteries. Journal of Power Sources, 195: 4554-4569.

Hamon Y, Douard A, Sabary F, et al. 2006. Influence of sputtering conditions on ionic conductivity of LiPON thin films. Solid State Ionics, 177:257-261.

Hayashi A, Hama S, Minami T, et al. 2003. Formation of superionic crystals from mechanically milled Li_2S-P_2S_5 glasses. Electrochemistry Communications, 5: 111-114.

Hovington P, Lagacé M, Guerfi A, et al. 2015. New lithium metal polymer solid state battery for an ultrahigh energy: nano C-$LiFePO_4$ versus nano $Li_{1.2}V_3O_8$. Nano Letters, 15: 2671-2678.

Inaguma Y, Itoh M. 1996. Influences of carrier concentration and site percolation on lithium ion conductivity in perovskite-type oxides. Solid State Ionics, 86-88: 257-260.

Kim J G, Son B, Mukherjee S, et al. 2015. A review of lithium and non-lithium based solid state batteries. Journal of Power Sources, 282: 299-322.

Liu W, Liu N, Sun J, et al. 2015. Ionic conductivity enhancement of polymer electrolytes with ceramic nanowire fillers. Nano Letters, 15: 2740-2745.

Mai L Q, Dong Y J, Xu L, et al. 2010. Single nanowire electrochemical devices. Nano Letters, 10: 4273-4278.

Manthiram A, Yu X W, Wang S F. 2017. Lithium battery chemistries enabled by solid-state electrolytes. Nature Reviews Materials, 2:16103.

Murugan R, Thangadurai V, Weppner W. 2007. Fast lithium ion conduction in garnet-type $Li_7La_3Zr_2O_{12}$. Angewandte Chemie International Edition, 46: 7778-7781.

Xu L, Tang S, Cheng Y, et al. 2018. Interfaces in solid-state lithium batteries. Joule, 2: 1991-2015.

Zhang H M, Liao X B, Guan Y P, et al. 2018. Lithiophilic-lithiophobic gradient interfacial layer for a highly stable lithium metal anode. Nature Communications, 9: 3729.

Zhang X, Liu T, Zhang S F, et al. 2017. Synergistic coupling between $Li_{6.75}La_3Zr_{1.75}Ta_{0.25}O_{12}$ and poly(vinylidene fluoride) induces high ionic conductivity, mechanical strength, and thermal stability of solid composite electrolytes. Journal of the American Chemical Society, 139: 13779-13785.

Zhou W D, Wang S F, Li Y T, et al. 2016. Plating a dendrite-free lithium anode with a polymer/ceramic/polymer sandwich electrolyte. Journal of the American Chemical Society, 138: 9385-9388.

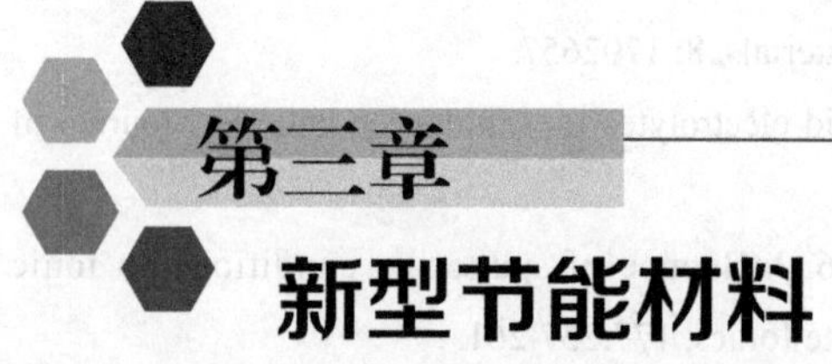

第一节 磁性材料在能源领域中的应用

一、研究概述

在能源领域中，电能是最普遍应用的能源，光伏、风力、核能、热能等最终都变成电能进行再应用。

19世纪，人类进入电气化时代，发电机、电动机、变压器等应运而生。从原理上考虑，磁与电相互依存，密不可分。例如，线圈切割磁力线可产生电，这是发电机的基本原理；导线中流过电流，周围可产生磁场，磁-电相互作用，可以将电能转变为机械能，这是电动机的基本原理。电能在长距离输运过程中需要将低电压升为高电压，使用时需将高电压降为低电压，都需要变压器；直流变交流也需要变压器。此外，开关电源、逆变器、磁悬浮列车、电磁弹射器等都离不开磁性材料。因此，磁性材料在生产电能和应用电能中有着重要的作用，尤其是在提高效率、降低能耗方面发挥着决定性的作用。

材料、信息和能源是当今社会的三大支柱。其中，材料是信息、能源的基础，而磁性材料属于基础性的功能材料，又融合于信息和能源领域。

二、磁性材料历史及现状

磁性材料大致上可分为两大类，即以磁滞回线为特性的材料、以磁与其他学科交叉耦合新性质为特性的材料。在能源领域中应用的主要是第一类，其中以软磁和永磁材料为主，本节将重点介绍。磁性材料典型的磁滞回线如图 3-1 所示。

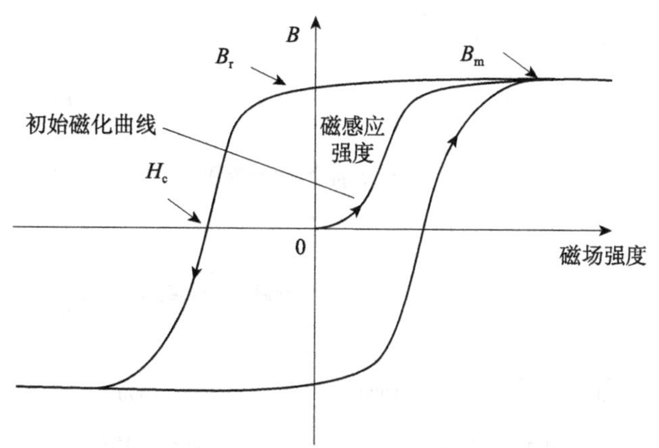

图 3-1　磁性材料典型的磁滞回线

H_c：矫顽力；B_r：剩磁；B_m：饱和磁化强度

对软磁材料通常要求矫顽力、剩磁尽可能小，对永磁材料的要求则刚好相反，要求二者尽可能大；软磁和永磁材料对饱和磁化强度的要求基本上一致，都希望尽可能高。

磁性材料的发展趋势为永磁材料矫顽力越来越高，软磁材料矫顽力越来越低，如图 3-2 所示。其中，钕铁硼稀土永磁材料性能最佳，被誉为“当代磁王”。软磁材料中，矫顽力最低的为非晶与纳米微晶软磁材料，软磁材料除要求矫顽力低外，尚需在所使用频段损耗要低。

三、能源领域磁性材料目前面临的重大科学问题及前沿方向

（一）永磁材料

永磁材料主要分为铁氧体永磁，如 $Sr(Ba)Fe_{12}O_{19}$；稀土永磁，如 $SmCo_5$、Sm_2Co_{17}、Sm-Fe-N、$Nd_2Fe_{14}B$；轻稀土双主相复合永磁三大类。铁氧体永磁价廉，化学稳定性佳，主要用于民用；稀土永磁性能好，磁能积比铁氧体约

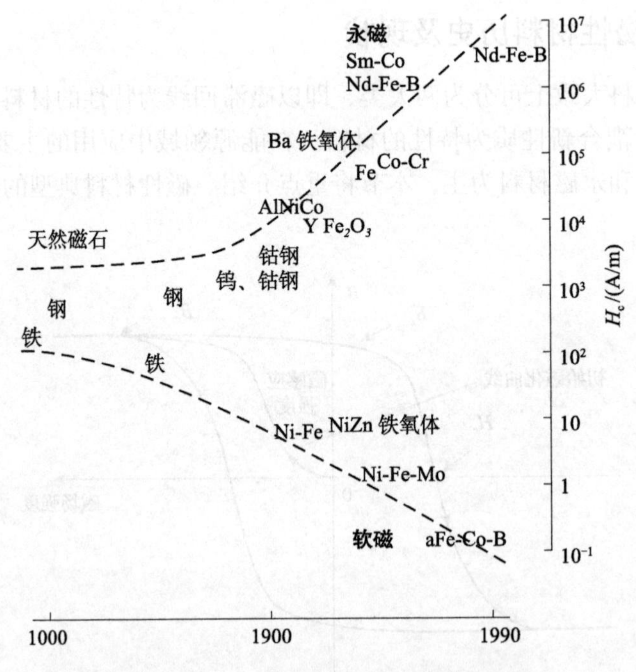

图 3-2 软磁、永磁材料矫顽力发展的总趋势

高 10 倍，但价格也高，主要应用于高科技领域等。

以永磁电机为例，与电励磁电机相比，永磁电机具有高效、节能、体积小等优点，二者对比如图 3-3 所示。

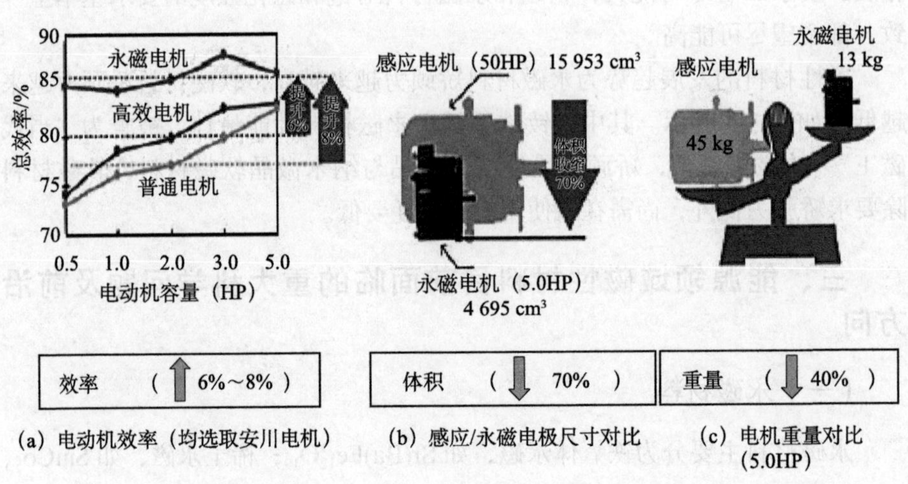

（a）电动机效率（均选取安川电机） （b）感应/永磁电极尺寸对比 （c）电机重量对比（5.0HP）

图 3-3 永磁电机与电励磁电机性能对比，5.0HP（Oliver G. et al.，2011）

稀土永磁钕铁硼的最高性能：磁能积 (BH)max 为 59.5MGOe[①]，达到理论值的 93%；剩磁为 1.55 T，达到理论值的 97%。目前国内主要开展以下研究。

第一，采用双合金工艺，降低重稀土元素 Dy、Tb 含量。

第二，以丰度高、价廉的稀土元素，如 La、Ce、Y 等取代 Nd，生产 (Ce，Nd)FeB/NdFeB 双主相稀土复合永磁体。

Ce 含量达稀土总量的 20%～30%，其磁能积可达 40 MGOe；矫顽力达 12 kOe，价格随轻稀土元素含量的增加而下降，以满足不同应用的需求。

轨道交通车辆牵引系统历经直流电机、交流异步电机和第三代永磁同步电机牵引系统三个发展阶段。永磁同步电机牵引系统具有多种优点：一是低传递损耗小，噪声低；二是转速平稳，过载能力强，可靠性和功率密度高；三是体积小，重量轻，效率高；四是电机采用全封闭结构，维护简单，无齿轮驱动，磨耗小，全寿命周期成本低；五是结构多样，转向架自由空间大，径向调节能力强。永磁同步电机牵引系统已经成为各发达国家竞相研究的技术热点。牵引系统能耗占轨道交通系统总能耗的 40%～50%。永磁同步电机牵引系统较传统的异步电机牵引系统能耗可降低约 11.5%。一台永磁电机可承载两台异步电机的运力并满足运行要求。我国轨道交通对钕铁硼的总需求约 4577 t。2015 年 1 月，我国首列永磁高铁试车。“永磁地铁”长沙地铁 1 号线正式投入载客运营。

2009～2020 年，每 3 年风电量约增加 1 倍，2009 年风电量为 59 213 MW，2020 年全球约为 1900 GW。3 MW 的风力发电机需 1.5 T 的稀土永磁来驱动，目前我国风电产量应占全球第一位。

（二）软磁材料

软磁材料在电机、变压器、开关电源、交直流转换器，以及电感元器件等应用中是不可或缺的功能材料，通常分为铁氧体、金属两大类。其中，铁氧体软磁材料主要用于较高频段直到微波、光波段，而金属软磁材料如硅钢片、铁-镍合金等主要用于低频段，尤其在电工中的应用，几乎是硅钢片的天下，但目前已有所变化。非晶与纳米微晶软磁材料已在电工变压器中得到部分应用，可以显著降低空载时的损耗。为了进一步降低损耗（能耗）、减少体积，近年来软磁性复合材料（soft magnetic composite materials，SMC）已开始进入电工领域。SMC 是金属软磁颗粒被绝缘黏合剂包裹形成的。随后

① 1 MGOe ≈ 7.96 KJ/m^3。

该材料可采用粉末冶金的工艺制备成所需的器件。软磁性复合材料的性能处于金属与铁氧体之间，如图 3-4 所示。

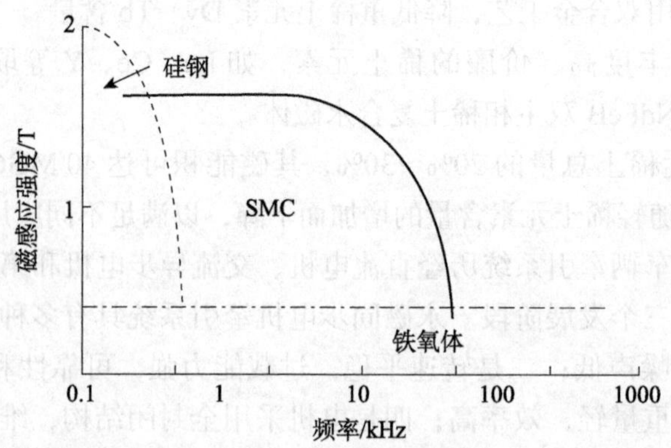

图 3-4 硅钢、SMC、铁氧体之磁通密度与频率的关系曲线对比

1936 年，日本报道了作为粉体磁性材料的 Fe-Si-Al 合金，其组成为 5%Al-10%Si-85%Fe。其合金性能如下：初始磁化率为 30 000，最大磁化率为 120 000。将其补磨成 10 μm 粉体，压成型后，该材料性能如下：矫顽力为 1.25 Oe，磁感应强度为 0.45 T，T_c=500 ℃。

磁粉芯的性能处于金属与铁氧体之间，除 Fe-Si-Al 合金外，可作为 SMC 的软磁材料品种很多，如纯 Fe 粉、Fe-Ni、Fe-Si、非晶与纳米微晶等磁粉，人们可以根据不同用途进行选择与研发。

铁粉芯的基本特性如下：①好的直流偏置特性；②较高的饱和磁感应强度；③好的磁导率频率特性线性度；④好的磁导率温度特性线性度。

软磁材料的发展趋势在低频段，0.5 kHz 之内的硅钢片依然具有优势；0.5～100 kHz 的 SMC 将具有优势，有望部分取代硅钢片与铁氧体；更高的频段则继续以铁氧体为主。

四、学科发展政策建议与措施

第一，加强轻稀土复合永磁材料的研发与生产。

第二，加强研发适合电机、电工用的软磁性复合材料，支持产业化生产。

第三，加强新型高效永磁电机的研发。

都有为（南京大学）

第四章

新能源材料发展的新概念、新应用、新机遇

第一节 原位资源利用中的能量与物质转换

一、研究概述

太空探索对揭示生命起源、保护地球安全、开拓探索疆域、开发太空资源、推动科技进步具有十分重要的意义，已成为人类当前面临的全球共同挑战，也是我国从航天大国走向航天强国、落实创新驱动发展战略的重大实践活动之一。近年来，太空探索的热度、广度和深度都得到了显著提升。2018年1月，在国际太空探索协调组的组织下，十四国航天局联合发布新版全球探索路线图，提出了重返月球、载人火星探索等重大里程碑任务（图4-1），通过太空探索，“拓展人类在太阳系的存在，更好地理解我们在宇宙中的位置”（International Space Exploration Coordination Group，2018）。美国历届政府均把太空探索作为国家航天计划的首要任务。美国特朗普政府重建了搁置25年的美国国家航天委员会，并于2017年12月11日签署1号太空政令，宣布美国宇航员将重返月球并最终前往火星。我国一直十分重视发展太空探索能力，在载人航天和探月工程等重大科技专项工程的基础上，将进一步实施“深空探测及空间飞行器在轨服务与维护系统”等科技创新重大项目。2016年以来，载人深空探索也受到了国际航天公司的广泛关注。SpaceX

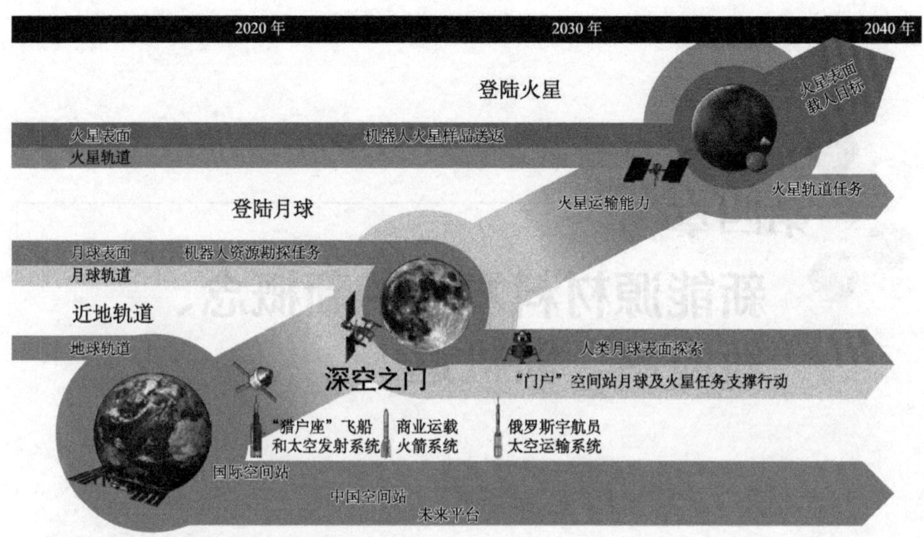

图 4-1　全球探索路线图（International Space Exploration Coordination Group，2018）

公司、洛克希德·马丁空间系统公司等航天公司也提出了载人火星探索计划。美国政府首开先河，批准私营公司开展登月活动，月球也已被纳入地球经济圈，也迅速催生了地月经济圈（cislunar econosphere）（图 4-2）、小行星采矿（asteroid mining）、月球采矿（moon mining）及太空制造（in-space manufacturing）等新领域。随着人类文明发展和科技进步，人类的探索疆域逐步向深空拓展。21 世纪，人类将有望实现地外移民，月球、小行星、火星等地外天体上将留下人类的足迹。

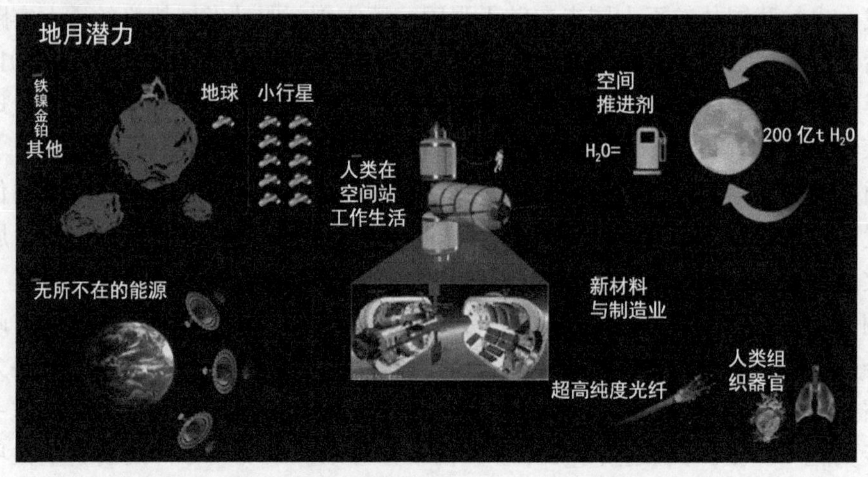

图 4-2　地月经济圈概念（Bergin，2018）

在载人深空探索活动中，地外生存是人类实现长期太空飞行（地球和月球轨道任务、地球和火星长期飞行任务）、地外长期居住（月球和火星基地）、地外移民的基本能力。地外生存过程面临基本的物质与能量需求。对于物质需求来说，为满足航天员基本生理需求，航天员须消耗水 27.6 kg/（d·人）、氧 1 kg/（d·人）、食物 1.5 kg/（d·人）、其他 3.5 kg/（d·人），共计 33.6 kg/（d·人）。对于 6 个航天员的火星任务，若不循环利用物资，其需携带的消耗品就高达 200 t，即使采用资源再生循环利用技术，需携带的消耗品仍达 38 t，发射质量超过 500 t（Rapp，2013）。对能源需求来说，对于一个小型地外前哨战，其电力和热能消耗一般为 10 千瓦级，因月夜长达 350 h，即使采用 500 W·h/kg 高比能再生燃料电池，储能质量也会超过 7 t（平均供电 10 kW）。从地球上携带资源来开展载人深空探索，任务成本代价极高，技术上也难以实现，因此，必须对飞行器废弃物原位资源和地外天体原位资源加以有效利用，才能大大减少从地球携带的物资，使载人深空探索任务具备可行性。

原位资源利用是利用太空可原位获取的资源转换为航天任务需要的各种产品的新方法。通过利用原位资源，可在其他星球上原位获取人类生存和活动所需的基本能源与物资，这将大大减少从地球的补给需求，降低太空探索的发射质量、成本和风险，使人类具备脱离地球的生存能力，真正实现可承受、可持续的太空探索。随着人类探索疆域的拓展，重返月球、载人火星探索等极具挑战性的航天任务逐步提上日程，在地外天体表面如何有效地实现原位资源的综合利用，成为实现人类太空疆域拓展亟须解决的首要难题。潜在的太空资源包括水、大气成分、太阳风注入的挥发物（氢、氦、碳、氮等）、金属和矿物质、太阳能、永久光照区和阴影区、太空自身的真空和微重力，甚至是人类探索活动产生的垃圾和废物，通过适当加工可以将这些原材料转化为有用的材料和产品。今天，飞行任务必须带上所有的推进剂、空气、食物、水和居住舱，以及为机组人员提供地球以外旅行所需的屏蔽。如果人类希望探索地球以外的空间，就必须在太空中找到和利用生产推进剂、维持生命以及支持系统和栖息地的建设所需的资源。原位资源利用的近期目标是降低载人登月和火星任务的成本，建立长期载人空间基地，并将能源或宝贵资源带回地球。原位资源利用主要包括四个应用领域：任务消耗品生产（推进剂、燃料电池反应物、生命维持消耗品和制造、建造用原料）、表面建造（辐射防护罩、着陆垫、墙壁、栖息地等）、制造和维修（备件、电线、桁架、集成系统等）、太空公用设施和电力设备（National Aeronautics and

Space Administration，2015）。

二、历史及现状

阿什（Ash）、道勒（Dowler）和瓦尔西（Varsi）于 1978 年最早提出了在火星上生产上升段燃料的设想，并创造了“原位燃料生产”这个词汇。近年来，随着技术的发展，利用原位资源有望生产更多的材料和产品，“原位燃料生产”被“原位资源利用”这个更具有广泛意义的词汇所代替，原位资源利用也逐渐成为当前的研究热点方向。2004 年，美国国家航空航天局（NASA）成立了原位资源利用能力路线图小组，该小组与其他 14 个能力路线图在 2005 年联合发布了路线图报告（NASA，2005），确定了资源提取、材料处理和输运、资源处理、原位制造、原位建造、原位消耗品生产和存储与分发、独特开发和认证能力等 7 个 ISRU 核心能力。美国 NASA 于 2015 年发布的《空间技术路线图》中将原位资源利用列为载人深空探索优先发展的首项技术和交叉技术领域的重要组成部分，重点发展原位资源采集（提取、输运、分离等）、生产（加热、萃取、催化、生物技术等）、应用（蓄热、辐射防护、建筑等）技术（National Aeronautics and Space Administration，2015），如图 4-3 所示。2016 年，我国由叶培建院士带领的载人深空探测中国学科发展战略工作组系统分析了实施载人深空探测所面临的关键科学和技术问题，认为原位资源利用技术属于有可能带来颠覆性、变革性的技术领域（中国科学院，2016）。2018 年 8 月，美国公布的 2020 财年预算提出重点支持以原位资源利用为核心的太空探索研发。

图 4-3　原位资源利用技术研究内容

欧美等国分别针对月球和火星资源利用开展了长期研究与技术攻关，开发了一系列原位资源利用的技术验证系统和原型机，规划了多个飞行试验计划。由于水资源在深空探测活动中的作用十分关键，而地面直接补给的造价又过于高昂，地外原位获取和转换利用水资源成为必然趋势。在经一系列探测任务，发现月球、火星和小行星上有水资源之后，随着地外天体探测技术的不断发展以及商业航天概念的不断推动，地外天体水资源获取和转换利用的研究热潮持续涌现，已成为近期航天活动的热点。由于地外天体的特殊极端环境，其水资源主要存在于土壤之中，且存在形式复杂（混合水、束缚水、深埋水等），不确定性（时变性、分层性、跨区域性等）特点显著，为其原位获取提出巨大挑战。美国 NASA 于 2005 年提出了资源勘探者（resource prospector，RP）任务，计划于 2022 年开展月球水冰等原位资源利用的探索实验。该任务包括“月球表层土壤和环境科学，以及氧气和挥发物提取”（Regolith and Environmental Science and Oxygen and Lunar Volatiles Extraction，RESOLVE）关键载荷，通过设备将月球表层土壤中的小分子成分（包括可能存在的水）提取并还原成氧气，利用氢气将月壤中的钛铁矿还原成水并最终转换为氧气（Sanders & Larson，2015），如图 4-4 所示。该原型机完成了地面测试。但目前主要载荷转至美国 NASA 新商业月球载荷服务计划支持的商业着陆器任务。俄罗斯针对月球探测规划了一系列“月球”（Luna）后续计划，并将在 2025 年发射代号为“月球 27 号”的“月球-资源”着陆器。欧洲空间局与俄罗斯联合，在“月球-资源”着陆器上安装

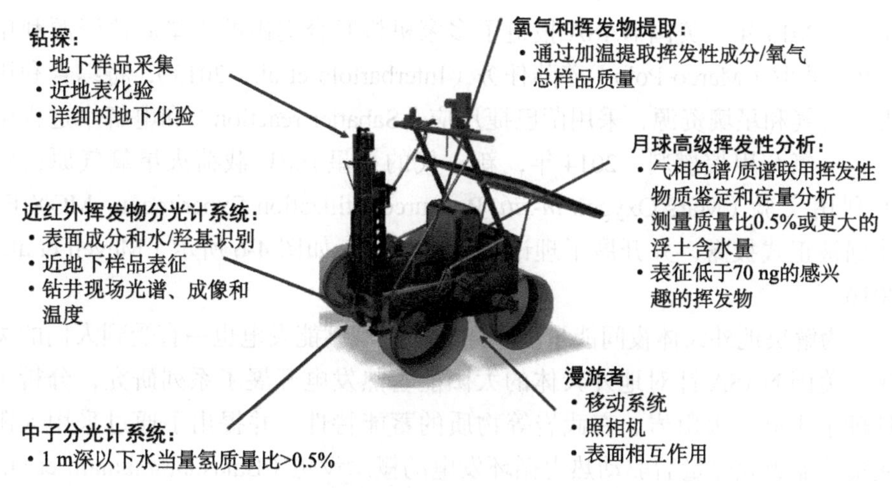

图 4-4　RP 任务及 RESOLVE 关键载荷

“资源观测和勘探商业开采和运输的原位勘探包”（The Package for Resource Observation and *in-situ* Prospecting for Exploration，Commercial exploitation and Transportation，PROSPECT）载荷，实现约 1 m 深度范围内提取和输运星壤，并从提取物中获取挥发物资源进行成分检测。针对月球和火星水提取任务需求，美国蜜蜂机器人（Honeybee Robotics）公司设计了一款移动式原位水提取装置（Zacny et al.，2012）。该装置采用钻取方式获取原始冻土并将其搬运至密封舱内，通过同位素进行加热处理，挥发的水蒸气通过冷凝管进行冷凝收集，汇集于储水罐内，用于月球、火星及近地小行星表面原位获取水资源。2018 年，我国钱学森实验室提出用直接太阳能光热法的钻取一体化装置实现地外原位水资源和其他挥发物的高效提取（姚伟等，2018）。

除水之外，另一个最重要的方面是从星壤中提取氧气，该过程包括化学还原、热解、酸处理或熔融星壤电解等方法（Schwandt et al.，2012）。美国洛克希德·马丁空间系统公司自 2005 年开始研制先驱者原位资源利用月球氧测试床系统（Precursor ISRU Lunar Oxygen Testbed，PILOT），美国 NASA 也开发了 ROxygen 系统，采用氢还原过程提取星壤中的氧，系统温度在 875～1050 ℃，氧气提取效率为 1%～2%。美国欧比泰克（Orbitec）公司开发了更加高效的碳热还原系统，通过太阳能聚光将星壤加热到 1800 ℃进行化学反应（Gerald & William，2013）。

在火星二氧化碳大气利用方面，美国提出的火星勘测 2011 登陆器（Mars Surveyor 2001 Lander，被取消）任务拟携带 MIP 载荷，尝试利用高温电解技术将二氧化碳还原成为氧气，验证从火星大气中制造氧气的能力，如图 4-5 所示。2013 年，美国 NASA 报道了多家机构联合提出的火星原位资源利用马可·波罗（Marco Polo）着陆任务（Interbartolo et al.，2013），将综合利用火星大气和星壤资源，采用萨巴捷反应（Sabatier reaction）和电解水过程生产氢、氧与甲烷燃料。2014 年，新一代的火星 ISRU 载荷火星氧气原位资源利用实验（Mars Oxygen *in-situ* Resource Utilization Experiment，MOXIE）计划被正式披露，并开展了理论及实验研究，如图 4-6 所示（Meyen et al.，2016）。

为解决地外天体夜间能量供应问题，星壤蓄能发电也一直受到人们的关注。美国 NASA 针对地外天体的太阳能蓄热发电开展了系列研究，分析了月球上干砂、火山岩、玄武岩等物质的蓄能特性，并提出了通过采用太阳能集热器吸收热量后驱动热力循环发电的概念设想（Balasubramaniam et al.，2011；Climent et al.，2014）。我国钱学森实验室为解决月球基地热电联供问题，开展了太阳能聚光增强星壤储能发电系统理论和实验研究（Lu et al., 2016）。

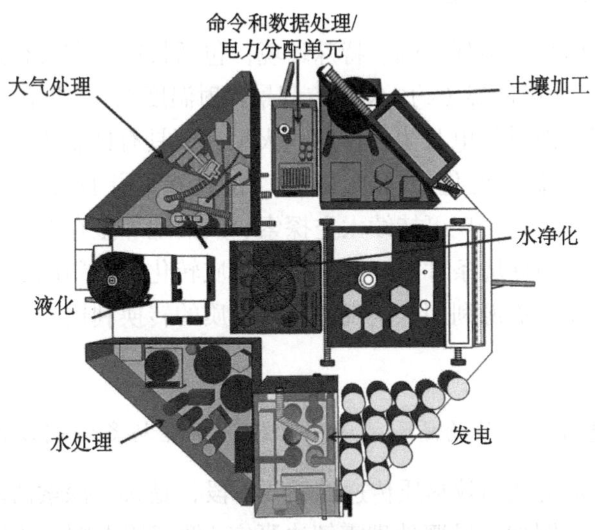

图 4-5 “马可·波罗”火星大气与星壤综合利用系统

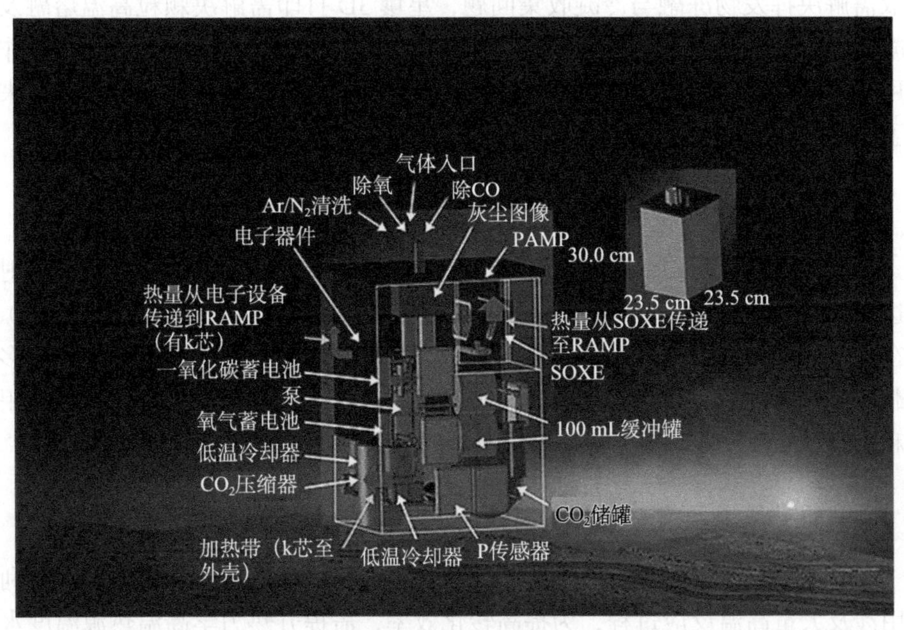

图 4-6 MOXIE 火星大气资源利用系统

三、目前面临的重大科学问题及前沿方向

（一）重大科学问题

太空探索面临一系列极端、特殊环境，包括长期深空飞行过程中面临的微重力（$<10^{-6}$）、极低温（约4 K宇宙背景辐射温度）、高真空、强辐射环境，在月球表面还需面对带电月尘的影响；火星表面具有低重力（0.38 g）、低气压（600 Pa）、高温差变化（140～300 K）的特点，并且受全球沙尘暴的影响。原位资源利用是实现可持续太空探索、人类地外生存的基础。在地外天体极端、特殊空间环境条件下，其能量和物质转化面临新问题、新挑战。在地外天体开展原位资源利用，实现能量与物质的转换利用，面临以下亟待解决的科学问题。

1. 低微重力下原位物质和能量多相输运机理与多场耦合机制

多相输运是实现高效热质传递的重要手段，是太空探索活动原位资源利用的重要途径。例如，星壤处理需解决真空/低气压下颗粒流输运问题，星壤氧提取和冶炼过程需解决颗粒熔融与高温化学反应的问题，地外水资源提取需解决挥发物冻融与冷凝收集问题，星壤3D打印需解决颗粒高温熔融及凝固机理等问题。空间弱引力条件使得表面张力、曳力、布朗力等次级力作用凸显，残余引力场和流场、温度场、浓度场强耦合，给多相输运过程带来了一系列新的科学问题。

2. 太空极端条件下太阳能全谱综合利用的热力学机制

与地面系统不同，在太空环境下，太阳能系统面临上下两个极端热辐射边界：超高温的太阳热辐射（约5800 K）和超低温的宇宙背景辐射（4 K）。在这两个极端辐射温度边界条件下，深入理解光子、电子、声子、离子等多载能子耦合过程，开发新材料，突破新机理，实现全谱段的太阳能综合利用和梯级转换，为太空探索活动提供所需的电能、化学能和热能。

3. 极端辐射和高温条件下高性能、长期稳定的新能源材料

由于太空环境面临高能辐射、强紫外辐射等极端辐射条件，原位资源利用涉及大量高温反应过程，为提高转化效率，需提升热力学高温热源温度水平。同时，由于太空探索任务的长期性和自主性，需要开发能够长期稳定可靠的新能源材料。一方面，需探索能够适应太空极端环境的新能源材料；另

一方面，需深入理解太空极端环境下的性能衰退和失效机理。

（二）前沿研究方向

（1）太空太阳能高效收集材料：需开发抗辐射自清洁聚光材料、高温稳定的高太阳吸收率光吸收材料、柔性高导光材料等。

（2）高效光热电能量转换材料：包括光伏材料，高热电优值、宽温区的半导体热电材料，高光谱转换效率和窄谱高发射率的热光伏转换材料，静态碱金属热发电等。

（3）实现原位资源转换的高效催化材料：包括进行二氧化碳、水等资源转换的光催化、电催化和热催化材料。

（4）恶劣环境下的安全储能材料：包括锂电热冲击下不会燃烧和热失控的高安全锂离子电池材料、低温环境下（−60 ℃）可安全工作的锂离子电池材料、高温环境下（200～450 ℃）可安全工作的锂离子电池材料等。

四、研究总结与展望

原位资源利用是实现人类地外生存的基础，在全球太空探索共同目标的驱动下，已成为当前航天科技的新领域、新方向。在地外天体极端、特殊环境条件下，原位能量和物质转化面临一系列新问题、新挑战，有望催生一批原创性、颠覆性的新技术，但当前国际上相关研究刚刚起步，研究基础薄弱，亟待加强。

结合我国深空探测和载人航天后续发展需求，加强原位资源利用领域的基础研究，将推动我国与欧美等航天强国同步甚至引领发展，可为我国航天强国建设提供坚实的科学技术基础。

更重要的是，地外天体原位资源利用的研究将推动能量与物质转换利用新概念、新技术、新材料的发展，有望催生可再生能源、资源再生利用领域的颠覆性技术和新能源材料发展，促进地球可持续发展。

五、学科发展政策建议与措施

我国深空探测和载人航天后续发展提出了原位资源利用的重大需求，原位资源利用面临的能量与物质转换利用的新挑战、新问题，将催生新能源领域的新学科方向。建议我国结合航天强国建设重大任务和可持续发展重大需求，加快推动原位资源利用能量与物质转换新学科方向的发展，形成以我国为主导的前沿学科方向。

张 策（中国空间技术研究院）、姚 伟（中国空间技术研究院）

本节参考文献

姚伟，王超，李啸天，等 . 2018. 一种地外天体表面水资源获取钻具装置及钻取方法 . 发明专利 : 201810162389.8.

中国科学院 . 2016. 中国学科发展战略·载人深空探测 . 北京：科学出版社 .

Balasubramaniam R, Gokoglu S, Sacksteder K, et al. 2011. Analysis of solar-heated thermal wadis to support extended-duration lunar exploration. Journal of Thermophysics & Heat Transfer, 25: 130-139.

Bergin C, 2018. ULA Laying the foundations for an econosphere in cisLunar space. https://www.kc4mcq.us/?=1629[2020-10-10].

Climent B, Torroba O, González-Cinca R, et al. 2014. Heat storage and electricity generation in the Moon during the lunar night. Acta Astronautica, 93, 352-358.

Gerald B S, William E L. 2013. Progress made in lunar *in situ* resource utilization under NASA's exploration technology and development program. Journal of Aerospace Engineering, 26: 5-17.

Interbartolo M A, Sanders G B, Oryshchyn L, et al. 2013. Prototype development of an integrated Mars atmosphere and soil-processing system. Journal of Aerospace Engineering, 26: 57-66.

International Space Exploration Coordination Group. 2018. The Global Exploration Roadmap.

Lu X C, Ma R, Wang C, et al. 2016. Performance analysis of a lunar based solar thermal power system with regolith thermal storage. Energy, 107: 227-233.

Meyen F E, Hecht M H, Hoffman J A. 2016. Thermodynamic model of Mars oxygen ISRU experiment (MOXIE). Acta Astronautica, 129: 82-87.

NASA. 2005. NASA capability roadmaps executive summary: 264-291.

National Aeronautics and Space Administration. 2015. NASA Technology Roadmaps.

Rapp D. 2013. Use of Extraterrestrial Resources for Human Space Missions to Moon or Mars. Berlin: Springer.

Sanders G B, Larson W E. 2015. Final review of analog field campaigns for *in situ* resource utilization technology and capability maturation. Advances in Space Research, 55: 2381-2404.

Schwandt C, Hamilton J A, Fray D J, et al. 2012. The production of oxygen and metal from lunar

regolith. Planetary & Space Science, 74: 49-56.
Zacny K, Chu P, Paulsen G, et al. 2012. Mobile *in-situ* water extractor (MISWE) for Mars, Moon, and Asteroids *in situ* resource utilization. AIAA SPACE 2012 Conf Expo, 1: 11-13.

第二节 仿生模拟光合作用（微光捕获 / 高效传能）

一、人工光合作用材料研究概述

（一）人工光合作用材料的研究意义

我们目前的能源主要来自化石燃料（石油、煤炭和天然气），其中石油预计将在未来 50～150 年耗尽，而且化石燃料的使用导致了许多环境问题，例如增加温室气体排放，最明显的是二氧化碳。2008 年，全球能源消耗总量为 15 TW。如果将这个数字对应于能量的燃烧率，那么预计到 2050 年，能源消耗总量将几乎翻倍。考虑到经济、环境和人类健康在内的重要因素，增加的能源数量必须来自可再生和可持续能源。在所有可再生能源（太阳能、风能、地热能、潮汐能、生物质能等）中，从基本的科学角度来看，太阳能是最可行、最具吸引力的替代能源。地缘政治和环境问题正在越来越多地推动能源的研究工作，而可再生绿色能源的发展是当今社会面临的最重要的科技挑战之一（Lewis & Nocera，2006；Kim et al.，2015；Esswein & Nocera，2007；Chu & Majumdar，2012）。如果我们要应对满足未来能源需求的挑战，开发无碳可再生燃料的高效技术至关重要。

氢气是一种高密度、无碳的能量载体，被认为是潜在的未来燃料，是经济和社会未来可持续发展的理想选择（Tollefson，2010；Pagliaro et al.，2010；Armaroli & Balzani，2011）。工业中使用的大多数氢气来自天然气、煤、石油和水。通过天然气的蒸汽重整，或者通过天然气的部分氧化和随后的煤重整，或者通过水的电解，可以大规模生产氢气。然而，这些常规制备方法受到副产物温室气体二氧化碳的排放或消耗电力的限制。显然，这不是绿色的制备方式，需要新的技术来实现可持续的大规模氢气生产，而人工光合作用以高效的方式产生清洁能源氢气而被广泛研究，为了实现氢气的大规模化制备，迫切需要探索提高氢气产率的制备技术。

理想情况下，氢气应使用可再生能源（如生物质能或太阳能）从水中通过分解水技术来制备。因此，人们非常希望开发稳定的、无碳和低成本的氢气制备方法，以支持新兴的氢经济。经过数十亿年的发展，大自然优化出了一个能够通过太阳能的转化提供可利用能量的系统。太阳每年提供大约 3×10^{24} J 的可用能量，约为当前能源需求的 10 000 倍，因此太阳能是未来经济和社会发展的可行能源。通过人工光合作用将太阳能直接转换为氢燃料已被认为是减轻能源危机和解决日益恶化的环境问题的理想途径（Xu et al.，2016；Oshima et al.，2015；Joya et al.，2013）。

受大自然的启发，利用太阳能开发可再生、无碳能源是满足未来能源需求的理想途径之一，充满挑战。高效的太阳能电池（即固态光伏器件）和太阳能驱动的水分解系统是利用太阳能的两种主要方式。虽然太阳能电池可以直接将太阳辐射转化为电能，但它们无法储存能量，太阳能驱动的水分解可以以燃料（氢气）和氧化剂（氧气）的形式存储能量。化学键中的能量储存是最有效的，因此，这种太阳能储存策略在过去三十年引起了人们的极大关注。氢气是一种环境友好型能源载体，因为它唯一的氧化产物是水。通过广泛使用直接在人工光合系统中产生或通过使用太阳能发电电解过程中产生的氢气作为燃料，可以减少对化石燃料的依赖，并相应地减少二氧化碳的排放。

利用太阳能实现高效水分解的关键是大幅降低水氧化所需的过电位，这可以利用催化剂来实现：①降低氧化半反应过程形成的最高能中间体的自由能；②增加产生的特定产物。这种催化剂促进水氧化所需的两个关键特征是：①底物水分子与有利于双电子氧化反应的位点结合，从而绕过能量上不利的单电子氧化（自由基）中间体；②这些位点非常接近，以便在过氧中间体中形成 O—O 键。第一个特征涉及与结合水的高效质子耦合电子转移反应。这可以通过能够配位水的过渡金属化合物提供。最佳催化剂应将能量密集型 O—H 离解步骤与 O—O 键形成的能量释放步骤结合，以协调顺序重排而不释放活性中间体。

目前，已经开发了基于贵金属（如钌、铱和铑）的均相和非均相催化剂。然而，这些由贵金属制成的催化剂可能会显著增加实际应用的成本。自然界将锰作为构建 Mn_4CaO_4 水氧化催化剂的关键金属离子之一，已人工合成各种锰络合物以模拟光系统Ⅱ（PSⅡ）的功能。至于水的还原，铂已被广泛用作有效的水还原催化剂。人们普遍认为，铂金必须由更便宜的材料取代，因为它价格昂贵，土壤丰度非常低。因此，由廉价的、地球上丰富的元素制成的高效水氧化催化剂和还原催化剂的开发在太阳能转换领域是一个非常重要的主题。

（二）人工光合作用的基本原理

在自然界中，光合作用对将太阳能转化为化学能起着至关重要的作用。化学能为地球上的生命提供燃料，并提供碳水化合物（食物）、氧气甚至化石燃料。在自然界中，植物和光合细菌通过利用太阳能，使二氧化碳与水被消耗，通过一系列光化学反应和暗反应产生氧气和碳水化合物。光合作用中有两个相关的光系统：光系统Ⅰ（PSⅠ）和光系统Ⅱ。当光系统Ⅰ被太阳光激发时，诱导电子转移过程以还原一系列电子受体，例如，叶绿醌和铁氧还原蛋白最终将辅因子烟酰胺腺嘌呤二核苷酸磷酸（$NADP^+$）还原为还原型烟酰胺腺嘌呤二核苷酸磷酸（NADPH），同时光系统Ⅰ被氧化。NADPH是卡尔文循环中二氧化碳转化为碳水化合物的电子和质子源。氧化光系统Ⅰ通过从光系统Ⅱ的几种电子继电器获得电子而再生，如质体醌、细胞色素 f 和质体蓝素。氧化的光系统Ⅱ介导另一个重要的反应——水氧化产生氧气，由 $CaMn_4O_4$ 簇催化。

模拟人工光合作用是利用太阳能在人工催化剂的作用下高效分解 H_2O，产生 H_2 和 O_2［式（4-1）］。光催化水分解反应分为两个半反应：H_2O 中 O^{2-} 被氧化，生成 O_2［式（4-2）］H^+ 还原，产生 H_2［式（4-3）］：

$$2H_2O \longrightarrow 2H_2 + O_2 \qquad \Delta E^0 = 1.23\ \text{V} \tag{4-1}$$

$$2H_2O \longrightarrow 4H^+ + 4e^- + O_2 \qquad E^0 = +1.23\ \text{V}\ vs.\ \text{NHE，pH=0} \tag{4-2}$$

$$4H^+ + 4e^- \longrightarrow 2H_2 \qquad E^0 = 0\ \text{V}\ vs.\ \text{NHE，pH=0} \tag{4-3}$$

来自太阳光的光子用于产生电荷分离态，驱使水氧化催化剂进入氧化水和释放电子以还原质子所需的高氧化态。H^+ 催化产氢的过程需要光敏剂与产氢催化剂的结合，光敏剂可以将太阳光辐射转变为激发电子，将其传输至催化剂催化中心，诱发产氢反应的发生。实现光分解水产生氧气的过程是在高能量的条件下，从两个水分子中移除 4 个质子和 4 个电子，形成 1 个新的 O—O 键，这个半反应被认为是实现整体水分解的障碍。

在标准条件下，氧化水所需的最小理论电位为 1.23 V。然而，在实践中，氧化水所需要的过电位取决于催化剂稳定最高能量中间体的能力。目前最好的贵金属催化剂实现了低至 0.32 V 的过电位。另外，在标准电解槽中必须克服其他因素，包括：①电池电阻（质子和电子电流电阻、电解质电导率、离子气体渗透性）；②不可逆过程（外来电荷极化、末端中间体的产生）。因此，实现水电解所需的实际电压通常为 1.85～2.05 V，相当于 0.5～0.7 V 的过电位。

原则上，可以利用丰富的太阳能资源来提供这种电位。实现该目标的方法之一是使用商业上可获得的光伏电池来为水电解供电，该技术已经被证明可以产生高达7%的太阳能产氢效率。但是，硅太阳能电池产生的最大电压仅为0.6 V左右，典型的电解槽工作电压约为2 V，因此，必须使用4个硅电池串联，才能产生所需电压以形成氢气。由于对4个硅电池的需求和电解槽的额外成本，使用这种技术从水中生产氢气对于大规模应用来说是不经济的，估计成本为28美元/kg（1 kg氢气在体积上大致相当于3.785 L）。

二、利用太阳能催化水分解的发展历史与现状

为了寻找环保节能的分解水方法，人们将目光转向了太阳能。很久之前，已有人发现了太阳能的巨大潜力。比如意大利光化学家贾科莫·恰米奇安（Giacomo Ciamician）在20世纪预测，太阳能可用于解决未来的能源危机。在1912年的科学文章中，他说：在干旱的土地上，将会出现没有烟雾、没有烟囱的工业殖民地；玻璃管的森林将延伸到平原上，玻璃建筑物将随处可见；其内部将发生迄今为止植物严密保护的秘密的光化学过程，但这将由人类工业掌握，知道如何让它们承载比自然更丰富的水果，因为大自然无须争分夺秒而人类是匆忙的（Ciamician，1912）。即使在遥远的未来，煤炭的供应完全耗尽，文明仍然不会受到威胁，因为只要有阳光普照，生命和文明就会持续下去！基于自然光合作用的原理，恰米奇安的愿景实现了，科学家创造了用于从阳光产生燃料的人工系统，可以被描述为“人工树叶”。这是一种光驱动系统，其中能量存储形式为氢气等，这些物质由水分解产生的质子的还原产生。

太阳能制备氢气的研究正在迅速扩大并吸引着不同学科的科学家，包括：①设计和合成分子光敏剂和催化剂并研究它们的结构-性质关系的化学家（Sakai et al.，2013；McNamara et al.，2012；Li et al.，2012；Wang et al.，2013；Kim et al.，2016）；②用新型电子结构构建半导体光催化剂的物理学家（Fujishima & Honda，1972；Hu et al.，2015；Kudo & Miseki，2009；Ma et al.，2014；Tachibana et al.，2012）；③建造具有新型结构和形态的独特光催化材料的材料科学家（Xiang et al.，2015；Tong et al.，2012；Liang et al.，2015；Yuan et al.，2016；Huang et al.，2016）。水分解是一种貌似简单的化学反应，在水被认为是氢和氧的化合物后，就引起了人们的注意。尽管研究人员在用于开发有效的太阳能产氢气转化系统的研究方面付出巨大的努力，但是在太阳辐射下的反应系统中，将整体水分解以产生氢气和氧气的

光催化作用仍然是一个巨大的挑战。文献中报道的可以将水分解为氢气和氧气［式（4-1）］的光催化材料还是很少的（Zou et al.，2001；Hara et al.，1998；Wang et al.，2012；Liao et al.，2014；Fujito et al.，2016）。水氧化催化剂已在两个领域广泛开发：固体金属氧化物催化表面和无机分子催化剂。太阳能水氧化的一个主要挑战是催化剂除需要稳定且具有高转换频率外，还必须是太阳能的固有收集器（集成催化 / 光转换）或易于结合到由大量（非贵重）材料制成的导电电极上（用于间接光转换）。

光子用于产生电荷分离，驱使氧化催化剂氧化水，同时释放电子以还原质子。正如富克鲁瓦（Fourcroy）和他同时代人所发现的那样，水氧化是更加困难的半反应，许多人认为它是该领域发展的瓶颈。因此，全球实施人工光合作用的主要挑战之一就是找到用于水氧化半反应的催化剂。实现光诱导的水分裂的主要困难之一是两个半反应是多电子过程。因此，利用太阳能来进行可再生燃料的制备，需要将几个基本但具有挑战性的光物理步骤与复杂的催化转化进行高效的协同。

因此，在人工光合系统中，必须设计一种有效的方法：①通过光敏剂吸收光子；②通过将电子转移到还原催化剂（通常是主要受体）上，形成电荷分离状态；③在还原催化剂处接受并积累两个连续电子，随后利用这些电子将两个氢质子还原为氢分子；④通过从氧化催化剂（通常是主要供体）转移电子；⑤从氧化催化剂一个接一个地转移四个连续电子后，氧化催化剂通过利用从两个水分子转移来的四个电子，产生一个氧气分子和四个氢质子，同时催化剂还原回到基态。

因此，人工光合系统的构建需要能将单电子转移耦合到驱动 4 电子-4 质子连续过程的催化单元。这些催化剂可以与半导体材料的表面相接触，通过两种不同的方法增强光电化学 H_2O 的分裂：①用于两个半反应的催化剂（H_2O 氧化和质子还原）可以被涂覆到具有适当能带结构的光捕获材料上，用于 H_2O 分裂；②催化剂可以单独连接到两个不同的半导体上，通过所谓的“硬线”连接。

在人工光合系统中，光敏剂可以首先附着到半导体上。通过从光激发的光敏剂到半导体的导带的电子转移，形成电荷分离状态。重要的是，这种电荷分离状态足够长，以使光敏剂从催化剂中提取电子，从而再生光敏剂和单电子氧化催化剂。在四次连续的电荷分离和伴随的电子迁移之后，催化剂通过从 H_2O 中提取电子来重新填充空穴，从而将其氧化成分子 O_2。从 H_2O 中提取的电子最终被输送到还原催化剂处，该催化剂介导质子（由 H_2O 氧化产生）还原成太阳能燃料（如 H_2）。产生的 H_2 可以直接用作能源，或者用于将 CO_2 还原成其他类型的更高复杂性的燃料，如 CH_4 或 CH_3OH。

为了研究 H_2 生成的反应机理，研究人员通常通过用适当的牺牲还原剂代替水氧化半反应来开发太阳能 H_2 生成系统。从太阳能光解水产氢的方法已经成功地以不同形式实现，包括在均相和多组分体系中使用半导体器件（Wang et al.，2009；Zhang et al.，2014；Yang et al.，2013；Chang et al.，2014；Wu et al.，2016）和光活性分子染料（Zhang et al.，2011，2009；Gong et al.，2011；Lazarides et al.，2009；McCormick et al.，2010）。由于莱恩（Lehn）和绍瓦热（Sauvage）观察到在 $[Ru(bpy)_3]^{2+}$ (bpy = 2,2'-联吡啶）作为光敏剂、$[Rh(bpy)_3]^{3+}$ 作为电子继电器、胶体 Pt 作为催化剂、三乙醇胺（TEOA）作为牺牲还原剂的体系中，光照可以驱动 H_2 的产生，因此均相和多组分光催化制氢系统引起了人们极大的关注（Lehn & Sauvage，1977；Kirch et al.，1979）。研究者已经构建了三种不同的光催化制氢系统，包括三组分、双组分和单组分光催化制氢系统。这些光催化制氢系统的关键组成部分如图 4-7 所示。图 4-7（a）所示的三组分光催化制氢系统是 20 世纪 70 年代末和 80 年代的研究热点（Lehn & Sauvage，1977；Kirch et al.，1979；Brown et al.，1979；Kiwi & Grätzel，1979；Chan et al.，1981）。在该系统中，电子继电器用于接受来自激发的光敏剂的电子并将它们转移到催化剂处，催化剂起电子受体的作用。与三组分光催化制氢系统不同，双组分光催化制氢系统在没有电子中继物质的情况下产生氢气（Kalyanasundaram et al.，1978；DeLaive et al.，1979；Krishnan & Sutin，1981；Krishnan et al.，1985；Brown et al.，1979）。研究者对这类体系的兴趣迅速增加。双组分光催化制氢系统的典型组成如图 4-7（b）所示，其中电子可以从激发的光敏剂上转移到催化剂上进行 H_2 析出反应，从而简化了 H_2 析出系统。然而，在这些三组分和双组分光催化制氢系统中，电子转移过程取决于许多难以控制的因素。为了控制电子转移过程，近年来研究者设计了将光捕获单元与催化中心结合的超分子光催化剂［图 4-7（c）］。在这种超分子光催化剂中，从光活化单元到催化中心的分子内电子转移可以得到潜在的控制，从而使得太阳能 H_2 产生达到更有效的电荷转移。有趣的是，三组分和双组分光催化制氢系统中的光敏剂组分被用作捕获光。此外，单组分体系中的超分子金属配合物被用作光催化剂，其不仅为光催化反应吸收光，而且还原质子产生 H_2。

在这些 H_2 催化系统中，第一步是光敏剂进行光子捕获，光敏剂的作用与光合色素类似（Zhang et al.，2011；Lazarides et al.，2009；McCormick et al.，

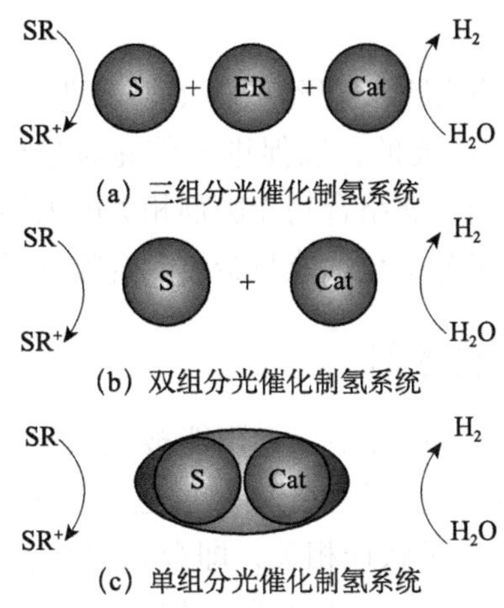

图 4-7　三种不同的光催化制氢系统

SR：牺牲剂；S：光敏剂；ER：电子继电器；Cat：催化剂

2010；Lehn & Sauvage，1977；Kirch et al.，1979）。光敏剂应该有效地吸收入射光子并将自身转换成激发态，该激发态可以向受体转移/从供体接受电子以产生电荷分离状态，产生质子还原反应所需的热力学驱动力。光敏剂对于有效的光捕获和激发电子产生与转移来说是必不可少的，这是决定光催化 H_2 生成系统的总效率的最重要因素之一。在过去的 40 多年中，研究人员已经构建了许多不同类型的光敏剂，其中包括不含金属的有机染料（Zhang et al.，2009）、金属络合物（Lehn & Sauvage，1977；Zhu et al.，2013；Zhang et al.，2014）和官能化的金属有机框架材料（Horiuchi et al.，2012），并应用于光催化制氢系统中。其中，无金属有机染料和金属配合物已被广泛研究并作为用于光催化 H_2 生产的光敏剂。然而，这些使用不含金属的有机染料作为光敏剂的光催化制氢系统通常由于有机染料的光降解而具有短寿命。与有机染料相比，金属配合物具有更高的稳定性，这是由于金属与其配体之间存在强耦合作用（Wenger，2009）。因此，高效金属络合物光敏剂的开发是光催化水分解领域的热门研究课题。光活性分子染料和半导体都可以诱导电子转移，并且有可能在未来用于 H_2O 裂解的人工装置中用作捕光光敏剂。

用于电子转移过程的受到最充分研究的分子光敏剂是多吡啶基钌（Ⅱ）、$[Ru(bpy)_3]^{2+}$ 型络合物。自 20 世纪 70 年代早期发现它们可以从激发态转移电子到牺牲性电子受体上以来，这些配合物已被多位研究人员研究过，由此开辟了新的研究途径，这些配合物已用于各种领域，还在光氧化还原催化领域受到了极大的关注。

研究者在人工光合作用领域已经取得了很大进展。我们认为，对该课题进行深入的研究是迫切需要的，以促进光催化领域的进一步发展。在本节中，我们系统地回顾了与金属络合物光敏剂相关的太阳能 H_2 生成领域近期和重大的科学进展，强调了光催化性能与金属配合物分子结构之间的关系。

（一）金属络合物光敏剂

在光催化 H_2 生产反应过程中，太阳能被光敏剂组分吸收，因此，光敏剂的作用是收获光并引发一系列电子转移反应。对于金属络合物光敏剂，光吸收与三种主要类型的电子跃迁相关，即金属中心、配体中心和金属-配体电荷转移（metal-to-ligand charge-transfer，MLCT）。金属络合物的 MLCT 状态在光催化反应中起关键作用。例如，在常用的八面体 d^6 金属多吡啶配合物中，最高占据分子轨道（HOMO）对应于金属定位的 t_{2g} 轨道，而最低未占据分子轨道（LUMO）是位于配体反键合 π^* 轨道。在可见光照射下，金属中心 t_{2g} 轨道中的一个电子被激发到配体中心的 π^* 轨道（Wenger，2009）。结果，金属配合物的氧化还原电位可以被戏剧性地改变。如式（4-4）和式（4-5）所示，激发态氧化和还原电位通常分别从基态减去或加上零-零激发能量（$E_{0\text{-}0}$）来估算。因此，激发的金属配合物是比它们的电子基态更好的氧化剂和更好的还原剂，还可以为电荷转移反应带来更多的热力学驱动力。金属络合物基于光诱导的氧化还原电位变化和激发态的长寿命的特点，吸引研究者深入研究了它作为光敏剂用于光催化 H_2 生产的性能。

$$E(S^+/S^*) = E(S^+/S) - E_{0\text{-}0} \tag{4-4}$$

$$E(S^*/S^-) = E(S/S^-) + E_{0\text{-}0} \tag{4-5}$$

任何光催化 H_2 生产系统的基本特征是光敏剂吸收光以产生电子激发态。激发的光敏剂可以通过电子给体或电子受体通过还原或氧化反应淬灭以回到基态。由光敏剂、催化剂和牺牲试剂组成的光催化体系中的典型反应机理如图 4-8 所示。受激光敏剂的还原淬灭涉及来自牺牲剂的电子转移，其产生还原形式的光敏剂（S^-）。然后该形式能够将电子转移到析氢催化剂上并返回其基态（S）。在氧化淬灭途径中，受激光敏剂被析氢催化剂氧化以形成氧化物质（S^+），然后在重复循环之前通过牺牲试剂还原氧化物质以恢复其原始状态。通过这些过程，催化剂可以驱动质子的还原以产生 H_2。

许多因素可以影响光敏剂对 H_2 产生的光催化性能。用于太阳能 H_2 产生

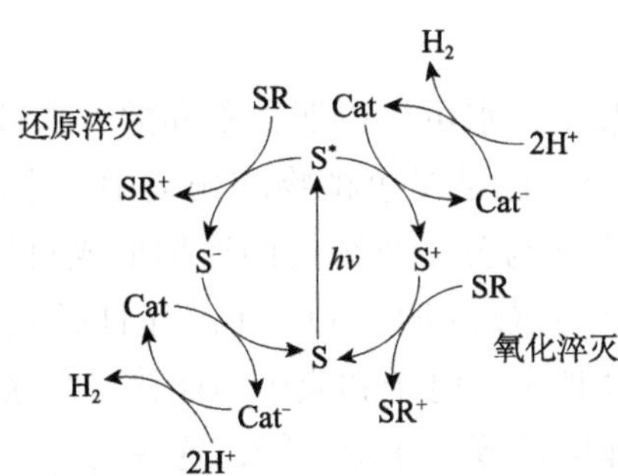

图 4-8　光催化 H_2 生产系统中两个不同淬灭路径的示意图

的高效金属络合物光敏剂的设计和构造的一些一般要求如下。①由于可见光的波长范围超过太阳光谱的 47%，因此金属络合物光敏剂应显示出非常宽的吸收范围，具有高摩尔消光系数（ε），特别是在可见光区域。②激发的光敏剂应该具有合适的氧化还原电位，用于从牺牲还原剂到催化剂的电子转移。对于还原淬灭过程，光敏剂、催化剂和牺牲剂的氧化还原电位应遵循以下趋势：$E(S/S^-) < E(Cat / Cat^-) < E(H^+ / H_2)$ 和 $E(S^* / S^-) > E(SR^+ / SR)$。如果激发的光敏剂通过氧化过程淬灭，相同关键组分的氧化还原电位应表现出以下趋势：$E(S^+/S^*) < E(Cat / Cat^-) < E(H^+ / H_2)$ 和 $E(SR^+ / SR) < E(S^+ / S)$。③光敏剂的激发态寿命的长度应足以进行电荷转移反应。④光敏剂的光稳定性应足够高，以确保其在太阳能 H_2 生成系统中长期运行。⑤光敏剂应在水和常用的有机溶剂中具有良好的溶解性，以确保在分子水平上发生光催化 H_2。

（二）水氧化催化剂

化石燃料的一个清洁替代品是使用阳光分解水获得的氢能。为了实现这一目标，研究人员仍需要充分地了解和控制几个关键的化学反应。其中之一是水催化氧化成分子氧，这也发生在绿色植物和藻类的光系统Ⅱ的放氧中心。

均相介质中的过渡金属水氧化催化剂为研究人员提供了一个极好的平台，可以研究和提取关键信息，准确描述复杂反应中涉及的不同步骤。在分子水平上提取的系统信息使研究人员能够理解控制该反应的因素，以及使系统脱轨以致引起分解的因素。因此，可以开发具有潜在技术应用的稳定且高效的水氧化催化剂。

过去的 5 年中，研究人员在设计新的过渡金属配合物方面取得了重要进展，不同的配体组和结构已被证明能够在过量牺牲氧化剂的情况下分解水产生分子氧。

根据转移的电子数量，水可以以四种不同的方式被氧化。四电子（$4e^-$）过程是较低的能量路径，也是绿色植物和藻类中光系统Ⅱ的放氧中心发生的过程。这种低能热力学与分子复杂性的增加形成对比。对于$4e^-$转移的情况，必须破坏来自两个H_2O的四个O—H键并且必须形成O—O键，因此该反应的催化剂必须处理多个电子和质子的迁移。在这方面，存在许多含有通式[M^{n+}-H_2]的水配体的过渡金属配合物，这些配合物可以可逆地迁移1个H^+和1个e^-并产生相应的[$M^{(n+1)+}$-OH]配合物。在过去30年中，经过深入研究，[Ru-OH_2]多吡啶配合物成为构成这类配合物的典型例子。[Ru-OH_2]多吡啶配合物可以失去质子和电子，并且在很窄的电位范围内很容易达到更高的氧化态，这要归功于氧代基团的σ-域和π-域赋形特征，可以稳定那些较高的氧化态。另外，一个质子和一个电子（H原子）的迁移阻止了库仑电荷的积累，否则该电荷会在简单的外球形电子转移过程中积累，并且会强烈地导致复合物不稳定。因此，质子耦合电子转移（proton coupled electron transfer，PCET）为氧化过程提供了低能量途径，确保了整个水氧化反应的相对快速反应动力学。在氧化还原平衡中，大自然在光合系统Ⅱ析氧催化剂（oxygen evolution catalyst-photosystem Ⅱ,OEC-PSⅡ）中也使用PCET来进入低能量途径。除迁移质子和电子之外，水氧化成分子氧需要形成O—O键。除OEC-PSⅡ之外，已有研究表明，亚氯酸盐歧化酶也可以进行这种反应。形式上，从机制角度来看，可以将过渡金属配合物促进的O—O键的形成考虑在内，考虑到游离水分子是否参与上述键的形成。从这个角度来看，存在两种可能性：水亲核攻击和两种金属配位单元的相互作用（interaction of two metal-oxygen moieties，I2M）。

1. 固态催化剂

目前正在评估各种固态材料用于水氧化催化的可能性，包括大量且多样的过渡金属氧化物族，可用作电催化剂和半导体金属氧化物，用于整合催化/光转化。最有效的电催化剂通常使用贵金属，如Pt、Ru和Ir，它们通常以外部施加的电位驱动。正在开发具有改进的稳定性和周转率的合金，但是可用性和成本的基本限制构成了这些系统的大规模应用的主要障碍。

最近的发展是使用具有催化活性的过渡金属氧化物相。一个有趣的例子是AB_2O_4尖晶石相，多年来已知其微米级粒度的颗粒具有催化活性。已发现Co_3O_4尖晶石的纳米级颗粒表现出相当高的催化水氧化速率（TOF = $1140s^{-1}$/

Co_3O_4）。已通过电沉积制备了其他活性固相。例如，在 Co^{2+} 的电氧化反应中形成无定形的、非导电的 Co：P：O 相，其在水氧化中是具有催化作用的。这些薄膜在 0.5 V 的过电位下产生 $1mA/cm^2$ 的电流密度。这种有希望的材料在溶解和 Co(Ⅱ) 离子损失方面是化学不稳定的，但是可以通过施加的电位在 HPO_4^{2-}存在下修复。这在可溶性 Co^{2+} 和 HPO_4^{2-}物质与不溶性但具有催化活性的 Co^{3+}-HPO_4^{2-}相之间建立了动态平衡。

已经开发了含有 Ti、Nb、Ta、W、Ga、In、Ge、Sn 和 Sb 的各种组合的金属氧化物，以努力利用光来实现水的有效电催化氧化。与 TiO_2 一样，这些材料中的许多材料（包括 $NaTaO_3$ 和 $SrTiO_3$）都具有较大的带隙，仅吸收一小部分太阳光谱。其他的（如 WO_3 和 Fe_2O_3）具有不适合的带边位置，因此在吸收光时产生的势能不能补充水的氧化和随后的质子还原所需的能量。另一类催化剂（包括 CdS 和 CdSe）具有理想带边位置的窄带隙，但在驱动水氧化所需的氧化电位下不稳定。因此，使用单一材料在能量吸收和催化电位之间达到最佳平衡是困难的。最近研究者感兴趣的材料包括具有紫外-蓝光响应的掺杂二元半导体。

克服这些困难的另一种方法是耦合适当的材料组合。例如 WO_3 和 Fe_2O_3 的不合适的带边位置可以通过将它们串联耦合到光伏器件上来克服，然后通过光伏器件产生分裂水所需的额外电位。通过将含有这些材料的水电解器与染料敏化太阳能电池相结合，已经实现了 3% 的氢生成效率。

将多种材料集成到形成多结器件的单个电极中也解决了这一问题，其中光电阳极包含一层水氧化光催化剂（如 $GaInP_2$），以及一层光伏材料（如 GaAs p-n），后者提供了完成电路所需的电位。

固态催化剂的其他发展包括在多结光电池上沉积掺杂的薄膜氧化物（$NiFeO_2$ 和 Fe_2O_3）。这些催化剂表现出更长的稳定性，并已与商用多结太阳能电池集成。然而，它们通常仍然需要大的过电位且使用高腐蚀性的碱性条件（pH = 14）。

总之，固体薄膜催化剂具有制备简单性和易于结合到装置中的潜在优点。当制备高比表面积纳米结构材料时，可以获得大密度的催化位点，即使对于每个位点表现出适度特定活性的材料，也会使总体转换数量得到显著改善。

2. 分子催化剂

分子催化剂提供了对原子结构和组成进行特定定制的可能，而没有受到延伸的固体晶格所施加的限制。这种催化剂有希望潜在地导致更有效的水氧化，因为其可以同时优化 O—H 键裂解和 O—O 键形成反应。它们可以由丰富（且便宜）的材料制成。

然而，“自下而上”设计和开发成功的分子催化剂是一项极具挑战性的任务。值得庆幸的是，大自然花费了30亿年的时间来优化水氧化系统，这是分子水氧化催化剂设计的优秀蓝图。这种最有效的分子水氧化催化剂是光系统Ⅱ中天然存在的水氧化复合物［water-oxidizing complex，WOC，也称为氧气放出中心（oxygen evolving complex，OEC）］。含氧光养型生物中 PSⅡ-WOC 的活性位点包含 $CaMn_4O_x$ 核心，其结构和催化特征在别处详细描述。除簇合物本身之外，通过光系统Ⅱ利用太阳能成功进行水氧化的关键步骤是通过导电酪氨酸残基（Y_z）与光吸收氧化色素分子 $P680^{0/+}$ 复合物相结合，从簇合物中有效迁移电子。

光系统Ⅱ的水氧化复合物激发了广泛的模型分子水氧化催化剂的构建。其中最成功的是过渡金属 Ru、Mn 和 Ir 类催化剂。如在 PSⅡ-WOC 中那样，这些金属的高氧化态通常通过引入给氧配体与氧和氮供体原子来稳定。因为它们最近已被详细评述，所以在此我们对这些分子催化剂的一些实例描述如下。

迄今，Ru 是水氧化分子催化剂中使用最广泛的过渡金属。Ru 络合物通常形成稳定的金属-配体键，这增加了参与水氧化的活性中间体的稳定性。然而如下所述，与金属配位球上的末端位置结合的强配体也可以通过阻止进一步反应使 Ru 基催化剂失活。此外，动力学惰性中间体，通常是 Ru^{2+} 物质，可能参与催化循环，这将影响催化剂周转速率。

均相分子水氧化催化剂的第一个例子是 $Ru^{Ⅲ}$ 二聚体，$[(bipy)_2(H_2O)Ru^{3+}\text{-}ORu^{3+}(H_2O)(bipy)_2]^{4+}$，其中 bipy=2,2-联吡啶。已经证明，使用 Ce(Ⅳ) 作为牺牲氧化剂，该复合物以每秒 0.0042 分子 O_2 的速率产生 O_2，平均进行 13 次转换（Gilbert et al.，1985）。该催化剂氧化水的机理仍然是争论的主题。一系列氧桥连 Ru 二聚体催化剂也随之被报道。这些催化剂是由胺基配体桥联的双核 Ru 配合物，例如 $[(terpy)_2(H_2O)Ru(dpp)Ru(H_2O)(terpy)_2]^{3+}$，其中 terpy = 1,2':6',2''−吡啶和 dpp = 2,4-（联吡啶）吡唑。据报道，这些配合物在氧化水方面比氧桥连二聚体具有更快的速率和更高的转换率。使用 Ce^{4+} 作为氧化剂，它们表现出每秒 0.014 分子 O_2 的析氧速率和 18.6 次的转换数（Sens et al.，2004）。增加的速率和稳定性似乎得到了胺桥的辅助，已经提出通过将两个钌中心紧密靠近来减缓催化剂分解并促进水氧化。由（萘啶基）吡啶基−哒嗪配体和四个轴向取代的吡啶配体支持的双核 Ru 复合物已显示，使用 Ce^{4+} 作为氧化剂能够以每秒 7.7×10^{-4} 分子 O_2 的速率进行水氧化催化，进行 3200 次转换（Zong & Thummel，2005）。

一类新的分子催化剂具有由两个苯基吡啶和两个水配体 [Ir(R_1R_2 苯基吡啶)$_2(H_2O)_2]^+$ 支持的单个 Ir^{3+} 中心。在均相溶液中，使用 Ce^{4+} 作为牺牲氧化剂，这些配合物的最大转换频率为每秒 1.5×10^{-3} 分子 O_2，转换数为 2500 次。通过改变苯基吡啶配体的侧链官能团，这些复合物的电化学性质可以被调整；笔者认为，这可以促进与光敏剂或电极表面的有效耦合。

几种 Mn 分子水氧化催化剂的制备受到 PSⅡ-WOC 的启发。文献中描述了多种二价和三价 Mn 配合物作为水氧化的“催化剂”。然而，很少有这些复合物可以获得水分解使氧气释放的明确证据，甚至更少的证据表明，其能够产生多次转换。最活跃的配合物是具有 $[Mn_2O_2]^{n+}$ 核的配合物。最值得注意的例子（Brimblecombe et al.，2009）是 $[Mn_2^{II}(mcbpen)_2(H_2O)_2]^{2+}$(mcbpen = *N*–甲基–*N*–羧甲基–*N*,*N*'–双（2–吡啶基–甲基）乙烷–1,2–二胺、$[(tpy)(H_2O)Mn(\mu\text{-}O)_2Mn(H_2O)(tpy)]^{3+}$ (tpy = 2,2':6',2''–三吡啶) 和 $[Mn_2(OAc)_2(bpmp)]^+$(bpmp = *N*,*N*-bis((6–甲基吡啶–2–基) 甲基)-*N*-2–吡啶基甲胺)。当使用牺牲的化学氧化剂如活化的过氧化物、次氯酸盐、Pb^{4+} 或 Ce^{4+} 时，这些化合物的最大转化率为每秒 1.6～2.7×10^{-3} 分子 O_2，并具有中等的转换数（通常 <20 次）。事实上，在没有来自牺牲氧化剂的电子转移时，这些配合物很少能够直接氧化 H_2O。在均相溶液中，$[(tpy(H_2O)Mn(\mu\text{-}O)_2Mn(H_2O)(tpy)]^{3+}$ 的活性取决于 $2e^-$O–原子供体氧化剂的存在。例如 HSO_5^-(oxone) 或 ClO^-使用一种电子氧化剂（如 Ce^{4+}），当复合物被支撑在黏土矿物的孔隙中时，观察到 17 次转换。在这种情况下，O_2 的释放速率对黏土层内催化剂浓度的平方的依赖性表明 H_2O 氧化可能涉及二聚体–二聚体相互作用。此外，催化剂分解涉及二聚体离解成单体，在黏土层内较慢。斯蒂林（Styring）、库尔茨（Kurz）及其同事（Beckmann et al.，2008）提供了 $[Mn_2(OAc)_2(bpmp)]^+$ 与 HSO_5^-或 $Pb(OAc)_4$ 组合使用时 H_2O 直接氧化的证据。岛崎优一（Shimazaki）等的早期研究发现，二锰（Ⅲ）–卟啉二聚体被间氯过苯甲酸（mCPBA）氧化成双核 $Mn^{5+}{=}O$ 络合物，其在加入酸时几乎按化学剂量地释放氧。然而，在这些研究中未证实催化转换（Shimazaki et al.，2004）。

分子催化剂通常与牺牲氧化剂结合使用。这些牺牲氧化剂有如下特点：①在反应过程中是无害的；②在催化反应过程中消耗掉；③在其操作中能量效率低（例如，Ce^{4+} 的氧化电位为 1.72 V *vs.* NHE）。这些氧化剂尽管可用于探测分子种类可能的催化活性，并阐明它们在溶液中的反应机理，但在水分解光电化学装置中使用这种化学氧化剂是不切实际的。

3. 光电化学电池的分子催化剂

为了在电化学装置中实现持续的水氧化，分子催化剂必须易于被电极表面氧化。只有少数报道描述了在电极表面通过直接电子转移氧化的分子催化剂，然后在均相溶液中继续催化水氧化。这些体系中的催化作用受催化剂向电极的扩散和催化剂与电极之间电子转移的效率的限制。为了克服这些局限并优化催化效果，需要将催化剂固定在电极表面。

催化中心与电子的有效耦合也是PSⅡ-WOC成功的关键。在该系统中，从WOC中迁移电子导致水氧化是通过一系列电子载体发生的，最终使用电子与水氧化期间形成的质子一起将二氧化碳转化为碳水化合物。水分解光电化学电池的目的类似于氧化水，产生氧、电子和质子，然后在对电极处重新结合以形成氢气。用于制氢的生物分子催化剂需要催化剂电耦合到导电表面。这种耦合有助于在水氧化期间有效地迁移和得到电子。装置配置需要促进气态产物的分离。

（三）水还原催化剂

1. 钴基催化剂

自20世纪70年代早期起，人们报道了许多钴配合物作为水还原催化剂，其中包括含有以下配体的体系：二烯-N_4和四烯-N_4大环、卟啉、双/多聚吡啶、η^5-环戊二烯、膦和二甲基乙二肟及相关衍生物。大多数这些钴水还原催化剂用于电催化产氢，其中质子源是有机酸而不是水。

1980年，费希尔（Fisher）和艾森伯格（Eisenberg）报道了两个早期的例子，它们使用均相Co大环配合物同时电催化还原CO_2和H_2O以产生CO与H_2。这些配合物能够在1.5～1.6 V *vs.* SCE的电位范围内还原CO_2和H_2O，电流效率高达93%。然而，总转换数（turnover number，TON）和TOF较低，并且反应存在较大的过电位。对照实验表明，任何一种催化剂对电解时CO和H_2的产生都是至关重要的，并且H_2O也是该系统的必要组分，因为当在CO_2存在下，在干燥的Me_2SO中进行电解时没有检测到CO或H_2。

1981年，克里希南（Krishnan）和萨廷（Sutin）报道了第一个使用钴联吡啶配合物作为催化剂、$Ru(bpy)_3^{2+}$作为光敏剂、抗坏血酸作为牺牲电子供体的光催化体系。在反应过程中，Co(Ⅰ) 联吡啶物种被认为是通过激光闪光光解的关键中间体。形成 H_2 的最佳 pH 为 5.0，这可能有利于形成氢化钴中间体。同位素实验证实，产生的氢源自水中的质子。

后来使用相同的催化剂进行 CO_2 和 H_2O 还原以产生 CO 与 H_2，但驱动力是可见光而不是电化学势。在该系统中，$Ru(bpy)_3^{2+}$ 用作光敏剂，胺［三乙胺（TEA）或三乙醇胺］是牺牲电子供体。该系统同时产生 CO 和 H_2，CO 和 H_2 的相对量与溶解的 CO_2 浓度成比例。

伯恩哈特（Bernhard）及其同事报道了一系列使用含 Ir 光敏剂、$Co(bpy)_3^{2+}$ 作为催化剂、三乙醇胺作为牺牲电子供体的可见光驱动的 H_2O 中 H_2 生成的系统。与经典的 $Ru(bpy)_3^{2+}$ 光敏剂相比，Ir 光敏剂具有更高的活性，H_2 生产的量子效率高达 37 倍。在光敏剂存在下实现了高达 9000 次的 TON，而在催化剂存在下可达到 74 次的 TON。

莱恩及其同事于 1983 年报道了一种基于钴肟的还原水催化剂的早期实例。该系统以 $Ru(bpy)_3^{2+}$ 作为光敏剂，三乙醇胺作为牺牲剂的供体。虽然在 pH ≈ 9.0 的 *N,N*-二甲基甲酰胺（DMF）/H_2O 混合物中以每小时 16 次转换的速率形成 H_2，但是 Co(Ⅱ) 二甲基乙二醛催化剂的不稳定性和空气敏感性在当时限制了该实验的进一步研究。

詹姆斯·埃斯彭森（James H. Espenson）最初在 1986 年报道了使用更稳定的 BF_2-环化的钴肟用于制氢，并且在过去的 10 年中由彼得斯（Peters）、阿尔泰罗（Artero）和丰塔克（Fontaceve）改良，用于电催化质子还原。阿尔泰罗和丰塔克也报道了基于相关的钴肟催化剂的光化学产生氢气的相关系统，但是在所有这些系统中，催化剂在非水介质中操作，其中添加酸或有机铵盐作为质子源。

2008 年，杜平武和艾森伯格报道了具有轴向吡啶配体的 H 桥联 Co(Ⅲ) 双（二甲基乙二醛）配合物作为 $MeCN/H_2O$ 的混合溶液中可见光驱动水还原的催化剂，并将该材料与光敏剂和三乙醇胺牺牲电子供体结合，构成光催化制氢系统。

在可见光驱动的氢气生产系统中成功应用 Co(Ⅲ) 二甲基乙二醛配合物后，艾森伯格及其同事发现了第一个仅基于地球富含元素用可见光驱动水生成氢气的均相系统，在使用有机染料（曙红 Y 或玫瑰红）作为光敏剂、Co(Ⅲ) 二甲基乙二醛配合物作为催化剂和三乙醇胺作为牺牲供体的情况下，在照射 12 h 内，在该系统获得 900 次的 TON。机理研究表明，通过重原子效应获得的那些有机染料中最低的三重态对电子转移有着非常重要的作用，以实现 Co(Ⅲ) 还原和随后的涉及 Co(Ⅱ)、Co(Ⅰ) 的析氢反应和 Co(Ⅲ) 氢化物中间体生成反应。荧光素没有重原子，因此导致三重态最小化，在类似系统中不产生氢。在进一步的研究中，当使用罗丹明染料（用 S 或 Se 取代黄原烯环中的 O）作为光敏剂时，实现了高达 9000 次的 TON。孙立成等研究了一种类似的系统，其中将曙红 Y 或玫瑰红作为光敏剂、BF_2-环化的钴肟作为催化剂、三乙醇胺作为牺牲供体，在 5 h 内达到的最高 TON 为 327 次。

2011 年，麦克纳马拉（McNamara）等报道用于水的光催化和电催化产氢的钴二硫杂环戊二烯络合物的例子。该配合物是一种非常活泼的催化剂，具有高达 2700 次的 TON（基于催化剂）和 880 mol H_2 /mol 催化剂 /h 的初始周转速率。当使用 $Ru(bpy)_3^{2+}$ 作为光敏剂和抗坏血酸作为牺牲剂的供体，并当电位保持在 1.0 V *vs.* SCE 时，该化合物在 MeCN / H_2O 混合溶液中也是一个活性电催化剂。斯图伯特（Stubbert）等发现，钴双（亚氨基吡啶）络合物也是一种高活性的电子催化剂，用于水性质子还原。在水溶液缓冲溶液中，在 1.4 V *vs.* SCE 恒定电位下电解获得的法拉第效率高达 87 % ± 10%。

水还原的另一个方法涉及建立分子光化学装置，用于直接从水中产生可见光驱动的氢气。阿尔泰罗和丰塔克于 2008 年报道了第一种钴基光化学分子器件［$Ru(bpy)_3^{2+}$ 衍生物和钴肟的组装］。在该体系中，铵盐（对氰基苯胺或三甲基氯化铵）作为质子源。后来他们用铱光敏剂代替 $Ru(bpy)_3^{2+}$，但实验仍在铵盐作为质子源存在下进行。后来报道了另外两个光化学装置的例子，用于直接使用水作为质子源的可见光驱动的氢生产。一种体系含有卟啉（作为光敏剂）和带有轴向吡啶配体（作为催化剂）的氢桥联钴肟；另一种体系是铱光敏剂和钴多吡啶络合物催化剂。然而，两种体系仅表现出中等的水分解产生氢转换水平（TON <30 次）。

除分子光化学装置外，还开发了基于 TiO_2 半导体的非均相光化学装置。拉卡达玛利（Lakadamyali）和莱斯纳（Reisner）报道了一种基于 TiO_2 的光化学装置，其由功能化的 $Ru(bpy)_3^{2+}$ 作为光敏剂锚定，由 [Co(Ⅲ)(dmgH)$_2$（吡啶基-4-磷酸酯）Cl] 作为催化剂。组装该系统非常容易，该系统在中性水（无有机溶剂）中、可见光照射下工作。尽管该系统的 TOF 似乎很低（$0.005\ s^{-1}$），但是改善的 TiO_2 半导体在约 400 mmol H_2 /（$h \cdot g_{TiO_2}$）的情况下表现出非常好的活性水平。

2. 镍基催化剂

镍是制造水还原催化剂的一种元素。大自然利用镍来制造 [NiFe] 氢化酶，用于质子还原和氢氧化。通过单晶 X 射线衍射测定 [NiFe] 氢化酶活性位点的结构。受自然界的启发，已开发出许多 [NiFe] 氢化酶模型配合物和镍基功能配合物。然而，很少有例子证明其可用于催化体系中的氢生产。劳赫富斯（Rauchfuss）小组报道了几种用于从酸中电催化质子还原的镍铁配合物，但到目前为止，当水是质子源时，没有将 [NiFe] 氢化酶的模型复合物用于产氢的报道。在本节中，我们仅介绍一些含有镍的功能模型配合物，用于从水中生产氢气。

在费希尔和艾森伯格首次报告均相 Ni(Ⅱ) 大环催化剂同时进行电催化 CO_2 和 H_2O 还原以生成 CO 与 H_2 后，科林（Collin）和索瓦热描述了两种相关的 Ni(Ⅱ) 大环配合物——$Ni(cyclam)^{2+}$ 和 $Ni_2(biscyclam)_2{}^{4+}$，用于相同的电催化 CO_2 和 H_2O 还原。在固定电位（−1.25 V *vs.* SCE）下，通过对比两种复合物的活性，他们提出了涉及镍氢化物中间体的反应机理。

杜波依斯（DuBois）等设计了一系列镍基分子催化剂，包括 $[Ni(P_2{}^{Ar}N_2{}^{Ar})_2](BF_4)_2$，其中每个配体含有两个膦配位点和两个非配位碱性位点（胺基），用于电催化产氢。研究人员认为，这些基本位点受 [FeFe] 氢化酶中类似的侧胺的启发，在实现高催化活性方面起着非常重要的作用。据报道，一种令人印象深刻的仅含有两个胺侧基的镍催化剂可用于铵盐的电催化质子还原。在相关工作中，阿尔泰罗及其同事报道了一种功能化的碳纳米管电极与杜波依斯的第一代镍催化剂一起用于电催化产氢。据报道，上述系统均未直接从水中形成 H_2（需要添加其他质子源），但其中一些在少量水（<0.5 mol/L）存在下起到了质子还原的作用。

麦克劳克林（McLaughlin）等报道了利用杜波依斯的镍（Ⅱ）基催化剂从水中光驱动产生氢的一个实例。在可见光照射下，该系统以曙红 Y 或 $Ru(bpy)_3{}^{2+}$ 作为光敏剂，以抗坏血酸作为牺牲电子供体，在混合的 MeCN∶H_2O (V∶V = 1∶1) 条件下产生氢气。在催化剂存在下，当曙红 Y 是光敏剂时，获得了高达 2700 次的 TON，这是迄今文献中报道的由无贵金属材料组成的 TON 最高的光催化体系。

3. 铁基催化剂

铁可能是自然界中最丰富的过渡金属之一，从生物学的角度来看，也是最重要的元素之一。所有氢化酶（[FeFe] 氢化酶、[NiFe] 氢化酶）含有的铁作为活性位点组分之一。氢化酶以每个位点每秒 9000 个分子的速率催化 H^+ 向 H_2 的相互转化。为了模拟氢化酶的功能，科学家正在合成和研究 [FeFe]

氢化酶、[NiFe] 氢化酶和仅有 [Fe] 氢化酶活性位点的模型复合物，已经研究了这些模型复合物中的一些，用于电催化质子产生，但是迄今很少有直接水还原的实例出现，并且所有这些已经发现的实例的电位都比平衡的热力学值更负，这是 pH 依赖性决定的。

最近，一类使用简单的羰基铁络合物进行光驱水还原的催化系统被报道。这些羰基铁催化剂包括 $[\{CpFe(CO)_2\}_2]$、$[Fe(CO)_5]$、$[Fe_2(CO)_9]$、$[(cot)Fe(CO)_3]$ 和 $[Fe_3(CO)_{12}]$。在这项工作中，$[Ir(bpy)(ppy)_2]$ PF_6 和 TEA 分别用作光敏剂与牺牲电子给体。在照射下，$[Fe_3(CO)_{12}]$ 体系以在 6 h 内产生的 H_2 的速率为 510 次的 TON 而被证明具有最高活性，进而提出一种羰基氢化铁作为关键反应中间体。后来通过添加不同的膦配体对同一体系进行改进，在 24 h 内产生高达 1500 次的 TON 和 13.4 光子转换效率，但它们在催化中的作用还未完全清楚。

三、面临的问题与前景

数十年来，金属络合物在分解水中的应用已经发展成为一个非常活跃且备受关注的研究课题。毫无疑问，这项研究的动力来自开发传统制氢技术的替代方法的愿望，以确保人类快速利用可更新的资源提供充足的能源供应。

许多研究表明，金属络合物由于独特的光物理和电化学性质，可以作为光催化 H_2 生产的高效光敏剂。虽然在用于光诱导 H_2 生成的金属络合物光敏剂的开发方面已经取得了很大进展，但是需要进一步优化它们的光物理和电化学性质以提高光催化体系的效率。一些主要问题如下。

（1）如何科学评价 H_2 生成系统的光催化性能？目前，在该领域的许多工作中，TON 值用于描述催化体系的性能。TON 反映了催化体系在工作条件下的稳定性。在相似反应条件下，具有相似类型结构或类似构造结构的光催化材料中的 TON 的比较可以提供关于反应的一些有用信息。该参数与照射时间有关，即较高的 TON 通常意味着较长的反应时间，较高的 TON 表示光催化 H_2 生成系统的耐久性较好。实际上，对于催化反应，应考虑反映催化效率的另一个参数，即 TOF，其基于活性位点的数量，找到控制高 TOF 的主要因素非常重要。除 TON 和 TOF 外，还应确定量子产率以评估光催化 H_2 生产系统的活性。

（2）进一步探索金属络合物光敏剂的结构-性质相关性。开发高效太阳能 H_2 生成系统所涉及的困难之一是确定金属络合物光敏剂的结构-性质相关性。许多因素可以影响金属络合物光敏剂的光催化活性，如激发态寿命、摩尔吸光系数、氧化还原电位和发光性质。尽管一些研究人员已经将注意力集中在这些因素与光催化性能之间的关系上，但由于其分子结构-性质关系尚不清楚，因此仍需进一步研究。显然，更好地理解金属络合物的结构-功能相关性有助于构建新型金属络合物作为用于光捕获的光敏剂和用于高效太阳能-氢生成的催化剂。

（3）设计和制造不含贵金属的金属配合物光敏剂。目前，人们已经使用贵金属配合物作为光敏剂构建了大多数高效均相 / 多组分光催化 H_2 生产系统。例如，$[Ru(bpy)_3]^{2+}$、$[Ir(ppy)_2(bpy)]^+$ 和 $[Pt(terpy)]^+$ 的衍生物作为光催化 H_2 进化的高效光敏剂已被广泛研究。其中，Ir(Ⅲ) 配合物是最有效的光敏剂，因为它们具有优异的激发态特性，帕克（Park）等观察到 Ir(Ⅲ) 光敏剂的最大 TON 为 17 000 次。然而，这些基于贵金属的光敏剂原料极其宝贵且稀缺，无法大规模用于光催化制氢。因此，非常希望开发出不含贵金属的金属络合物光敏剂。近年来，已经设计和开发了许多基于地球含量丰富的金属元素的新型金属络合物光敏剂，用于从水中光催化制氢。例如，铝原卟啉和 Cu(Ⅰ) 配合物首次被用作光催化制氢的有效光敏剂。不幸的是，这些光催化体系具有低活性和短寿命。将来，可以预期更智能的设计和合成用于光催化制氢的其他不含贵金属的复合光敏剂。例如新型锌卟啉配合物，作为用于染料敏化太阳能电池的高效光敏剂，其有望成为用于光催化制氢的贵金属配合物的合适替代品。

（4）基于金属络合物光敏剂的光催化整体水分解体系的构建。在均相 / 多组分光催化 H_2 生产系统领域的大多数工作集中在水还原半反应系统上，需要额外的牺牲还原剂（如 TEOA、TEA 或 H_2A）。对于实际应用，水还原半反应系统的光催化 H_2 产生研究是没有意义的，因为它需要高成本的牺牲还原剂。原则上，在可见光和近红外光照射下，光催化整体水分解是可能的，因为 H_2/H_2O 和 H_2O/O_2 半反应的电位差仅为 1.23 V。然而，光催化整体水分解仍然具有挑战性。在均相和非均相光催化体系中，成功开发光催化整体水分解体系的最大瓶颈是水氧化半反应，这在热力学上非常苛刻。一些具有低量子产率的均相光催化水氧化体系已建成。未来，更多的项目将聚焦水氧化研究和整体水分解系统光催化剂研究。

（5）开发基于金属络合物光敏剂的光电化学电池用于水分解。在半反应体系中获得的高量子产率，意味着半反应体系的必要组分（光敏剂和催化剂）已经在光催化整体水分解中存在，至少在概念验证水平上缺一不可。如果电荷-复合反应足够慢，光催化水分解将变得类似。金属络合物光敏剂的电荷复合在敏化半导体系统中得到了有效的抑制，因此，金属络合物光敏剂和半导体的组合对开发高效的光催化整体水分解非常重要。

催化剂是实现高效能量转换和储存的关键因素之一。在过去几十年中，使用由过渡金属元素（Co、Ni 和 Fe）制成的材料进行水分解取得了很大进展。据报道，使用钴肟水还原催化剂从水中生产氢气已达到 9000 次的 TON，并且已经发现使用钴基水氧化催化剂生产氧气已高达 1000 次的 TON。对于镍基材料，产氢已经达到超过 2700 次的 TON，并且已开发出用于水氧化的氧化镍膜。铁可能是制造水分解催化剂的研究者最感兴趣的元素之一。在自然界中，氢化酶和单加氧酶是两类含铁蛋白质，它们分别使用铁的氧化还原能力来高效活化氢和氧。据报道，对于铁基催化剂，水中产氢活性高达 1500 次的 TON，水中产氧超过 1000 次的 TON。对于所有基于 Co、Ni 和 Fe 的水还原催化剂，提出金属氢化物为作用机理中的关键反应中间体，并已通过光谱检测了其中一些中间体。在水氧化反应中，已经提出 Co(Ⅳ) 和 Ni(Ⅲ) 类分别在使用 Co-水氧化催化剂和 Ni-水氧化催化剂的析氧过程中作为关键中间体。虽然没有建立基于 Fe-水氧化催化剂的水氧化反应中的关键中间体，但有关通过氧化酶使 O_2 活化的大量文献指出了在这类反应中 Fe═O 的重要性。

除上述水分解的进展外，合成化学家和物理化学家基于土壤中的丰富元素 Co、Ni 和 Fe 来设计经济上可行的分子与非均相催化剂仍然是一个至关重要的挑战。显而易见的是，在该领域将开展更多的工作以扩展基于这些元素的催化剂的活性、效率和稳健性。对于上述用于还原水和水氧化的大多数分子催化剂，仍存在许多常见问题。第一，稳定性是一个关键问题，据报道，催化过程中分子催化剂的寿命从几分钟到几天不等，催化剂通常在反应过程中分解。实际应用中要求催化剂的寿命很长，以容纳数百万次到数十亿次的转换数。第二，空气敏感性是另一个主要问题。对于水还原催化剂，这些分子催化剂的还原形式，特别是在有机介质中通常是对空气敏感的，大气中的 O_2 能够快速氧化它们。理想的还原水催化剂应该比氧化更快地促进产氢，以避免分解或失活。第三，耐水性是许多报道的产生 H_2 的催化剂的重要问题。虽然报道的一些分子催化剂在有机溶剂中表现良好，但很明显，水是水分解的首选溶剂。迄今，许多在有机溶剂中可发挥良好作用的分子催化剂在仅有少量水存在的环境下效率会迅速下降。第四，电催化分解水的工作电位也非常重要。理想的工作电位应尽可能接近水还原和水氧化的两个半反应的热力学势，如果工作电位过大，则过程中会损失一些能量（即工作电位与所需最小电位之间的差异），半反应会失效。如何将过电位降低到水分解的最小值是设计更好的催化剂的挑战。对于非均相催化剂，相关的问题是效率。

然而，对于光驱动的半反应，过电位和反应障碍的要求会发生变化。在 H_2 的光生成中通过从激发态光敏剂（PS^*）到催化剂的电子转移（氧化淬灭）或通过从牺牲供体到 PS^* 的电子转移（还原淬灭），PS^* 将经历电子转移淬灭。每个步骤的可行性由催化剂、牺牲供体和 PS^* 的氧化还原电位决定，氧化和还原淬灭路径的相对速率将由特定的淬灭速率常数与淬灭剂浓度控制。在诸如这些的系统中，用于产生 H_2 的催化剂的过电位不是直接因素。相反，它有助于确定可遵循哪种光化学路径。

在本节所述的研究中，我们主要关注分子系统在析氧电催化反应中的应用，即质子还原为 H_2 或水中 O^{2-} 氧化为 O_2。这些研究的主要目标是开发每个半反应的组分，使其具有活性且高效和稳定。处理水分解半反应的研究中的实际能量储存不是或不应该是主要的研究目标，并且在某些情况下涉及光和牺牲电子供体或受体，甚至不能实现净能量储存。实际上，对于质子还原为 H_2，诸如叔胺的牺牲供体的氧化可能提供产生氢所需的大部分潜在能量，并且对于水中 O^{2-} 氧化成 O_2，电子受体如过硫酸盐是化学上足够强的氧化剂，因此不需要光来驱动反应。虽然牺牲供体 / 受体的消耗似乎是半反应系统的主要缺点，但通过这些水分解半反应的研究，可以开发出通过水分解存储太阳能的实际系统。一旦在稳定性和活性方面充分开发了用于进行两个半反应的系统，它们将需要通过导电膜连接。该系统还需要允许质子从一个半反应室移动到另一个半反应室，以便不会发生显著的质子梯度。

已经证明，将固体层沉积到导电电极上是研究分子的电催化潜力的有效手段。然而，由于固体层的不稳定性，这种方法似乎在光电化学装置中具有有限的用途。将催化剂掺杂到聚合物膜中比将固体层沉积在导电电极上会提供更大的稳定性。控制电极表面催化剂浓度的最新进展表明，该技术可成功应用于能够产生高电流密度的系统的开发中。然而，我们注意到，对于 Ru 配合物，在高催化剂负载下竞争的双分子分解反应通常会降低其催化效率。相反，增加层状材料中 Mn 催化剂的负载量可以提高氧气的产生速率。因此，报道的将 Mn 催化剂加入这些聚合物层中可在较高催化剂浓度下取得实际成功。最近在光催化组件方面的研究工作证明了这种方法的潜力，该组件利用生物激发的立方烷簇作为水氧化催化剂。

分子催化剂与电极表面的共价连接代表了开发高效太阳能水分解装置的另一种有希望的方法。在这方面，最显著的例子为通过共价连接的 Ru 催化剂成功地催化水氧化，每个催化剂催化产生了多达三个的氧分子。此外，已经有研究成功实现了几个潜在催化簇的光敏化，在光电化学电池中实现有效、持续、光驱动的分子水氧化所需的各个过程。然而，迄今电极结合的分子催化剂的高效光驱动水氧化仍未得到报道。鉴于分子催化剂的最新进展以及准分子系统的报道，该系统将吸附在 TiO_2 纳米颗粒上的 Ru 染料与 IrO_2 催化剂结合，以实现光驱动的水氧化。我们预期，电极结合水氧化催化剂与光敏化电极将在不久的将来实现。

于振涛（南京大学）

本节参考文献

Armaroli N, Balzani V. 2011. The hydrogen issue. ChemSusChem, 4: 21-36.

Beckmann K, Uchtenhagen H, Berggren G, et al. 2008. Formation of stoichiometrically ^{18}O-labelled oxygen from the oxidation of ^{18}O-enriched water mediated by a dinuclear manganese complex—a mass spectrometry and EPR study. Energy & Environmental Science, 1: 668-676.

Berardi S, Drouet S, Francàs L, et al. 2014. Molecular artificial photosynthesis. Chemical Society Reviews, 43: 7501-7519.

Brimblecombe R, Dismukes G C, Swiegers G F, et al. 2009. Molecular water-oxidationcatalysts for photoelectrochemical cells. Dalton Transactions, 9374-9384.

Brown G M, Brunschwig B S, Creutz C, et al. 1979. Homogeneous catalysis of the photoreduction of water by visible light. Mediation by a tris(2,2'-bipyridine)ruthenium(Ⅱ)-cobalt(Ⅱ) macrocycle system. Journal of the American Chemical Society, 101: 1298-1300.

Brown G M, Chan S F, Creutz C, et al. 1979. Mechanism of the formation of dihydrogen from the photoinduced reactions of tris(bipyridine)ruthenium(Ⅱ) with tris(bipyridine)rhodium(Ⅲ). Journal of the American Chemical Society, 101: 7638-7640.

Chan S F, Chou M, Creutz C, et al. 1981. Mechanism of the formation of dihydrogen from the photoinduced reactions of poly(pyridine)ruthenium(Ⅱ) and poly(pyridine)rhodium(Ⅲ) complexes. Journal of the American Chemical Society, 103: 369-379.

Chang K, Mei Z, Wang T, et al. 2014. MoS_2/graphene cocatalyst for efficient photocatalytic H_2 evolution under visible light irradiation. ACS Nano, 8: 7078-7087.

Chu S, Majumdar A. 2012. Opportunities and challenges for a sustainable energy future. Nature, 488: 294-303.

Ciamician G. 1912. The photochemistry of the future. Science, 36: 385-394.

DeLaive P J, Sullivan B P, Meyer T J, et al. 1979. Applications of light-induced electron-transfer reactions. Coupling of hydrogen generation with photoreduction of ruthenium(Ⅱ) complexes by triethylamine. Journal of the American Chemical Society, 101: 4007-4008.

Esswein A J, Nocera D G. 2007. Hydrogen production by molecular photocatalysis. Chemical Reviews, 107: 4022-4047.

Fujishima A, Honda K. 1972. Electrochemical photolysis of water at a semiconductor electrode. Nature, 238: 37-38.

Fujito H, Kunioku H, Kato D, et al. 2016. Layered perovskite oxychloride Bi_4NbO_8Cl: a stable visible light responsive photocatalyst for water splitting. Journal of the American Chemical Society, 138: 2082-2085.

Gilbert J A, Eggleston D S, Murphy W R, et al. 1985.Structure and redox properties of the water-oxidation catalyst $[(bpy)_2(OH_2)RuORu(OH_2)(bpy)_2]^{4+}$. Journal of the American Chemical Society, 107: 3855-3864.

Gong L M, Wang J, Li H, et al. 2011. Acriflavine-cobaloxime-triethanolamine homogeneous photocatalytic system for water splitting and the multiple effects of cobaloxime and triethanolamine. Catalysis Communications, 12: 1099-1103.

Hara M, Kondo T, Komoda M, et al. 1998. Cu_2O as a photocatalyst for overall water splitting under visible light irradiation. Chemical Communications, (3): 357-358.

Horiuchi Y, Toyao T, Saito M, et al. 2012. Visible-light-promoted photocatalytic hydrogen production by using an amino-functionalized Ti(Ⅳ) metal-organic framework. Journal of Physical Chemistry C, 116: 20848-20853.

Housecroft C E, Constable E C. 2015. The emergence of copper(Ⅰ)-based dye sensitized solar cells. Chemical Society Reviews, 44: 8386-8398.

Hu J Q, Liu A L, Jin H L, et al. 2015. A versatile strategy for shish-kebab-like multi-heterostructured chalcogenides and enhanced photocatalytic hydrogen evolution. Journal of the American Chemical Society, 137: 11004-11010.

Huang Z F, Song J J, Li K, et al. 2016. Hollow cobalt-based bimetallic sulfide polyhedra for efficient all-pH-value electrochemical and photocatalytic hydrogen evolution. Journal of the American Chemical Society, 138: 1359-1365.

Joya K S, Joya Y F, Ocakoglu K, et al. 2013. Water-splitting catalysis and solar fuel devices: artificial leaves on the move. Angewandte Chemie International Edition, 52: 10426-10437.

Kalyanasundaram K, Kiwi J, Grätzel M. 1978. Hydrogen evolution from water by visible light, a homogeneous three component test system for redox catalysis. Helvetica Chimica Acta, 61: 2720-2730.

Kärkäs M D, Verho O, Johnston E V, et al. 2014. Artificial photosynthesis: molecular systems for catalytic water oxidation. Chemical Reviews, 114: 11863-12001.

Kim D, Sakimoto K K, Hong D C, et al. 2015. Artificial photosynthesis for sustainable fuel and chemical production. Angewandte Chemie International Edition, 54: 3259-3266.

Kim D, Whang D R, Park S Y. 2016. Self-healing of molecular catalyst and photosensitizer on metal-organic framework: robust molecular system for photocatalytic H_2 evolution from water. Journal of the American Chemical Society, 138: 8698-8701.

Kirch M, Lehn J M, Sauvage J P. 1979. Metabolites of microorganisms. The aspochalasins A, B, C, and D. Helvetica Chimica Acta, 62: 1345-1384.

Krishnan C V, Brunschwig B S, Creutz C, et al. 1985. Homogeneous catalysis of the photoreduction of water. 6. Mediation by polypyridine complexes of ruthenium(Ⅱ) and cobalt(Ⅱ) in alkaline media. Journal of the American Chemical Society, 107: 2005-2015.

Krishnan C V, Sutin N. 1981. Homogeneous catalysis of the photoreduction of water by visible light. 2. Mediation by a tris(2,2'-bipyridine)ruthenium(Ⅱ)-cobalt(Ⅱ) bipyridine system. Journal of the American Chemical Society, 103: 2141-2142.

Kudo A, Miseki Y. 2009. Heterogeneous photocatalyst materials for water splitting. Chemical Society Reviews, 38: 253-278.

Lazarides T, McCormick T, Du P W, et al. 2009. Making hydrogen from water using a homogeneous system without noble metals. Journal of the American Chemical Society, 131: 9192-9194.

Lehn J M, Sauvage J P. 1977. Chemical storage of light energy-catalytic generation of hydrogen by visible-light or sunlight-irradiation of neutral aqueous-solutions. New Journal of Chemistry, 1: 449-451.

Lewis N S, Nocera D G. 2006. Powering the planet: chemical challenges in solar energy utilization. Proceedings of the National Academy of Sciences USA, 103: 15729-15735.

Li X Q, Wang M, Zheng D H, et al. 2012. Photocatalytic H_2 production in aqueous solution with host-guest inclusions formed by insertion of an FeFe-hydrogenase mimic and an organic dye into cyclodextrins. Energy & Environmental Science, 5: 8220-8224.

Li X, Yu J G, Jaroniec M. 2016. Hierarchical photocatalysts. Chemical Society Reviews, 45: 2603-2636.

Liang Q H, Li Z, Yu X L, et al. 2015. Macroscopic 3D porous graphitic carbon nitride monolith for enhanced photocatalytic hydrogen evolution. Advanced Materials, 27: 4634-4639.

Liao L B, Zhang Q H, Su Z H, et al. 2014. Efficient solar water-splitting using a nanocrystalline CoO photocatalyst. Nature Nanotechnology, 9: 69-73.

Ma Y, Wang X L, Jia Y S, et al. 2014. Titanium dioxide-based nanomaterials for photocatalytic fuel generations. Chemical Reviews, 114: 9987-10043.

McCormick T M, Calitree B D, Orchard A, et al. 2010. Reductive side of water splitting in artificial photosynthesis: new homogeneous photosystems of great activity and mechanistic insight. Journal of the American Chemical Society, 132: 15480-15483.

McNamara W R, Han Z J, Yin C J, et al. 2012. Cobalt-dithiolene complexes for the photocatalytic and electrocatalytic reduction of protons in aqueous solutions. Proceedings of the National Academy of Sciences USA, 109: 15594-15599.

Oshima T, Lu D L, Ishitani O, et al. 2015. Intercalation of highly dispersed metal nanoclusters into a layered metal oxide for photocatalytic overall water splitting. Angewandte Chemie International Edition, 54: 2698-2702.

Pagliaro M, Konstandopoulos A G, Ciriminna R, et al. 2010. Solar hydrogen: fuel of the near future. Energy & Environmental Science, 3: 279-287.

Prier C K, Rankic D A, MacMillan D W C. 2013. Visible light photoredox catalysis with transition metal complexes: applications in organic synthesis. Chemical Reviews, 113: 5322-5363.

Ran J R, Zhang J, Yu J G, et al. 2014. Earth-abundant cocatalysts for semiconductor-based photocatalytic water splitting. Chemical Society Reviews, 43: 7787-7812.

Sakai T, Mersch D, Reisner E. 2013. Photocatalytic hydrogen evolution with a hydrogenase in a mediator-free system under high levels of oxygen. Angewandte Chemie International Edition, 52: 12313-12316.

Sens C, Romero I, Rodriguez M, et al. 2004. A new Ru complex capable of catalytically oxidizing water to molecular dioxygen. Journal of the Amerrican Chemical Society, 126: 7798-7799.

Shimazaki Y, Nagano T, Takesue H, et al. 2004. Characterization of a dinuclear Mn^{V}=O complex and its efficient evolution of O_2 in the presence of water. Angewandte Chemie International Edition, 43: 98-100.

Swierk J R, Mallouk T E. 2013. Design and development of photoanodes for water-splitting dye-sensitized photoelectrochemical cells. Chemical Society Reviews, 42: 2357-2387.

Tachibana Y, Vayssieres L, Durrant J R. 2012. Artificial photosynthesis for solar water-splitting. Nature Photonics, 6: 511-518.

Tollefson J. 2010. Hydrogen vehicles: fuel of the future? Nature, 464: 1262-1264.

Tong H, Ouyang S X, Bi Y P, et al. 2012. Nano-photocatalytic materials: possibilities and challenges. Advanced Materials, 24: 229-251.

Ueno Y, Ohawara M. 1979. Desulfurizative stannylation of propargylic or allylic sulfides via an S_H' process. Journal of the American Chemical Society, 101: 1893-1894.

Wang F, Liang W J, Jian J X, et al. 2013. Highly luminescent and ultrastable $CsPbBr_3$ perovskite quantum dots incorporated into a silica/alumina monolith. Angewandte Chemie International Edition, 52: 8134-8138.

Wang H L, Zhang L S, Chen Z G, et al. 2014. Semiconductor heterojunction photocatalysts: design, construction, and photocatalytic performances. Chemical Society Reviews, 43: 5234-5244.

Wang M, Chen L, Sun L C. 2012. Recent progress in electrochemical hydrogen production with earth-abundant metal complexes as catalysts. Energy & Environmental Science, 5: 6763-6778.

Wang X C, Maeda K, Thomas A, et al. 2009. A metal-free polymeric photocatalyst for hydrogen production from water under visible light. Nature Materials, 8: 76-80.

Wang X, Xu Q, Li M R, et al. 2012. Photocatalytic overall water splitting promoted by an α-β phase junction on Ga_2O_3. Angewandte Chemie International Edition, 51: 13089-13092.

Wenger O S. 2009. Long-range electron transfer in artificial systems with d^6 and d^8 metal photosensitizers. Coordination Chemistry Reviews, 253: 1439-1457.

Willkomm J, Orchard K L, Reynal A, et al. 2016. Dye-sensitised semiconductors modified with molecular catalysts for light-driven H_2 production. Chemical Society Reviews, 45: 9-23.

Wu B H, Liu D Y, Mubeen S, et al. 2016. Anisotropic growth of TiO_2 onto gold nanorods for plasmon-enhanced hydrogen production from water reduction. Journal of the American Chemical Society, 138: 1114-1117.

Xiang Q J, Cheng B, Yu J G. 2015. Graphene-based photocatalysts for solar-fuel generation. Angewandte Chemie International Edition, 54: 11350-11366.

Xu Y, Kraft M, Xu R. 2016. Metal-free carbonaceous electrocatalysts and photocatalysts for water splitting. Chemical Society Reviews, 45: 3039-3052.

Yang J H, Wang D E, Han H X, et al. 2013. Roles of cocatalysts in photocatalysis and photoelectrocatalysis. Accounts of Chemical Research, 46: 1900-1909.

Yuan Y J, Ye Z J, Lu H W, et al. 2016. Engineering coexposed {001} and {101} facets in oxygen-deficient TiO_2 nanocrystals for enhanced CO_2 photoreduction under visible light. ACS Catalysis, 6: 532-541.

Zhang P, Zhang J J, Gong J L. 2014. Tantalum-based semiconductors for solar water splitting. Chemical Society Reviews, 43: 4395-4422.

Zhang W, Hong J D, Zheng J W, et al. 2011. Nickel-thiolate complex catalyst assembled in one step in water for solar H_2 production. Journal of the American Chemical Society, 133: 20680-20683.

Zhang X H, Yu L J, Zhuang C S, et al. 2014. Highly asymmetric phthalocyanine as a sensitizer of graphitic carbon nitride for extremely efficient photocatalytic H_2 production under near-infrared light. ACS Catalysis, 4: 162-170.

Zhang X J, Jin Z L, Li Y X, et al. 2009. Efficient photocatalytic hydrogen evolution from water without an electron mediator over Pt-rose bengal catalysts. Journal of Physical Chemistry C, 113: 2630-2635.

Zhao J Z, Wu W H, Sun J F, et al. 2013. Triplet photosensitizers: from molecular design to applications. Chemical Society Reviews, 42: 5323-5351.

Zhu M S, Li Z, Xiao B, et al. 2013. Surfactant assistance in improvement of photocatalytic hydrogen production with the porphyrin noncovalently functionalized graphene nanocomposite. ACS Applied Materials & Interfaces, 5: 1732-1740.

Zong R, Thummel R P.2005. A new family of Ru complexes for water oxidation. Journal of the American Chemical Society, 127: 12802-12803.

Zou Z, Ye J, Sayama K, et al. 2001. Direct splitting of water under visible light irradiation with an oxide semiconductor photocatalyst. Nature, 414: 625-627.

第三节　基于微生物的“半”人工光合成材料

一、基于微生物的“半”人工光合成材料研究概述

（一）基于微生物的“半”人工光合成材料研究的基本原理

自然界通过光合作用储存太阳能，即利用太阳能将水和二氧化碳转化为碳水化合物与氧气。利用人造材料模拟自然光合作用也能实现太阳能到化学能的转化，但人工光合成技术目前主要以氢气的形式将太阳能储存下来（Li et al.，2013；Yao et al.，2018；陈雅静等，2019），难以通过固定二氧化碳实现完整的卡尔文循环（Cook et al.，2010）。人工光合系统具有更强的可设计性和更大的太阳辐射波长利用范围，自然光合系统具有更强的催化选择性，因此，近年来一些研究人员开始尝试通过人工光合系统的功能组分（光吸收层、光电极）与自然的催化体系协同复合来实现从太阳能到化学能的转化，为实现自然光合作用与人工光合作用的耦合提供了一种新的策略，这种新的太阳能-化学能转化模式通常被称为“半”人工光合作用。

在详细介绍“半”人工光合作用前，首先需要介绍一下光合作用。光合作用按时间的先后顺序主要分为3个物理化学过程，即光能捕获、电荷产生与分离、催化反应。在自然光合作用中，光系统Ⅰ和光系统Ⅱ通过叶绿素捕获光能并将电子泵至激发态，在光系统Ⅱ的锰、钙-氧基团簇上形成空穴，发生水氧化反应产氧（Herek et al.，2002）。激发态电子传递到光系统Ⅰ后，

与光系统Ⅰ端的空穴结合，光系统Ⅰ上的激发态电子则参与后续的暗反应过程，这就是所谓的Z-scheme（Barber，2009）。在理想条件下，这些过程的电荷分离量子效率接近100%。然而自然光合作用的效率在约20%的标准太阳光强度下就会达到光饱和，更强的光照会导致生物光损伤（Zhu et al.，2008，2010），光损伤的修复消耗能量，同时降低太阳能-生物质的转化效率。自然光合作用的催化反应是在多个酶共同参与下的多种代谢途径的综合表现，可以利用二氧化碳、氮气、水等小分子选择性地合成大分子产物。但是自然光合作用的生物质转化效率一般不超过6%（农作物为1%～2%，大多数植物为0.1%）（Blankenship et al.，2011），因为自然界的光合生物优先进行光合生长而不是进行高能代谢产物的合成。当下兴起的合成生物学手段为提高生物质转化效率提供了一种合理有效的光合生物改造途径（Jagadevan et al.，2018；Kong et al.，2019；Kung et al.，2012）。此外，仿生无机纳米材料也可用于改造光合微生物（Xiong et al.，2013，2015；Jiang et al.，2015；Léonard et al.，2010；Ko et al.，2013；熊威等，2019）。

在人工光合作用方面，光捕获过程始于半导体材料的光吸收。相比于植物的叶绿素仅能选择性地捕获太阳光谱中的部分光（Cheng et al.，2018），这些半导体材料不仅可以实现宽太阳光谱吸收，还可以设计成叠层结构以实现太阳光谱的分段互补吸收。此外，半导体材料通过掺杂和异质结等手段可以实现高效的电荷分离，以及导电层的电荷传输。相比于自然光合系统，人工光合系统构成更加简易，更易于以模块化方式进行替换和改进。实际上，利用水和二氧化碳产生燃料的人工光合实验装置已经实现了远高于普通光合生物的太阳能转化效率（Khaselev & Turner，1998；Zhou et al.，2016；Verlage et al.，2015；Ager et al.，2015；Jia et al.，2016）。然而，上述人工光合系统需要使用昂贵的高纯度半导体材料，而且这些半导体材料在电解质溶液中长时间工作容易被分解或腐蚀，也没有自修复机制来缓解寿命问题，因此这种人工光合系统尚不具备大规模使用的可行性。基于目前人工光合系统的快速进展以及未来的前景，研究人员仍在努力探索和开发高效且价格合理的催化剂，以选择性地生成各种复杂的碳基和氮基化合物。同时，通过稳定反应中间体（Azcarate et al.，2016；Rosen et al.，2011；Gong et al.，2017；Kim et al.，2015）和控制界面反应物浓度（Hall et al.，2015；Ma et al.，2016）来模拟酶的工程策略已显现出一定的前景，但尚未实现高效率和高选择性的基于二氧化碳的C—C偶联还原反应。此外，虽然人工光合体系已经可以模拟自然光合作用的Z-scheme，但是几乎不能复制自然光合作用中跨膜pH梯度的产生或维持以驱动三磷酸腺苷（ATP）的合成（图 4-9）（Su & Vayssieres，2016）。

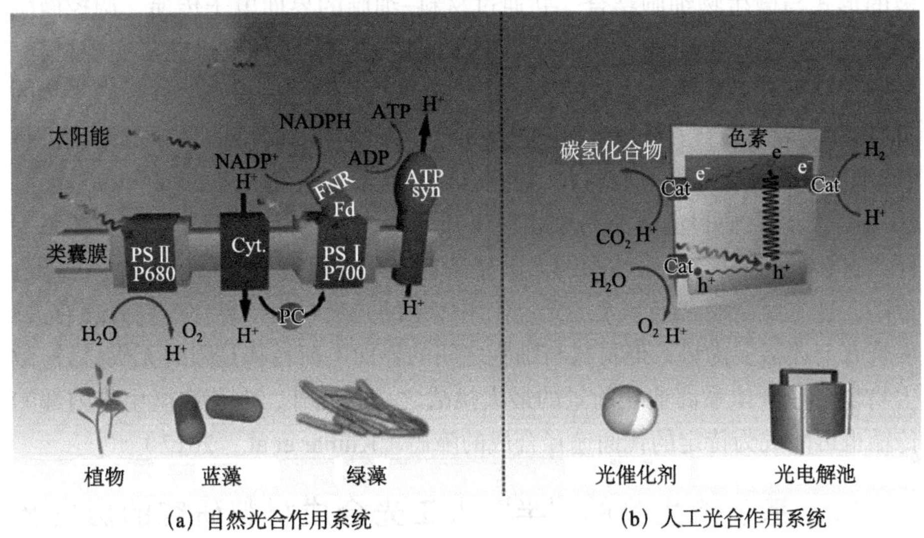

图 4-9　自然光合作用系统和人工光合作用系统示意图（Su & Vayssieres，2016）

ADP：腺苷二磷酸

（二）基于微生物的“半”人工光合成材料研究的意义

鉴于上述两种系统的局限性，“半”人工光合作用系统寻求整合自然光合系统（催化反应选择性高）和人工光合作用系统（光吸收效率高）各自的优势，旨在实现系统的优势互补。在此背景下产生了以无机材料-酶复合和无机材料-微生物复合形式构建的两类“半”人工光合作用系统，每种系统都具有其自身的优势和局限性。一方面，酶是人工催化剂设计的灵感来源；当酶与光电极或光吸收剂复合时，它们可以接近 100% 的选择性高速转换电荷，催化在动力学上难以实现的一系列简单产物的合成（Armstrong & Hirst，2011）。然而，分离酶的大量纯化过程（导致低可扩展性）及其固有的不稳定性（特别是在离体环境中）使得它们在太阳能转换中缺乏可持续的商业应用价值。另一方面，无机材料-微生物复合系统可以利用无机材料作为光吸收剂，微生物细胞作为“催化剂”，引导光能或电能通过微生物的代谢途径产生多种人工光合系统或材料-酶复合系统无法合成的复杂产物。酶和微生物的本质区别在于酶是蛋白质大分子，而微生物是生命体。无机材料-微生物复合的“半”人工光合系统中的无机材料主要以光催化材料为主，微生物主要以蓝藻和少数具有特殊代谢功能的细菌为主，无机材料以纳米颗粒或光电极的形式与微生物细胞结合，并通过材料-细胞的界面电子传递，调控微生物细胞的代谢产物合成。此外，借助合成生物学手段可以进一步实现材料-细胞复合光合系统产物生成的多样性与选择性。除这些优点之外，微生物的自我复制特性赋予了材料-细胞复合光合体系潜在的应用前景。虽然无机材料-细胞复合光合体系具有材料-酶复合体系所不具备的优势（Sakimoto et al., 2017），但是它们目前还是一个相对未知的系统。与材料-酶复合光合体系一样，材料-细胞复合光合体系的太阳光谱利用率（相比于非光合细胞）通常也很高，这主要是与“半”人工光合作用体系的光吸收由半导体组件处理有关。尽管有上述诸多优势，但材料与细胞之间的界面电荷转移是阻碍无机材料-微生物复合光合体系商业化前景的最大挑战之一。此外，电荷在微生物内部的传输也可能成为特定的代谢途径合成的障碍（Kumar et al., 2017）。

二、基于微生物的“半”人工光合成材料研究的历史及现状

针对自然光合作用和人工光合作用的研究都已经开展多年，但是任何一种途径都不能实现理想的太阳能-化学能转化和存储。在过去的 10 年里，“半”人工光合作用已经发展成为一个新兴的、涉及多学科的研究方向。其中，基于无机材料-酶复合的“半”人工光合体系是最早发展起来的，国内外都有课题组开展相关工作；基于无机材料-微生物复合的“半”人工光合体系的研究是最近几年才逐渐兴起的，美国目前处于领先地位。虽然国外在基于微生物的“半”人工光合作用领域已有多篇相关综述报道，但是国内对于这个领域的研究尚处于萌芽阶段（Kornienko et al., 2018；Sakimoto et al., 2018；Xu et al., 2018；Lee et al., 2019）。因此，本节主要针对基于微生物的“半”人工光合作用的最新进展进行系统的归纳和总结，并分析其主要的问题，展望未来这个领域的发展趋势，希望能为国内的研究人员提供一些启发和帮助。

（一）基于无机材料-微生物复合的“半”人工水氧化

在光合作用过程中，水氧化生成氧气的反应是化学反应。水是自然和人工光合作用的最终电子供体，并且水在自然界中大量存在，这使其成为生产太阳能燃料的最合适基质（Lewis & Nocera, 2006）。因此，理解水氧化反应的机制，进而提高其效率，对实现高度功能化以及高效率的自然 / 人工光合

作用是至关重要的。光系统Ⅱ是自然界中唯一能够氧化水的酶，它同时也能有效地吸收光、分离光激发的电荷并将它们引导至所需的终点。实际上，这种酶负责所有大气中的氧气产生。该功能使光系统Ⅱ成为探究水氧化反应机理的理想模型系统［图 4-10（a）］，因为在人工光合系统中，每一步反应的效率都是要追求极致的。但是光系统Ⅱ对光非常敏感，每隔约 15 min 在体

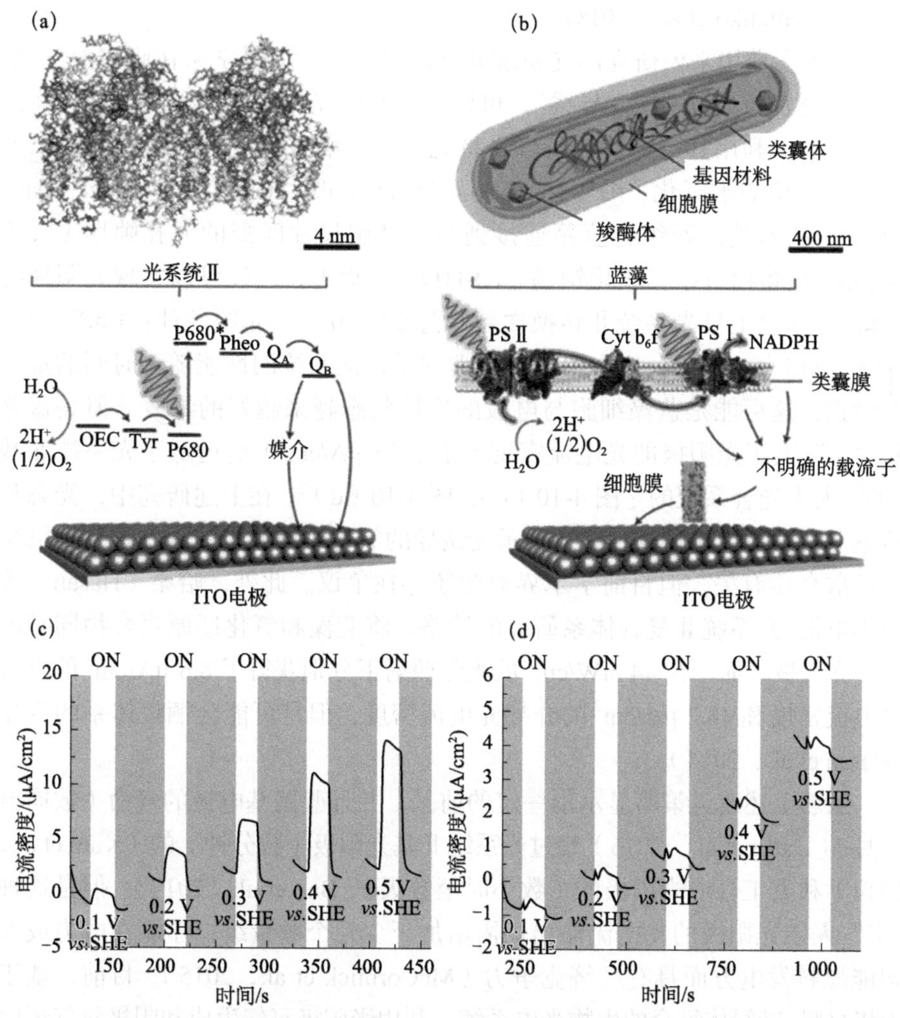

图 4-10 （a）IO-ITO| 光系统Ⅱ“半”人工光阳极示意图（Kornienko et al.，2018）；（b）IO-ITO| 蓝藻“半”人工光阳极示意图（Kornienko et al.，2018）；（c）IO-ITO|PS Ⅱ“半”人工光阳极在无外生介质和断续光照射下的多电位阶跃计时电流密度曲线（Zhang et al.，2018）；（d）IO-ITO| 蓝藻“半”人工光阳极在无外加导电介质和断续光照射下的多电位阶跃计时电流密度曲线（Zhang et al.，2018）。ON：光照

内修复一次，所以基于光系统Ⅱ的电极也具有类似的运行寿命（Kato et al.，2014）。目前，除与电极复合构成基于光系统Ⅱ的“半”人工光阳极之外，光系统Ⅱ也可以和光催化剂复合，实现“半”人工全水分解（Wang et al.，2014）。但是由于光系统Ⅱ存在提取过程烦琐、离体稳定性差和光吸收限制等问题，基于光系统Ⅱ的“半”人工光合作用系统更适合用来做模型研究，而不适合商业化应用（Kornienko et al.，2018）。

绝大多数用来做研究的光系统Ⅱ都取自蓝藻，但科学家在研究中发现，一些蓝藻表面具有“纳米导线”，可以将细胞内光合作用产生的电子传递到细胞表面。利用其电荷传输特性，将光合蓝藻贴在电极上培养繁殖可以实现自我维持的水氧化［图 4-10（b）］（Zhang et al.，2018；McCormick et al.，2015）。莱斯纳课题组将蓝藻连接到与其细胞尺寸匹配的大孔隙度（大于 10 μm）反蛋白石结构氧化铟锡（IO-ITO）电极上，已经可以实现光阳极电流持续 5d 以上且光系统Ⅱ转换次数超过 20 000 次（光照条件：1 mW/cm^2，λ=685 nm）（Zhang et al.，2018）。有趣的是，该系统的性能随着时间的推移而增强，这可能是蓝藻细胞与电极的连接状态越来越好的缘故。但是蓝藻基“半”人工光阳极的光电流密度（小于 15 μA/cm^2）仍远低于光系统Ⅱ基“半”人工光合系统的［图 4-10（c）、图 4-10（d）］。在上述研究中，光态和暗态下的循环伏安测试对比还显示光诱导的氧化还原波可能与一种传递电荷的扩散介质有关，但目前学术界对此还存在争议。此外，哈桑（Hasan）等利用电极-光系统Ⅱ复合体系研究的策略，将蓝藻和氧化还原聚合物同时固定在光电极表面，在 44 mW/cm^2 的光强照射下分别获得了 8.6 μA/cm^2 的直接光电流密度和 48.2 μA/cm^2 的介导光电流密度，但是可能会牺牲体系的寿命（Hasan et al.，2014）。

虽然上述改进策略显示出一定的前景，并且蓝藻基电极的寿命（数周至数月）（Darus et al.，2016）超过光系统Ⅱ基光阳极（数分钟）的（Kato et al.，2014）和人工合成光阳极的（数小时至数天）（Bae et al.，2017），但是这种“半”人工光阳极的光电流输出仍需增加至少几个数量级，才能在电力或太阳能燃料发电方面具有经济竞争力（McCormick et al.，2015）。目前，基于无机材料-叶绿体复合的生物光电系统，利用光阳极氧气生成和阴极氧气还原的催化环路，可以实现高达 500 μA/cm^2 的介导光电流密度（Bhardwaj et al.，1981）。因为蓝藻与叶绿体在生物进化上具有同源性，并且具有相似的结构和功能，所以使用蓝藻基“半”人工光阳极也有望实现 500 μA/cm^2 的光电流输出，但是该设计策略目前尚未在基于蓝藻的“半”人工光合系统中实现，这可能与用来构建该体系的蓝藻特性有关。蓝藻自身形成胞外电子的能力是构建该体系的基础。提高“半”人工光阳极的界面电荷转移率是将光电流密度提高到实用水平的关键。目前，已经在共聚焦荧光显微镜（Pirbadian et al., 2014）、电化学阻抗谱（He & Mansfeld, 2009）、表面增强拉曼/红外光谱（Millo et al., 2011；Busalmen et al., 2008）和纳米电极（Jiang et al., 2010；Ding et al., 2016）等一些检测平台上建立起许多技术，可以为“半”人工光阳极的界面电荷转移过程提供丰富的机理信息。例如，光谱检测扩散介质的分泌或膜结合细胞色素中的氧化还原变化，可以揭示它们在细胞-电极的电荷转移机制中的作用（Kornienko et al., 2018）。蓝藻基“半”人工光阳极的另一个挑战是有效地引导通过细胞产生的电子仅流向水氧化途径，而不是流向生物质积累和代谢副反应的途径。可通过仔细选择培养条件，或者使用特定的代谢化学抑制剂，甚至通过基因工程的手段，如将蓝藻光合机构中的氨基酸替换，提高光化学量子产率和水氧化催化率（Vinyard et al., 2014），这些途径都有可能实现有效的电子传递路径调控。

基于无机材料-微生物复合的“半”人工水氧化反应，目前主要依靠蓝藻基“半”人工光阳极实现，其基本原理是蓝藻细胞内的光系统Ⅱ发生水氧化反应时产生的电子流到细胞膜上，然后通过所谓的“纳米导线”传递到电极表面，最后在外电路中形成电流。相比于光系统Ⅱ基“半”人工光阳极，蓝藻基“半”人工光阳极具有更高的稳定性、持久性以及更简便的制备过程，但是效率和光电流是它的劣势。就目前来看，基于电极-蓝藻复合的“半”人工水氧化距离实际应用还很遥远。蓝藻细胞自身的光合作用效率，以及蓝藻细胞与电极界面的电荷转移效率是未来优化和改进蓝藻基“半”人工光阳极的关键。当然，其他一些化学反应动力学因素也是很重要的方面。其中，蓝藻细胞自身的光合作用效率可以通过合成生物学的手段和仿生无机纳米材料来改进（Jagadevan et al., 2018；Xiong et al., 2013；Jiang et al., 2015），蓝藻细胞与电极界面的电荷转移效率则可以通过对接触界面的调控来改善。此外，该体系可以为研究细胞向无机半导体材料表面的电荷转移提供支持。

（二）基于无机材料-微生物复合的“半”人工光合还原

自然光合作用包括光反应和暗反应两个阶段，在光反应阶段主要发生水氧化和电子传递，在暗反应阶段主要进行以二氧化碳还原为核心的卡尔文循环。光反应为暗反应提供电子和能量来源，所以暗反应对于整个系统来说是还原反应。在“半”人工光合作用中，还原反应比水氧化反应具有更复杂的形式和更多变的产物，是整个体系的终端反应，也是“半”人工光合作用领域研究的焦点。目前，已经开发了基于无机材料-酶复合和无机材料-微生物复合的“半”人工光合还原系统，研究“半”人工光合还原系统的目标主要是理解和改进光合产氢、固碳和固氮的反应过程。基于无机材料-光系统Ⅰ复合的“半”人工光合还原系统是最早开始研究的（Millsaps et al.，2001；Evans et al.，2004；Grimme et al.，2008），目的是实现“半”人工光合产氢。随着氢酶提取和保存技术的进步，又出现了基于无机材料-氢化酶复合的“半”人工光合体系（Tran et al.，2010；Brown et al.，2012，2010；Wilker et al.，2017；Caputo et al.，2014；Hutton et al.，2016）。在此基础之上，将氢化酶换成二氧化碳还原酶或者固氮酶，也实现了“半”人工光合固碳或固氮（Woolerton et al.，2010；Brown et al.，2016a，2016b；Hickey et al.，2018；Milton et al.，2016）。但是基于无机材料-酶复合的“半”人工光合系统存在固有的弊端，一是酶的提取和分离过程十分烦琐；二是酶离体后的稳定性和持久性差等；三是材料与酶的复合可能影响酶的催化活性，而单一酶的催化难以实现复杂产物的合成。因此，科学家开始寻求具有相关还原反应代谢途径的微生物细胞与人工合成的无机半导体材料的结合，构建基于无机材料-微生物复合的“半”人工光合还原系统。目前基于无机材料-微生物复合的“半”人工光合还原主要可以通过光催化剂-微生物复合体系和电极-微生物复合体系来实现。

1. 光催化剂-微生物复合体系

在最初的光催化剂-微生物复合的“半”人工光合体系研究中，杨培东课题组利用热醋穆尔氏菌的自我防御机制，将有生物毒性的镉离子（Cd^{2+}）在细胞培养的过程中沉积到细菌细胞表面，形成CdS纳米颗粒［图4-11（a）］（Sakimoto et al.，2016）。在光照的条件下，来自CdS的光生电子穿过细胞膜通过伍德-林格尔（Wood-Lundahl）途径成功地参与二氧化碳转化为乙酸的反应［图4-11（b）］，量子产率高达85%（光照条件：435～485 nm，5×10^{13} photon/cm^2 LED）（Sakimoto et al.，2016）。受到先前量子点酶研究工作的启发，杨培东课题组采用泵浦探针瞬态吸收光谱对该系统的机理进行研究（Kornienko et al.，2016）。通过建立CO_2光化学固定效率和瞬态吸收动力学的相关性与热醋穆尔氏菌氢化酶表达量的函数关系，在不同时间尺度上证明直接的电荷转移途径依赖于显性的双氢化酶介导。为了消除对牺牲剂的依赖性，杨培东课题组又将 CdS-热醋穆尔氏菌复合物与二氧化钛-锰-酞菁水氧化光催化剂在反应体系中共悬浮，也实现了对 CO_2 的固定，但是 CH_3COOH 产率不高（Sakimoto et al.，2016）。未来需要用无毒的光吸收剂来替代 CdS，通过选择透过性膜或区室化来防止 O_2 和活性氧物质伤害微生物，以及改善该系统在高太阳光强下的工作效率（Kato et al.，2014）。为了降低光催化材料与细菌细胞结合后产生的毒性，杨培东课题组在之前工作的基础上，将具有生物相容性的 Au 纳米团簇引入热醋穆尔氏菌细胞内部［图 4-11（c）］（Zhang et al.，2018），依然起到 CdS 的光催化作用，并且其在产生光生电子的同时，具有抑制活性氧自由基产生的功能［图 4-11（d）］，从而使得 CO_2 固定反应

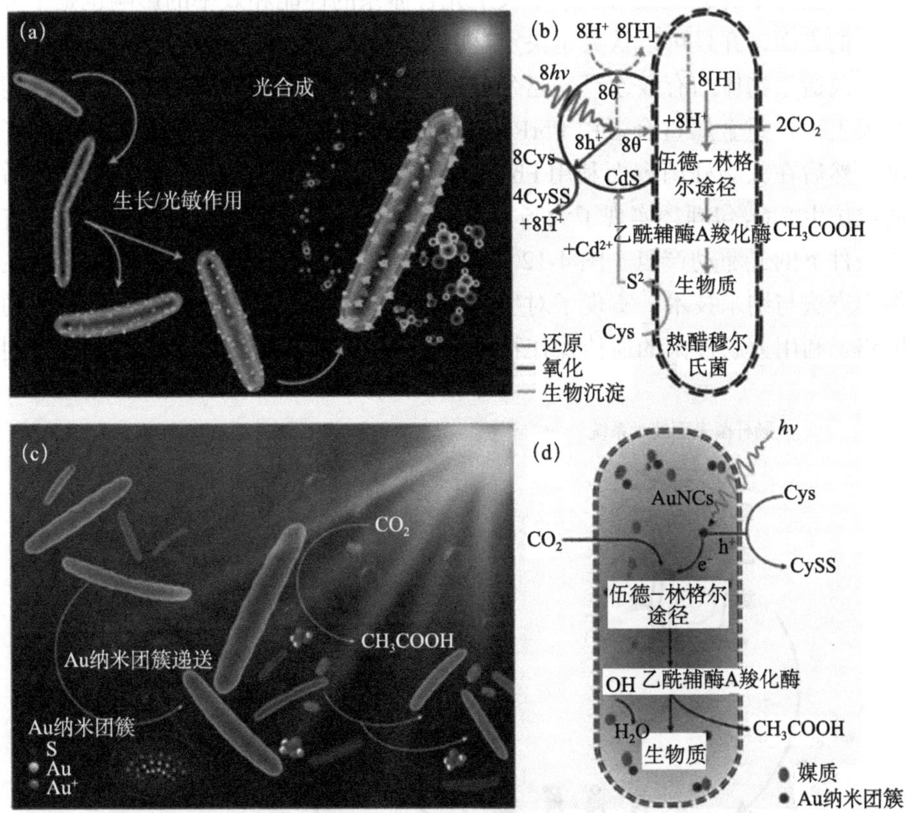

图 4-11　(a) CdS- 热醋穆尔氏菌复合体及其自光敏化太阳能转化；(b) CdS- 热醋穆尔氏菌复合物中太阳能 - 化学能转化的可能途径（Sakimoto et al.，2016）；(c) Au 纳米团簇在热醋穆尔氏菌内部复合及其自光敏化太阳能转化；(d) 热醋穆尔氏菌 -Au 纳米团簇复合体中太阳能 - 化学能转化的可能途径（Zhang et al.，2018）

可以维持至少 4 d。但是这个体系依然存在寿命问题，当细胞分裂形成新的细胞时，很难保证 Au 纳米团簇还在细胞中。以上的研究说明，当微生物体系确定时，光催化剂的选择和设计对整个“半”人工光合体系的性能有至关重要的影响。在选择和设计光催化剂时，一是要根据微生物细胞的生理特性选择生物毒性小的光催化剂，二是所选择的光催化剂能够以纳米材料的形式在细胞表面结合或者进入细胞内部。

通过以上的研究工作，可以总结出光催化剂-微生物复合的“半”人工光合作用的基本原理是：光催化剂产生的光生电子被微生物利用，参与微生物的代谢反应，从而选择性合成某种代谢产物。因此，除了材料的选择和设计，微生物本身的性质对“半”人工光合体系的性能和效率的影响也是十分重要的方面，并且可以认为是决定性因素。当下兴起的合成生物学手段就是一种改造生物体的有效途径。赵劲课题组通过合成生物学手段构建了一种能在膜上表达重金属螯合蛋白 PbrR 和在细胞内部表达 [NiFe] 氢化酶的大肠杆菌，然后在大肠杆菌表面利用 PbrR 结合镉离子，形成 CdS 纳米颗粒，最后通过仿生 SiO_2 包埋，实现了大肠杆菌的聚集，并诱导了大肠杆菌聚集体在有氧条件下的光驱动产氢（图 4-12）(Wei et al.，2018)。这项工作利用合成生物学方法与纳米技术，实现了对生物体的定向改造，为生物改造提供了新的思路。利用大肠杆菌的遗传操作特性，可以通过基因工程改造建立起一系列

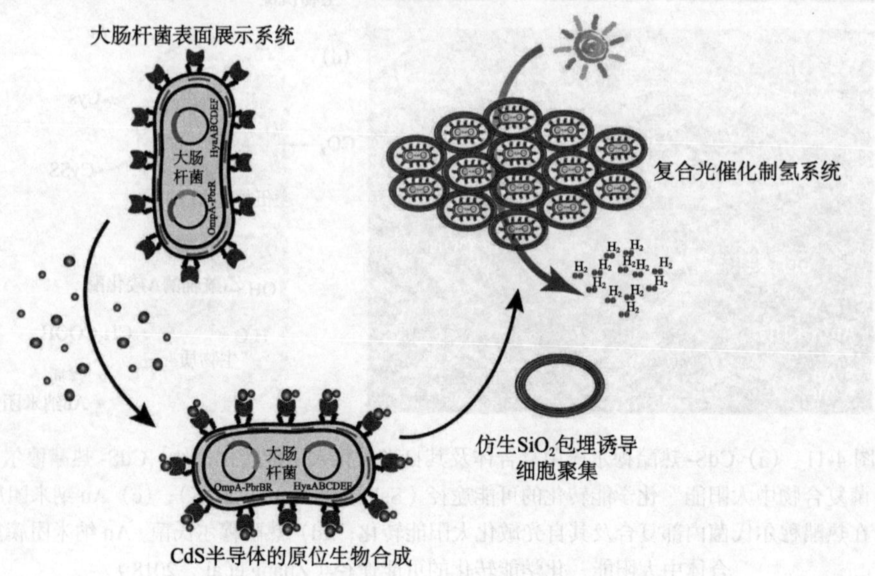

图 4-12　表面展示的生物复合技术诱导空气环境中的光驱动产氢（Wei et al.，2018）

生物催化反应途径，但是使用无机光催化纳米材料与细菌结合，容易对细胞产生毒害作用，从而限制其催化反应效率。最近，郭俊凌等使用简单的多酚化学将InP纳米颗粒与转基因的酿酒酵母（*Saccharomyces cerevisiae*）细胞表面复合，以生成一种生物无机杂化体（Guo et al.，2018）。当白光照射在利用太阳能的酿酒酵母细胞上时，它们会捕获光生电子并将其用于再生化合物NADPH，这是一种在莽草酸合成过程中分解的辅助因子，因此它可以再次参与合成，从而促进莽草酸的生物合成（Guo et al.，2018）。该研究首次将光催化剂与微生物结合，用于非能源物质的合成，拓展了“半”人工光合作用的应用领域，也为生物制造提供了新思路。

2. 电极-微生物复合体系

与基于光催化剂-微生物复合的“半”人工光合还原体系相比，基于电极-微生物复合的“半”人工光合还原体系具有更好的可调性和更强的应用前景，并且对电极的修饰与改造比对微生物细胞表面的材料修饰更加具有操控性。该体系的最初发展得益于从微生物电合成化学池和半导体-电极界面的基础研究中获得的理解和经验（Rabaey & Rozendal，2010）。杨培东课题组首先研究了希瓦氏菌（*Shewanella oneidensis*）MR-1在硅纳米线阵列上的识别（Jeong et al.，2013），并且用NaCl诱导了细菌在纳米线上的自组装（Sakimoto et al.，2014）。随着基于材料-酶复合的“半”人工光合作用研究的深入，为了克服基于材料-酶的“半”人工光合体系的缺点，基于电极-微生物的“半”人工光合体系应运而生。这种系统的独特能力是通过微生物细胞内部的代谢途径引导电极表面产生的还原物种（电子、氧化还原介质、H_2）转化为复杂的CO_2和N_2衍生产物（Claassens et al.，2016）。根据还原物种的形式，这个体系可以分为集成式和分散式。集成式，就是电极与微生物细胞是紧密结合的，还原物种的形式为电子，电极产生的电子直接传递给微生物来参与CO_2还原途径。在分散式中，电极只是插入微生物细胞的分散液中，并不和细胞界面连接，还原物种的形式为H_2或其他氧化还原介质，电极产生的电子先被用来产生H_2或者被氧化还原介质利用，然后再被微生物利用。因此，基于电极-微生物复合的“半”人工光合作用的基本原理可归纳为电极表面产生的还原物种进入微生物细胞，参与其内部代谢途径，转化为复杂的代谢产物。

对于集成式系统，实现电子在电极和细胞之间直接的传递需要两者之间具有良好的接触，因此高比表面积的电极对提高电流密度就显得十分重要。

杨培东课题组在国际上率先用 Si 纳米线阵列作为光阴极与厌氧细菌 *Spormusa ovata*（*S. ovata*）的细胞复合，用 TiO_2 纳米阵列作为光阳极，可以将 CO_2 选择性地还原为 CH_3COOH（图 4-13）（Liu et al.，2015），并且该系统的法拉第效率达到 90%。这种表面具有纳米线阵列的光阴极由于具有独特的几何结构，可将 O_2 从其固定厌氧细菌的纳米线通道中排除，从而使整个系统可以在大气条件下工作，并在 100 mW/cm^2 的光照强度下，与 TiO_2 光阳极配对时，以 0.3 mA/cm^2 的光电流密度驱动无辅助介质的 CO_2 还原。微生物“催化剂”的自我复制性质使得该系统的操作寿命为 200 h，并且可通过基因工程改造使大肠杆菌将 CH_3COOH 进一步升级为一系列更复杂的化学产物。

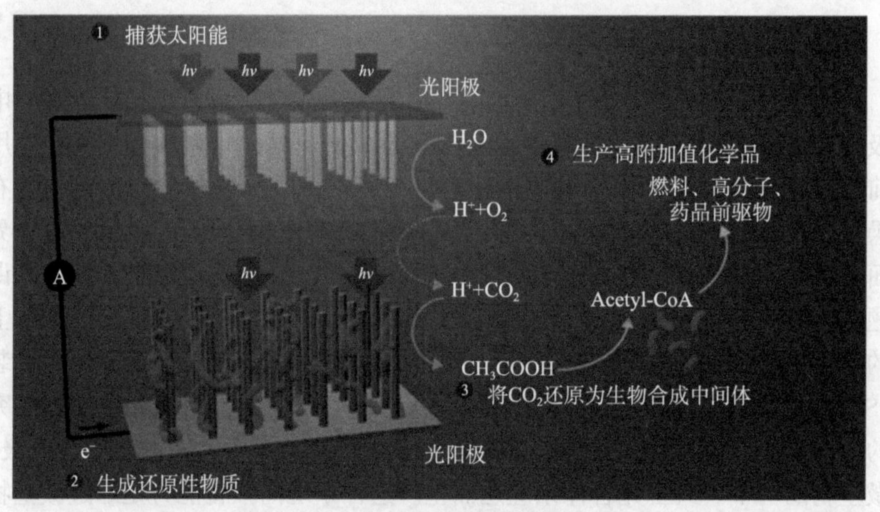

图 4-13　Si 纳米线阵列电极 –*S. ovata* 细菌集成式复合的“半”人工光合作用系统（Liu et al.，2015）

在分散式系统中，诺塞拉课题组通过将硅基“人造树叶”（阴极）（Torella et al.，2015）和 InP-TiO_2（阳极）（Nichols et al.，2015）串联插入富养罗尔斯通氏菌的细胞培养液中，在阴极产生 H_2，通过细菌的 H_2 代谢过程还原 CO_2 产生丁醇和异丙醇（图 4-14）。但是在这样的系统中，阳极会产生 O_2，所以容易在系统中积累活性氧自由基，并且电极催化剂材料容易在反应的过程中被腐蚀并释放出对微生物有毒害的金属离子。为了改进这个系统，诺塞拉课题组又开发了不会在溶液中产生显著活性氧物质或浸出金属离子的 Co-P 阴极-CoPi 阳极催化体系（Liu et al.，2016），这是“半”人工光合系统向商业化应用迈出的有希望的一步。该电化学系统与 Ge-GaAs-GaInP_2 三结光伏电

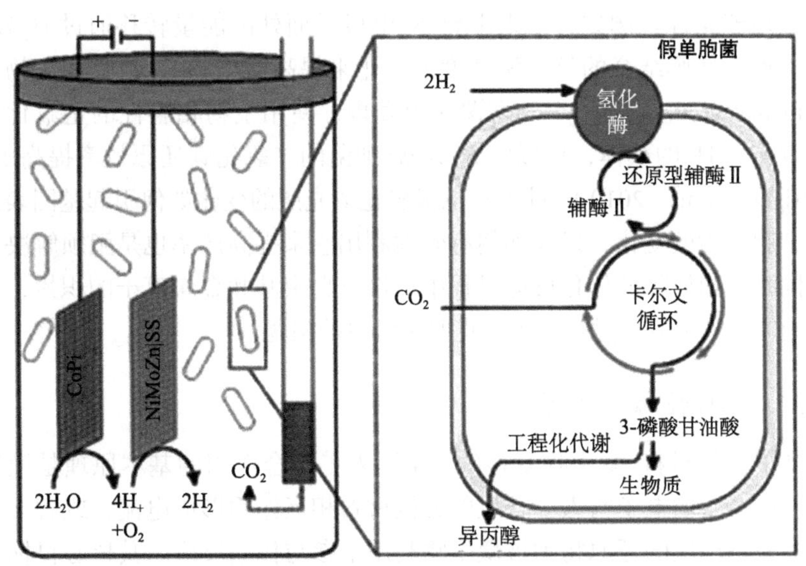

图 4-14　基于富养罗尔斯通氏菌的分散式“半”人工光合系统（Nichols et al.，2015）

池模型光伏电源和富养罗尔斯通氏菌相结合，直接利用 H_2 作为能量和还原物种输入，实现了持续 5 d 的太阳能驱动的 CO_2 还原，并且太阳能-化学能转化效率可达到 6%（Liu et al.，2016，2018）。上述 Co-P 阴极-CoPi 阳极催化体系被进一步用于自养黄色杆菌（*Xanthobacter autotrophicus*），可以驱动 N_2 向 NH_3 的电化学固定（Liu et al.，2017）。这些自养黄色杆菌甚至可以作为肥料直接添加到土壤中，从而避免了产品提取和纯化的需要。鉴于固氮酶的氧敏感性，利用微生物作为保护性支架是使 N_2 固定系统寿命延长的简便策略（Milton et al.，2017）。

虽然目前的研究进展是有应用前景的，但该系统关键的限制是基于微生物的“半”人工光合系统的体积产品产率，而目前最优的系统能达到 CH_3COO^-的产率约为 1.3 g/（L·d）。为了使反应器和产物的分离成本最低，扩大规模必须要提高产量。但是电化学池占用空间大、细胞与电极的低效连接，以及单个物种持续生长的要求是基于微生物的“半”人工光合系统推向商业化应用的关键障碍。对微生物电合成等类似体系的研究已经表明，电荷转移（直接、间接或两者兼有）过程依赖于应变、环境、电极和时间（Marshall et al.，2013；Siegert et al.，2014；Zhang et al.，2013）。因此，首先需要深入理解界面化学和能量转移，然后通过代谢工程提高产品生成速率和增加化学多样性，最后将环境耐受性扩展到无污染的实验室之外的大规模培养条件。例如，如果半导体-电极界面处的能量转移通过 H_2 发生，则可以通过使膜结合的氢化酶密度最大化来提高微生物 H_2 吸收和产物产率（Kornienko et al.，2016）。刘翀课题组发现，使用生物相容性的全氟化碳纳米乳剂作为 H_2 的载体，可以将 *S. ovata* 细菌的二氧化碳还原效率提高 190%（Rodrigues et al.，2019）。此外，电极极化或光照的变化如何引起基因表达和其他细胞行为的变化，以及如何有效地利用它来提高产率也是亟须解决的问题。同时，如何降低催化反应过程中活性氧自由基和金属离子的积累，也是逐步走向大规模应用过程中需要克服和解决的问题。

3. 无机材料-微生物界面

基于无机材料-微生物复合的“半”人工光合作用的基本原理是微生物光合作用产生的电子与人工合成的无机材料相互作用形成电流，或者无机材料在光照下产生的还原物种被微生物利用来参与代谢反应。其核心目标是无机材料与微生物催化体系的优势互补，无机材料与微生物的界面相互作用，决定二者间的能量和电荷传递，因此深入理解无机材料-微生物的界面对未来提高微生物基“半”人工光合系统的太阳能-化学能转化效率至关重要。在无机材料-微生物界面的相互作用中，有两个基础性问题：一是无机材料与微生物之间的能量和电荷转移问题；二是无机材料与微生物之间的相容性问题（图 4-15）（Sakimoto et al.，2018）。

目前实现无机材料与微生物之间的电子转移主要有两种模式：直接转移和介导转移（Sakimoto et al.，2018）。直接转移模式主要对应于光催化剂-微

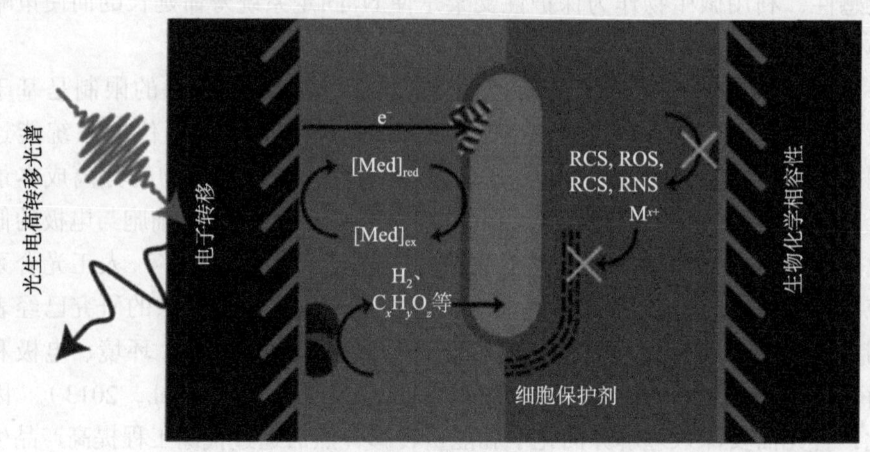

图 4-15　太阳能－化学能转化中的无机材料－微生物界面示意图（Sakimoto et al.，2018）

生物和集成式电极-微生物“半”人工光合体系，不需要外加氧化还原穿梭电对来传递电子，但是需要无机材料与微生物的紧密结合，这种模式对无机材料-微生物的界面通常有特殊的要求，需要无机材料-微生物具有良好的匹配性。介导转移模式主要对应于分散式的电极-微生物“半”人工光合体系，利用可溶性的氧化还原穿梭电对将电极-溶液界面处产生的电子转化为还原物种从而提供给微生物（Sakimoto et al.，2017；Liu et al.，2016，2017）。这些还原物种继而将 NAD（P）$^+$ 还原为 NAD（P）H，作为通用的生物电子供体，也可以产生 ATP 供生物催化反应使用。目前的介导转移系统可采用产氢电催化剂与 H_2 氧化、CO_2 还原或 N_2 还原细菌串联配对（Liu et al.，2016，2017；Marshall et al.，2013），利用 H_2 作为还原物种，实现“半”人工光合还原。这些简易的系统具有高达约 10% 的太阳能-生物质能转化效率（Liu et al.，2016），比典型的植物太阳能-生物质能转化效率高一个数量级（Ort et al.，2015）。其他的氧化还原电对介质，如甲酸盐（Li et al.，2012；Yishai et al.，2016）、紫罗碱和吩嗪（Moscoviz et al.，2016），也有一定的应用前景，但是它们对微生物的毒害性较大。单纯从电子转移效率来分析，直接转移模式比介导转移模式更有优势，因为氧化还原穿梭电对在传递电子的过程中不可避免地存在能量损失，但是无机材料与微生物的直接接触对彼此的相容性是很大的考验，细胞的生物活性以及无机光催化剂的活性都可能受到影响。此外，无机光催化剂的选择、电极的形貌结构和表面修饰都会对无机材料-微生物界面电子转移产生重要影响，这三个方面也是目前改进和优化系统的主要关注点。

考虑到无机材料与微生物直接接触可能发生的不利影响，如何在无机材料的生物相容性和微生物的化学相容性之间找到平衡点，一直以来都是这个领域的热点与难点。首先，材料的结构和形貌会影响其与微生物的结合，进而影响微生物的活性，如将电极表面功能化或纳米化会有利于微生物细胞在电极表面的吸附生长，而且二者接触面积的增大也会增加电流密度（Xie et al.，2011；Mink et al.，2012）。其次，任何一种含金属元素的催化剂在溶液中使用时，都会不可避免地溶解产生金属离子，而绝大多数的金属离子对于微生物来说都是具有毒性的。当微生物自身的活性下降后，内部的生物催化反应活性也可能受到影响，从而使整个系统的性能下降。但也有微生物对金属离子的响应是积极的，比如有些微生物本身就具备对某些重金属离子的抵抗作用（Cunningham & Lundie，1993；Wang et al.，2017），将此类金属元素以单质或化合物的形式沉积在表面，能够促进材料-微生物的功能协同。此外，无机半导体材料在发生催化反应时，时常伴随着活性氧、活性氯和活性氮自由基的产生（Sakimoto et al.，2018），它们会对微生物细胞的活性产生毒害作用，因此需要在材料选择和设计时尽量避免此类情况的发生，并同时研究微生物对这些自由基活性物种的响应机制。为了保护微生物，除了在无机半导体材料端做改进，通过化学和材料的手段对微生物细胞进行表面改进也是常用的方法，如二氧化硅单细胞包裹微生物（Xiong et al.，2013；Yang et al.，2009）、水凝胶包裹微生物（Lee & Mooney，2012）、金属有机框架材料包裹微生物（Liang et al.，2016）等。

三、基于微生物的“半”人工光合成材料研究目前面临的重大科学问题及前沿方向

表 4-1 列出了近几年来基于无机材料–微生物复合的“半”人工光合作用的主要代表性工作。“半”人工光合作用研究在近 10 年取得了迅猛的发展，其初期的焦点主要集中在基于材料–酶复合的“半”人工光合作用，除了一些国外的课题组，国内李灿院士团队在该领域也做出了具有代表性的工作（Wang et al.，2016，2014；Li et al.，2017）。基于无机材料–微生物复合的“半”人工光合系统研究则是在最近几年开始逐渐开展起来的，目前处于领先地位的是美国的几个课题组，以诺塞拉和杨培东为代表。虽然他们已在这个领域做出了开拓性的工作，并且也在《科学》、《自然》（*Nature*）子刊以及《美国国家科学院院刊》（*PNAS*）上发表了多篇论文（Kornienko et al.，2018；Sakimoto et al.，2016；Zhang et al.，2018；Guo et al.，2018；Torella et al.，2015；Liu et al.，2016，2017），但是这个领域依然处于刚刚兴起的阶段（Kornienko et al.，2018；Sakimoto et al.，2018）。国内的麦立强课题组和美国的田博之课题组也开始关注这个领域中材料与生物的界面问题（Xu et al.，2018；Lee & Tian，2019）。目前人们能够通过无机材料–微生物复合的“半”人工光合体系实现二氧化碳固定，生成乙酸、乙醇、莽草酸等小分子含碳有机物，但是其效率、规模和成本尚未达到商业化应用的水平，还有很多科学和工程问题亟待解决。

表 4-1　基于无机材料 – 微生物复合的“半”人工光合作用系统

微生物	材料体系	混合方法	功能	参考文献
集胞藻	IO-ITO 电极	整体混合	H_2O 氧化、生物光伏	Zhang et al.，2018
热醋穆尔氏菌	CdS	纳米涂覆	CO_2 还原	Sakimoto et al.，2016
热醋穆尔氏菌	CdS-TiO_2-酞菁锰	纳米涂覆	CO_2 还原	Kornienko et al.，2016
热醋穆尔氏菌	Au 纳米团簇	引入纳米团簇	CO_2 还原	Zhang et al.，2018

续表

微生物	材料体系	混合方法	功能	参考文献
大肠杆菌	CdS/SiO_2	纳米涂层与多细胞封装	产氢	Wei et al., 2018
酿酒酵母	InP 纳米颗粒	纳米涂覆	CO_2 还原	Guo et al., 2018
厌氧细菌 *S. ovata*	Si 纳米线阵列电极	整体混合	CO_2 还原	Liu et al., 2015
假单胞菌	NiMoZn-CoPi	分散混合	CO_2 还原	Nichols et al., 2015
假单胞菌	CoP-CoPi	分散混合	CO_2 还原	Liu et al., 2016
自养黄色杆菌	CoP-CoPi	分散混合	固氮	Liu et al., 2017
厌氧细菌 *S. ovata*	CoP-CoPi	分散混合	CO_2 还原	Rodrigues et al., 2019

微生物基的“半”人工光合系统中的一个核心问题就是材料-生物的界面相互作用，关键是能量与电荷在材料与微生物界面处的转移。材料与微生物的界面相互作用不仅影响微生物自身的性质，还会对材料性质产生影响。这种相互作用直接影响电荷在界面处的转移，因此提高界面电荷转移速率对提高微生物基“半”人工光合系统的太阳能-化学能转化效率至关重要。

四、研究总结与展望

由于材料与微生物的界面相互作用主要取决于微生物的表面特性、合成材料的性质、微生物细胞与材料的结合形式，因此改进二者的相互作用可以从以下三个方面入手。

（1）对微生物本身进行改造。在微生物改造方面，合成生物学发挥着越来越重要的作用（Jagadevan et al.，2018；Kong et al.，2019；Kung et al.，2012）。虽然合成生物学一直致力于增强细胞的体内功能（如合成回路和代谢途径的发展），但生物和非生物组分之间的界面是其将来要解决的关键领域。为此，应用合成生物学的工具可以从头设计具有精细功能单元的多组分生物系统。该方法原则上可以使微生物适应从电极到工程代谢循环的高强度电荷通量。除合成生物手段外，通过仿生无机纳米材料对微生物进行功能化改造也是一种有效的途径（Xiong et al.，2013，2015；Ko et al.，2013；熊威等，2019）。

（2）根据微生物的特性和最终的产物需求选择合适的材料体系。目前的“半”人工光合体系，主要集中在合成小分子含碳有机物和氨等方面。如何进一步发挥生物体酶促反应的专一性，得到结构更复杂和价值更高的产物是未来探索的一个重要方向，这方面需要基于对相关微生物代谢途径和分子机制的理解，结合化学与材料的手段构建合适的“半”人工光合体系。

（3）改变材料-微生物的复合形式。未来的机遇还在于将合成材料与生物系统整合在一起，以产生能量转换和催化循环的新途径（Xu et al.，2018）。不同的复合形式会产生不同的材料-微生物相互作用，而不同的相互作用对微生物的代谢活动可能产生不同的影响。由于材料-微生物的界面是“半”人工光合体系的核心，因此材料-生物界面的原位表征技术是探索材料-微生物复合形式的重要基础。

五、学科发展政策建议与措施

“半”人工光合作用是自然光合作用和人工光合作用交叉互补的一个体系，它的出现是自然光合作用和人工光合作用研究发展到一定阶段而产生的，未来的发展不仅取决于两个领域的研究水平和认知程度，更取决于二者的融合程度。虽然基于无机材料-微生物复合的“半”人工光合作用目前尚处于概念验证阶段，但是过去几年的研究已经为基于无机材料-微生物的“半”人工光合作用未来的发展奠定了坚实的基础，未来这个领域在现有的基础上还有进一步完善和认识的空间。特别是微生物基的“半”人工光合体系可以作为模型体系来探索一些在纯自然光合作用或纯人工光合作用中无法用实验进行研究的问题，在新的微生物体系、材料体系与结构、材料-微生物结合方式等方面更加具有广阔的探索前景。此外，在一些特殊环境条件（如海洋或太空航行）下，“半”人工光合系统可能存在更多的应用途径，但必须充分考虑资源管理和材料可回收性。在“半”人工光合体系迈向商业化应用的过程中，目标产物与生物催化体系的分离也是其需要解决的问题。

熊　威（南昌大学）、邹志刚（南京大学）

本节参考文献

陈雅静，李旭兵，佟振合，等 . 2019. 人工光合成制氢 . 化学进展，31: 38-49.

熊威，唐睿康，马为民，等 . 2019. 仿生无机纳米材料改造生物体的研究进展 . 无机化学学报，35: 1-24.

Ager J W, Shaner M R, Walczak K A, et al. 2015. Experimental demonstrations of spontaneous, solar-driven photoelectrochemical water splitting. Energy & Environmental Science, 8: 2811-2824.

Armstrong F A, Hirst J. 2011. Reversibility and efficiency in electrocatalytic energy conversion and lessons from enzymes. Proceedings of the National Academy of Science USA, 108: 14049-14054.

Azcarate I, Costentin C, Robert M, et al. 2016. Through-space charge interaction substituent effects in molecular catalysis leading to the design of the most efficient catalyst of CO_2-to-CO electrochemical conversion. Journal of the American Chemical Society, 138: 16639-16644.

Bae D, Seger B, Vesborg P C K, et al. 2017. Strategies for stable water splitting via protected photoelectrodes. Chemical Society Reviews, 46: 1933-1954.

Barber J. 2009. Photosynthetic energy conversion: natural and artificial. Chemical Society Reviews, 38: 185-196.

Bhardwaj R, Pan R L, Gross E L. 1981. Solar energy conversion by chloroplast photoelectrochemical cells. Nature, 289: 396-398.

Blankenship R E, Tiede D M, Barber J, et al. 2011. Comparing photosynthetic and photovoltaic efficiencies and recognizing the potential for improvement. Science, 332: 805-809.

Brown K A, Dayal S, Ai X, et al. 2010. Controlled assembly of hydrogenase-CdTe nanocrystal hybrids for solar hydrogen production. Journal of the American Chemical Society, 132: 9672-9680.

Brown K A, Harris D F, Wilker M B, et al. 2016. Light-driven dinitrogen reduction catalyzed by a CdS: nitrogenase MoFe protein biohybrid. Science, 352: 448-450.

Brown K A, Wilker M B, Boehm M, et al. 2012. Characterization of photochemical processes for H_2 production by CdS nanorod-[FeFe] hydrogenase complexes. Journal of the American Chemical Society, 134: 5627-5636.

Brown K A, Wilker M B, Boehm M, et al. 2016. Photocatalytic regeneration of nicotinamide cofactors by quantum dot-enzyme biohybrid complexes. ACS Catalysis, 6: 2201-2204.

Busalmen J P, Esteve-Núñez A, Berná A, et al. 2008. C-type cytochromes wire electricity-producing bacteria to electrodes. Angewandte Chemie International Edition, 47: 4874-4877.

Caputo C A, Gross M A, Lau V W, et al. 2014. Photocatalytic hydrogen production using polymeric carbon nitride with a hydrogenase and a bioinspired synthetic Ni catalyst. Angewandte Chemie International Editions, 53: 11538-11542.

Cheng W H, Richter M H, May M M, et al. 2018. Monolithic photoelectrochemical device for direct water splitting with 19% efficiency. ACS Energy Letters, 3: 1795-1800.

Claassens N J, Sousa D Z, Santos V A P M D, et al. 2016. Harnessing the power of microbial autotrophy. Nature Reviews Microbiology, 14: 692-706.

Cook T R, Dogutan D K, Reece S Y, et al. 2010. Solar energy supply and storage for the legacy and nonlegacy worlds. Chemical Reviews, 110: 6474-6502.

Cunningham D P, Lundie L L. 1993. Precipitation of cadmium by Clostridium thermoaceticum. Applied & Environmental Microbiology, 59: 7-14.

Darus L, Ledezma P, Keller J, et al. 2016. Marine phototrophic consortia transfer electrons to electrodes in response to reductive stress. Photosynthesis Research, 127: 347-354.

Ding M N, Shiu H Y, Li S L, et al. 2016. Nanoelectronic investigation reveals the electrochemical basis of electrical conductivity in shewanella and geobacter. ACS Nano, 10: 9919-9926.

Evans B R, O'Neill H M, Hutchens S A, et al. 2004. Enhanced photocatalytic hydrogen evolution by covalent attachment of plastocyanin to photosystem Ⅰ. Nano Letters, 4: 1815-1819.

Gong M, Cao Z, Liu W, et al. 2017. Supramolecular porphyrin cages assembled at molecular-materials interfaces for electrocatalytic CO reduction. ACS Central Science, 3: 1032-1040.

Grimme R A, Lubner C E, Bryant D A, et al. 2008. Photosystem Ⅰ/molecular wire/metal nanoparticle bioconjugates for the photocatalytic production of H_2. Journal of the American Chemical Society, 130: 6308-6309.

Guo J L, Suástegui M, Sakimoto K K, et al. 2018. Light-driven fine chemical production in yeast biohybrids. Science, 362: 813-816.

Hall A S, Yoon Y, Wuttig A, et al. 2015. Mesostructure-induced selectivity in CO_2 reduction catalysis. Journal of the American Chemical Society, 137: 14834-14837.

Hasan K, Yildiz H B, Sperling E, et al. 2014. Photo-electrochemical communication between cyanobacteria (*Leptolyngbia* sp.) and osmium redox polymer modified electrodes. Physical Chemistry Chemical Physics, 16: 24676-24680.

He Z, Mansfeld F. 2009. Exploring the use of electrochemical impedance spectroscopy (EIS) in microbial fuel cell studies. Energy & Environmental Science, 2: 215-219.

Herek J L, Wohlleben W, Cogdell R J, et al. 2002. Quantum control of energy flow in light harvesting. Nature, 417: 533-535.

Hickey D P, Lim K, Cai R, et al. 2018. Pyrene hydrogel for promoting direct bioelectrochemistry: ATP-independent electroenzymatic reduction of N_2. Chemical Science, 9: 5172-5177.

Hutton G A M, Reuillard B, Martindale B C M, et al. 2016. Carbon dots as versatile photosensitizers for solar-driven catalysis with redox enzymes. Journal of the American Chemical Society, 138: 16722-16730.

Jagadevan S, Banerjee A, Banerjee C, et al. 2018. Recent developments in synthetic biology and metabolic engineering in microalgae towards biofuel production. Biotechnology for Biofuels, 11: 185.

Jeong H E, Kim I, Karam P, et al. 2013. Bacterial recognition of silicon nanowire arrays. Nano

Letters, 13: 2864-2869.

Jia J Y, Seitz L C, Benck J D, et al. 2016. Solar water splitting by photovoltaic-electrolysis with a solar-to-hydrogen efficiency over 30%. Nature Communications, 7: 13237.

Jiang N, Yang X Y, Deng Z, et al. 2015. A stable, reusable, and highly active photosynthetic bioreactor by bio-interfacing an individual cyanobacterium with a mesoporous bilayer nanoshell. Small, 11: 2003-2010.

Jiang X C, Hu J S, Fitzgerald L A, et al. 2010. Probing electron transfer mechanisms in *Shewanella oneidensis* MR-1 using a nanoelectrode platform and single-cell imaging. Proceedings of the National Academy of Sciences USA, 107: 16806-16810.

Kato M, Zhang J Z, Paul N, et al. 2014. Protein film photoelectrochemistry of the water oxidation enzyme photosystem Ⅱ. Chemical Society Reviews, 43: 6485-6497.

Khaselev O, Turner J A. 1998. A monolithic photovoltaic-photoelectrochemical device for hydrogen production via water splitting. Science, 280: 425-427.

Kim C, Jeon H S, Eom T, et al. 2015. Achieving selective and efficient electrocatalytic activity for CO_2 reduction using immobilized silver nanoparticles. Journal of the American Chemical Society, 137: 13844-13850.

Ko E H, Yoon Y, Park J H, et al. 2013. Bioinspired, cytocompatible mineralization of silica-titania composites: thermoprotective nanoshell formation for individual chlorella cells. Angewandte Chemie International Edition, 52: 12279-12282.

Kong F T, Yamaoka Y, Ohama T, et al. 2019. Molecular genetic tools and emerging synthetic biology strategies to increase cellular oil content in *Chlamydomonas reinhardtii*. Plant & Cell Physiology, 60: 1184-1196.

Kornienko N, Sakimoto K K, Herlihy D M, et al. 2016. Spectroscopic elucidation of energy transfer in hybrid inorganic-biological organisms for solar-to-chemical production. Proceedings of the National Academy of Sciences USA, 113: 11750-11755.

Kornienko N, Zhang J Z, Sakimoto K K, et al. 2018. Interfacing nature's catalytic machinery with synthetic materials for semi-artificial photosynthesis. Nature Nanotechnology, 13: 890-899.

Kumar A, Hsu L H H, Kavanagh P, et al. 2017. The ins and outs of microorganism-electrode electron transfer reactions. Nature Reviews Chemistry, 1: 24.

Kung Y, Runguphan W, Keasling J D. 2012. From fields to fuels: recent advances in the microbial production of biofuels. ACS Synthetic Biology, 1: 498-513.

Lee K Y, Mooney D J. 2012. Alginate: properties and biomedical applications. Progress in Polymer Science, 37: 106-126.

Lee Y V, Tian B Z. 2019. Learning from solar energy conversion: biointerfaces for artificial photosynthesis and biological modulation. Nano Letters, 19: 2189-2197.

Léonard A, Rooke J C, Meunier C F, et al. 2010. Cyanobacteria immobilised in porous silica gels: exploring biocompatible synthesis routes for the development of photobioreactors. Energy & Environmental Science, 3: 370-377.

Lewis N S, Nocera D G. 2006. Powering the planet: chemical challenges in solar energy utilization. Proceedings of the National Academy of Sciences USA, 103: 15729-15735.

Li H, Opgenorth P H, Wernick D G, et al. 2012. Integrated electromicrobial conversion of CO_2 to higher alcohols. Science, 335: 1596.

Li X B, Tung C H, Wu L Z. 2018. Semiconducting quantum dots for artificial photosynthesis. Nature Reviews Chemistry, 2: 160-173.

Li Z S, Feng J Y, Yan S C, et al. 2015. Solar fuel production: strategies and new opportunities with nanostructures. Nano Today, 10: 468-486.

Li Z S, Luo W J, Zhang M L, et al. 2013. Photoelectrochemical cells for solar hydrogen production: current state of promising photoelectrodes, methods to improve their properties, and outlook. Energy & Environmental Science, 6: 347-370.

Li Z, Wang W Y, Ding C M, et al. 2017. Biomimetic electron transport via multiredox shuttles from photosystem Ⅱ to a photoelectrochemical cell for solar water splitting. Energy & Environmental Science, 10: 765-771.

Liang K, Richardson J J, Cui J W, et al. 2016. Metal-organic framework coatings as cytoprotective exoskeletons for living cells. Advanced Materials, 28: 7910-7914.

Liu C, Colón B C, Ziesack M, et al. 2016. Water splitting-biosynthetic system with CO_2 reduction efficiencies exceeding photosynthesis. Science, 352: 1210-1213.

Liu C, Colón B E, Silver P A, et al. 2018. Solar-powered CO_2 reduction by a hybrid biological | inorganic system. Journal of Photochemistry & Photobiology A: Chemistry, 358: 411-415.

Liu C, Gallagher J J, Sakimoto K K, et al. 2015. Nanowire-bacteria hybrids for unassisted solar carbon dioxide fixation to value-added chemicals. Nano Letters, 15: 3634-3639.

Liu C, Sakimoto K K, Colón B C, et al. 2017. Ambient nitrogen reduction cycle using a hybrid inorganic-biological system. Proceedings of the National Academy of Sciences USA, 114: 6450-6455.

Ma S C, Sadakiyo M, Luo R, et al. 2016. One-step electrosynthesis of ethylene and ethanol from CO_2 in an alkaline electrolyzer. Journal of Power Sources, 301: 219-228.

Marshall C W, Ross D E, Fichot E B, et al. 2013. Long-term operation of microbial electrosynthesis systems improves acetate production by autotrophic microbiomes. Environmental Science & Technology, 47: 6023-6029.

McCormick A J, Bombelli P, Bradley R W, et al. 2015. Biophotovoltaics: oxygenic photosynthetic organisms in the world of bioelectrochemical systems. Energy & Environmental Science, 8:

1092-1109.

Millo D, Harnisch F, Patil S A, et al. 2011. *In situ* spectroelectrochemical investigation of electrocatalytic microbial biofilms by surface-enhanced resonance raman spectroscopy. Angewandte Chemie International Edition, 50: 2625-2627.

Millsaps J F, Bruce B D, Lee J W, et al. 2001. Nanoscale photosynthesis: photocatalytic production of hydrogen by platinized photosystem Ⅰ reaction centers. Photochemistry & Photobiology, 73: 630-635.

Milton R D, Abdellaoui S, Khadka N, et al. 2016. Nitrogenase bioelectrocatalysis: heterogeneous ammonia and hydrogen production by MoFe protein. Energy & Environmental Science, 9: 2550-2554.

Milton R D, Cai R, Sahin S, et al. 2017. The *in vivo* potential-regulated protective protein of nitrogenase in azotobacter vinelandii supports aerobic bioelectrochemical dinitrogen reduction *in vitro*. Journal of the American Chemical Society, 139: 9044-9052.

Mink J E, Rojas J P, Logan B E, et al. 2012. Vertically grown multiwalled carbon nanotube anode and nickel silicide integrated high performance microsized (1.25 μL) microbial fuel cell. Nano Letters, 12: 791-795.

Moscoviz R, Toledo-Alarcón J, Trably E, et al. 2016. Electro-fermentation: how to drive fermentation using electrochemical systems. Trends in Biotechnology, 34: 856-865.

Nichols E M, Gallagher J J, Liu C, et al. 2015. Hybrid bioinorganic approach to solar-to-chemical conversion. Proceedings of the National Academy of Sciences USA, 112: 11461-11466.

Nocera D G. 2017. Solar fuels and solar chemicals industry. Accounts of Chemical Research, 50: 616-619.

Ort D R, Merchant S S, Alric J, et al. 2015. Redesigning photosynthesis to sustainably meet global food and bioenergy demand. Proceedings of the National Academy of Sciences USA, 112: 8529-8536.

Pirbadian S, Barchinger S E, Leung K M, et al. 2014. *Shewanella oneidensis* MR-1 nanowires are outer membrane and periplasmic extensions of the extracellular electron transport components. Proceedings of the National Academy of Sciences USA, 111: 12883-12888.

Rabaey K, Rozendal R A. 2010. Microbial electrosynthesis-revisiting the electrical route for microbial production. Nature Reviews Microbiology, 8: 706-716.

Rodrigues R M, Guan X, Iniguez J A, et al. 2019. Perfluorocarbon nanoemulsion promotes the delivery of reducing equivalents for electricity-driven microbial CO_2 reduction. Nature Catalysis, 2: 407-414.

Rosen B A, Salehi-Khojin A, Thorson M R, et al. 2011. Ionic liquid-mediated selective conversion of CO_2 to CO at low overpotentials. Science, 334: 643-644.

Sakimoto K K, Kornienko N, Cestellos-Blanco S, et al. 2018. Physical biology of the materials-microorganism interface. Journal of the American Chemical Society, 140: 1978-1985.

Sakimoto K K, Kornienko N, Yang P D. 2017. Cyborgian material design for solar fuel production: the emerging photosynthetic biohybrid systems. Accounts of Chemical Research, 50: 476-481.

Sakimoto K K, Liu C, Lim J, et al. 2014. Salt-induced self-assembly of bacteria on nanowire arrays. Nano Letters, 14: 5471-5476.

Sakimoto K K, Wong A B, Yang P D. 2016. Self-photosensitization of nonphotosynthetic bacteria for solar-to-chemical production. Science, 351: 74-77.

Sakimoto K K, Zhang S J, Yang P D. 2016. Cysteine-cystine photoregeneration for oxygenic photosynthesis of acetic acid from CO_2 by a tandem inorganic-biological hybrid system. Nano Letters, 16: 5883-5887.

Siegert M, Yates M D, Call D F, et al. 2014. Comparison of nonprecious metal cathode materials for methane production by electromethanogenesis. ACS Sustainable Chemistry & Engineering, 2: 910-917.

Su J Z, Vayssieres L. 2016. A place in the sun for artificial photosynthesis? ACS Energy Letters, 1: 121-135.

Torella J P, Gagliardi C J, Chen J S, et al. 2015. Efficient solar-to-fuels production from a hybrid microbial-water-splitting catalyst system. Proceedings of the National Academy of Sciences USA, 112: 2337-2342.

Tran P D, Artero V, Fontecave M. 2010. Water electrolysis and photoelectrolysis on electrodes engineered using biological and bio-inspired molecular systems. Energy & Environmental Science, 3: 727-747.

Tu W G, Zhou Y, Zou Z G. 2014. Photocatalytic conversion of CO_2 into renewable hydrocarbon fuels: state-of-the-art accomplishment, challenges, and prospects. Advanced Materials, 26: 4607-4626.

Verlage E, Hu S, Liu R, et al. 2015. A monolithically integrated, intrinsically safe, 10% efficient, solar-driven water-splitting system based on active, stable earth-abundant electrocatalysts in conjunction with tandem Ⅲ-Ⅴ light absorbers protected by amorphous TiO_2 films. Energy & Environmental Science, 8: 3166-3172.

Vinyard D J, Gimpel J, Ananyev G M, et al. 2014. Engineered photosystem Ⅱ reaction centers optimize photochemistry versus photoprotection at different solar intensities. Journal of the American Chemical Society, 136: 4048-4055.

Wang B, Zeng C P, Chu K H, et al. 2017. Enhanced biological hydrogen production from *Escherichia coli* with surface precipitated cadmium sulfide nanoparticles. Advanced Energy

Materials, 7: 1700611.

Wang W Y, Chen J, Li C, et al. 2014. Achieving solar overall water splitting with hybrid photosystems of photosystem Ⅱ and artificial photocatalysts. Nature Communications, 5: 4647.

Wang W Y, Wang H, Zhu Q J, et al. 2016. Spatially separated photosystem Ⅱ and a silicon photoelectrochemical cell for overall water splitting: a natural-artificial photosynthetic hybrid. Angewandte Chemie International Edition, 55: 9229-9233.

Wei W, Sun P Q, Li Z, et al. 2018. A surface-display biohybrid approach to light-driven hydrogen production in air. Science Advances, 4: 9253.

Wilker M B, Utterback J K, Greene S, et al. 2017. Role of surface-capping ligands in photoexcited electron transfer between CdS nanorods and [FeFe] hydrogenase and the subsequent H_2 generation. Journal of Physical Chemistry C, 122: 741-750.

Woolerton T W, Sheard S, Reisner E, et al. 2010. Efficient and clean photoreduction of CO_2 to CO by enzyme-modified TiO_2 nanoparticles using visible light. Journal of the American Chemical Society, 132: 2132-2133.

Xie X, Hu L B, Pasta M, et al. 2011. Three-dimensional carbon nanotube-textile anode for high-performance microbial fuel cells. Nano Letters, 11: 291-296.

Xiong W, Yang Z, Zhai H L, et al. 2013. Alleviation of high light-induced photoinhibition in cyanobacteria by artificially conferred biosilica shells. Chemical Communications, 49: 7525-7527.

Xiong W, Zhao X H, Zhu G X, et al. 2015. Silicification-induced cell aggregation for the sustainable production of H_2 under aerobic conditions. Angewandte Chemie International Edition, 54: 11961-11965.

Xu L, Zhao Y L, Owusu K A, et al. 2018. Recent advances in nanowire-biosystem interfaces: from chemical conversion, energy production to electrophysiology. Chem, 4: 1538-1559.

Yang S H, Lee K B, Kong B, et al. 2009. Biomimetic encapsulation of individual cells with silica. Angewandte Chemie International Edition, 48: 9160-9163.

Yao T T, An X R, Han H X, et al. 2018. Photoelectrocatalytic materials for solar water splitting. Advanced Energy Materials, 8: 1800210.

Yishai O, Lindner S N, de la Cruz J G, et al. 2016. The formate bio-economy. Current Opinion in Chemical Biology, 35: 1-9.

Zhang H, Liu H, Tian Z Q, et al. 2018. Bacteria photosensitized by intracellular gold nanoclusters for solar fuel production. Nature Nanotechnology, 13: 900-905.

Zhang J Z, Bombelli P, Sokol K P, et al. 2018. Photoelectrochemistry of photosystem Ⅱ *in vitro* vs *in vivo*. Journal of the American Chemical Society, 140: 6-9.

Zhang T, Nie H R, Bain T S, et al. 2013. Improved cathode materials for microbial electrosynthesis. Energy & Environmental Science, 6: 217-224.

Zhou X H, Liu R, Sun K, et al. 2016. Solar-driven reduction of 1 atm of CO_2 to formate at 10% energy-conversion efficiency by use of a TiO_2-protected Ⅲ - Ⅴ tandem photoanode in conjunction with a bipolar membrane and a Pd/C cathode. ACS Energy Letters, 1: 764-770.

Zhu X G, Long S P, Ort D R. 2008. What is the maximum efficiency with which photosynthesis can convert solar energy into biomass? Current Opinion in Biotechnology, 19: 153-159.

Zhu X G, Long S P, Ort D R. 2010. Improving photosynthetic efficiency for greater yield. Annual Review of Plant Biology, 61: 235-261.

第四节　水伏科学与技术

一、水伏科学与技术研究概述

近年来，以我国学者为代表的研究者在石墨烯、碳纳米管薄膜等低维材料中发现，水与低维材料通过表界面相互作用可以直接输出电能。类比于光伏技术，我们称这类生电现象为水伏效应。水伏效应使得人们能够用纳米材料从水流、水波、雨滴落下、蒸发、湿度变化等水运动和循环过程中直接捕获能量，为水能利用提供了全新的方式。

（一）研究背景

水覆盖了约 71% 的地球表面，约占人体重量的 70%，是地球表面能量循环和生命活动的基本载体。太阳辐射到地表的能量的年平均功率达 84 PW（1 PW=10^{15} W），其中 70% 以上被水动态吸纳，这比人类目前年平均能量消耗功率（0.018 PW）高三个数量级。水以热能、动能的形式存储所吸收的热量，再以蒸发、凝结、降雨、风浪等方式把存储的太阳能转化成机械能、热能等多种形式。更为重要的是，水通过蒸发、对流等过程，无时无刻不转移着环境中的巨大能量。

从水的储能和换能过程中提取对人类有用的能量一直是人类的追求。人类利用水力机械的历史可追溯到古希腊时期，并随电磁学的诞生发展了水电技术。传统技术主要利用水流的动能产生有用的机械能，并间接发电。如今，全球 17% 的电力生产仍来自水力发电。水力发电原则上是一种可持续的能源形式，因为其资源丰富且可再生，但它通常需要建设大型设施（如大坝和水库），可能会对当地社区和环境造成负面影响。水力发电的另一种方式是利用盐水的渗透。这种发电方式需要流动形式的水，可通过设计特定的表面形貌和化学性质以最大限度地获得渗透能。虽然对这种发电方式的研究已经在学界引起了相当大的关注，但其在原理上依赖于水的盐度梯度，相对较高的成本和较低的效率阻碍了其发展。

（二）水伏效应及水伏学

近年来，以我国学者为代表的研究者发现，通过碳纳米管、石墨烯等低维碳材料与水直接相互作用，可以将水中更丰富的热能和机械能转化为有用的电能输出（Anon，2018；郭万林，2018；郭万林和张助华，2018；Zhang et al.，2018）。例如，石墨烯可通过双电层的边界运动将拖动和下落水滴的能量直接转化为电能（拖曳势）、把波动能转化为电能（波动势）。后来更是发现廉价的炭黑等纳米结构材料可通过大气环境下无所不在的水的自然蒸发持续产生伏级的电能。蒸发发电带来的最大优势是它不需要任何机械输入。在环境蒸发条件下，1 cm 大小的炭黑片可稳定输出 1 V 的电压。类比于光伏、压电等能量转换效应，这类通过材料与水作用直接转化水能为电能的现象被称为水伏效应（郭万林和张助华，2018）。2018 年 12 月的《自然-纳米技术》以封面亮点标题的形式提出了“水伏学”一词（Anon，2018），并指出其是从水中获取电能的全新途径。水伏效应的理论与技术研究目前仍处于初期研究阶段，但其所展示的发展潜力和独特应用前景已透出水伏科学技术的曙光。

二、水伏科学与技术研究的历史及现状

早期对水与低维材料结构相互作用的研究发现，水分子流经碳纳米管可通过声子和库仑拖曳效应在碳纳米管中产生电信号（Kral & Shapiro，2001；Ghosh et al.，2003；Cohen，2003），其原理通常为量子力学现象的二级效应，因此信号微弱，缺乏提升空间和应用潜力。

在 2014 年发表于《自然-纳米技术》等的系列论文中，南京航空航天大学纳米科学研究所的郭万林领导的多学科研究团队率先在国际上发现，液滴在涂覆单层石墨烯的固体表面运动会产生与液滴运动速度成正比的拖曳液滴发电的拖曳势（Yin et al.，2014a），液面沿涂覆石墨烯的固体表面上下波动会产生与波动速度成正比的波动发电的波动势（Yin et al.，2014b），并总结出双电层边界运动发电的新的动电效应理论，实现了基于拖曳势的传感、云雾雨滴能量收集等新生电现象（图 4-16）。该系列研究结果正式开启了水伏效应的研究热潮。2017 年进一步可靠、稳定地实现了大气环境下水的自然蒸发在廉价碳纳米材料薄膜中产生持续的伏级电压或输出直流电流（Xue et al.，

2017），数平方厘米的薄膜产生的电能已经能够直接驱动液晶显示器。

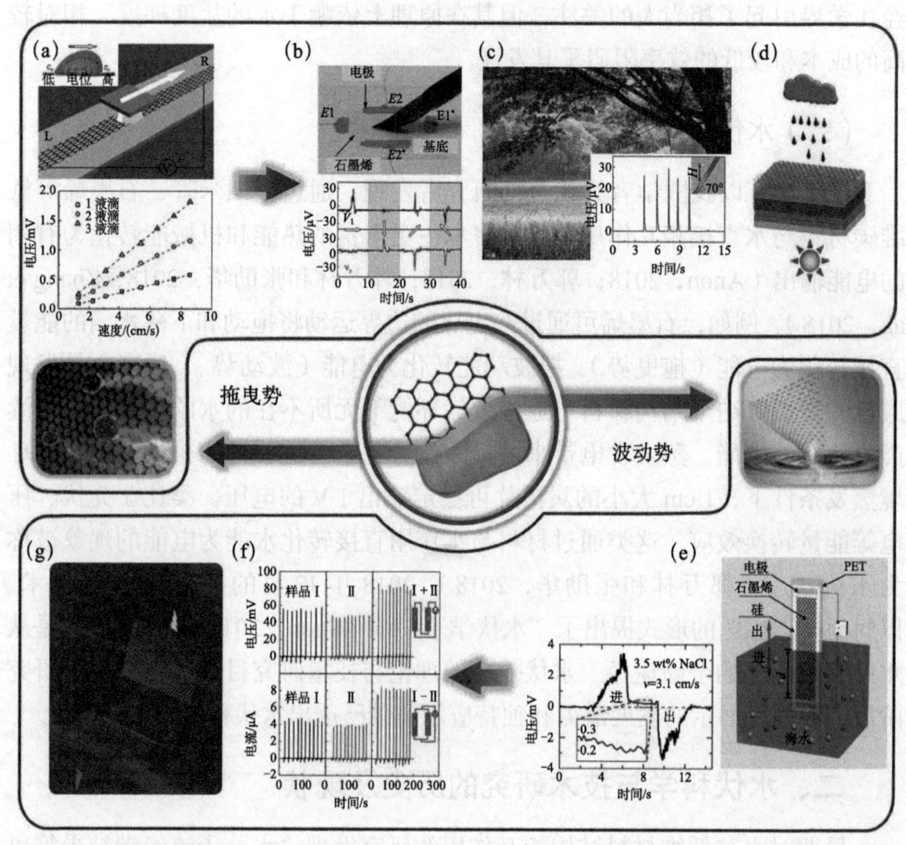

图 4-16 在石墨烯中发现的水伏效应：拖曳势与波动势

（a）上图为石墨烯拖曳势的实验装置示意图，其内插图为液滴运动产生电位差的原理图；下图表示同时拖动 1～3 液滴运动时的电压速度关系。（b）石墨烯作为书写传感器件的应用。当在四电极石墨烯器件表面书画时，相对电极间产生的拖曳势可以感知笔画方向。（c）拖曳势对雨滴能的利用：离子液滴溅落在倾斜的石墨烯表面时产生的电压是拖动该液滴产生电压的数倍。（d）石墨烯捕获雨滴能量以及与太阳能电池结合形成全天候供电系统。（e）右图为石墨烯波动势的实验装置图；左图表示石墨烯样品插入溶液过程中的典型电压信号。（f）两个石墨烯样品串联放大电压、并联放大电流的信号图。（g）石墨烯样品在海水波浪中发电的示意图

在上述发现之后，国际上迅速跟进，如 Kwak 等将石墨烯覆层置于 PTFE 基底上以增强静电摩擦效应（Kwak et al.，2016），发现 0.1 mL 的液滴可以产生 100 mV 以上的电压，0.6 mL 的液滴可以产生 0.4 V 电压，3 个 0.6 mL 的液滴串联可以得到 1.1 V 电压。在模拟降雨的条件下，这类石墨烯器件可以把不同的降雨强度转化为幅度不等的电脉冲（Zhang et al.，2016）。曲良体研

究组制备了内部含氧基团呈梯度分布的氧化石墨烯膜，发现水分子吸附能引起内部离子的重新分布，产生几十到几百 mV 的电位差（Zhao et al.，2015；Xue et al.，2016）。2018 年，华中理工大学的周军课题组在湿度发电方面也取得了喜人的进展，把水在实验室大气环境下的自然蒸发产生的电流提高了两个数量级以上（Xue et al.，2018）。

此外，水与材料相互作用的研究还可拓展至利用纳米结构吸水-脱水导致的变形形成的机械能输出，如美国得克萨斯大学达拉斯分校的鲍曼（Baughman）研究组利用碳纳米管纤维吸水膨胀的效应制作了可高速旋转的碳纳米管转子用作人工肌肉（Foroughi et al.，2011）。美国哥伦比亚大学的陈曦等利用枯草芽孢杆菌涂覆的聚酰亚胺胶带，在相对湿度从 10% 变化到 90% 时，利用细菌的吸水膨胀使胶带长度有数倍的伸长，并将其用于驱动发电机（Chen et al.，2015；Cavusoglu et al.，2017）。

目前已实现的水伏装置可以构成非常简洁的清洁能源供应系统和智慧能源系统。例如水蒸发无处不在，不受天气、时空的影响，其水伏系统可结合风能、太阳能、废热等显著提高水蒸发发电量，在理论上具有比光伏技术更大的发展空间。水伏系统天然容易与流体环境和生物环境结合：利用液滴和液面的拖曳势，可以直观地测量液体流向、流速乃至离子浓度等流体信息（Anon，2018；郭万林和张助华，2018）；利用碳纳米管纤维可在血流中产生准伏级电信号（Xu et al.，2017）；与光电等微纳传感器件结合可以构成自供能、自驱动传感系统等（Zhong et al.，2017）。现有的电子器件大多在固体结构或固体-固体界面工作，水伏系统在原理上工作于固-液或固-液-气界面，是对传统电子器件工作模式的发展，具有巨大的发展空间。

三、水伏科学与技术研究目前面临的重大科学问题及前沿方向

当前人们对水伏效应的生电机理认识还处于经验式推测的阶段，对水伏效应涉及的固液界面力-电-热相互作用规律和影响因素缺乏量子层次的深入理解。这也导致了对不同水伏生电现象缺少统一认识，水伏能量转换的基本物理规律不清晰。由于水伏效应依赖固液界面的相互作用，体系复杂，对牛顿力学和量子力学等传统理论均提出了挑战。另外，水与功能材料的界面包含力、电、热、流等耦合，描述这类耦合并最终认识水伏效应的物理本质需要拓展纳米尺度多场耦合物理力学理论框架。因此，探索水伏学的物理力学规律和建立水伏学的物理力学理论体系是目前面临的关键科学问题。

水伏学的研究还可进一步启发人们认识人类智慧的起源——大脑，并以此为基础发展类脑人工智能技术。神经信号主要为电信号，具体表现为溶液中的电位变化，其产生和传递涉及离子传输、离子定向扩散等。传统神经科学理论认为，这些以离子的分布和迁移为主要载体的电信号在神经元之间的传递是人脑实现智能化运行的基础。然而，纯水流经带电腔道可产生电压这一事实暗示生物通道中的电信号产生与传递可能并不限于离子，可能存在更多机制。不同于电脑依赖于晶体管，大脑运行则主要依赖于液体和软物质，而人们对液体的微纳观水伏效应认识远滞后于固体。对应于针对固体的量子力学，发展针对固-液的新物理力学理论无论对水伏效应研究发展还是对脑科学认识的进步都不可或缺。

四、研究总结与展望

迄今，人们已发掘和设计了很多材料结构以研究水伏效应，以期将水体中储存的巨大能量直接转化为电能，但总的来说，能量密度低和转化效率低仍是制约其应用的首要问题。现有报道的水伏装置，输出功率一般介于亚微瓦到毫瓦量级，功率密度一般小于 10 W/m^2，这还远远不能与现有的电力设备相提并论（三峡电站单个水轮机组的输出功率就高达 700 多 MW，其对水体机械能的转化效率更可高达 90% 以上）。因此，发展水伏技术仍需要从材料、理论等多个方面同时努力，协同突破。例如，蒸发发电材料需人工设计更佳的三维构型和表面基团修饰，以更有效地诱导电荷转移、增加通量和降低内阻；动电效应器件中，可以集成气液多种界面和大量平行微纳通道以提升转化效率和输出功率。在这些方面，实验和理论都进展很快，已经逐渐有了利用水伏器件直接驱动商业化电子装置的示范。例如，几个串接的蒸发发电装置可以持续地驱动小型薄膜型液晶显示屏，湿度驱动的氧化石墨烯膜可以点亮发光二极管设备等。可以预期，不久之后水伏能量捕获将是现有绿色能源体系的有力补充。

我们可以设想这样的基于水伏原理的能量综合利用系统（图 4-17）：在广阔的浪涌海域利用波动势发电，或在江河湖海等蒸发旺盛地区安装膜状或网状的蒸发发电“渔场”；在获取电力的同时，还可获得经蒸发过滤的净水，对远陆地海域有着重要的实用价值。当蒸汽经过大气输运形成降雨时，可以利用拖曳势发电装置直接收集雨滴能量；降雨在地面汇集过程中可以利用流动势装置发电。这种系统直接将环境中的潜热等低品质能源转化为高品质电能，不仅以环境无损的方式提供了电能，而且兼有净水生产、气温调节的功效，极具发展潜力。

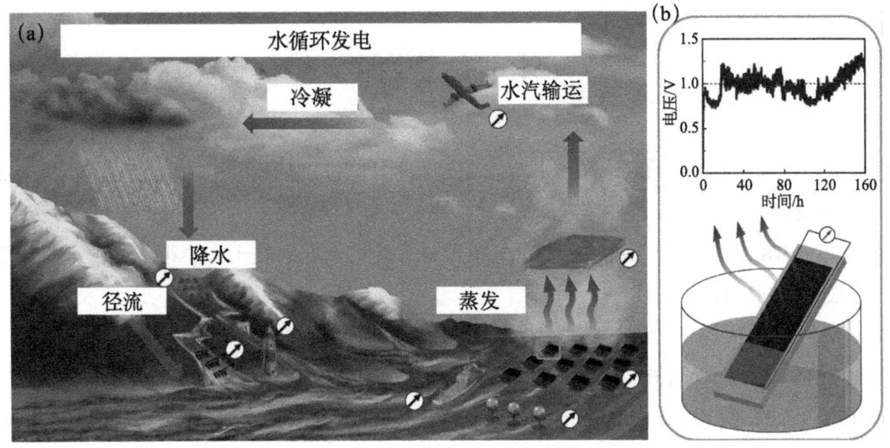

图 4-17　基于水伏原理的能量综合利用系统示意图

（a）地球水循环的五个过程：蒸发、水汽输运、冷凝、降水、径流。电流表符号指示相应的水能可通过适当的水伏技术转化为电能（见文中详细介绍）。（b）对应（a）图中的蒸发发电放大图，其中下图为蒸发发电的实验装置示意图，实验中将载有炭黑薄膜的载玻片斜插在器皿中，炭黑薄膜部分浸入水中；上图为水蒸发在炭黑两端产生的电压与时间的关系，说明蒸发发电是一个持续稳定的过程

五、学科发展政策建议与措施

水伏效应涉及低维材料的微观结构设计和宏观性能的优化，以及水在受限环境中复杂的动力学和电荷转移过程，最根本的是处于液固界面处不同状态的原子 / 分子间的场耦合、伴随的电荷动态重布和能量转移，以及不同特征尺度下液固耦合的协同作用。这是汇聚多尺度、跨学科、多场耦合等特征的前沿问题。

我国学者不仅对水伏效应概念的提出和研究的发起发挥了主导作用，在后续研究中也一直引领着该领域的发展，使得水伏科学与技术成为当前我国宝贵的原始创新研究阵地。这为我国形成既有原始概念创新又有自主知识产权的颠覆性技术提供了重大机遇。组织国内的优势力量，在国家专项基金等的强力资助下，通过 5～10 年的持续研究，水伏效应不仅有望产生新的清洁能源供应，而且可以为水能利用、胶体科学、人脑科学等领域的进步提供基础性的科学和技术支撑。

郭万林（南京航空航天大学）

本节参考文献

丁天朋，刘抗，李嘉，等 . 2018. 水蒸发驱动柔性自支撑复合发电碳膜 . 科学通报，63: 2846-2852.

郭万林，2018. 水伏科学与技术的曙光 . 中国科学院第十九次院士大会第六届学术年会学术报告汇编，1: 608-618.

郭万林，张助华 . 2018. 水伏科学与技术的召唤 . 科学通报，63: 2804-2805.

Anon. 2018. More power from water. Nature Nanotechnology, 13: 1087-1087.

Cavusoglu A H, Chen X, Gentine P, et al. 2017. Potential for natural evaporation as a reliable renewable energy resource. Nature Communications, 8: 617.

Chen X, Goodnight D, Gao Z H, et al. 2015. Scaling up nanoscale water-driven energy conversion into evaporation-driven engines and generators. Nature Communications, 6: 7346.

Cohen A E. 2003. Carbon nanotubes provide a charge. Science, 300: 1235-1236.

Foroughi J, Spinks G M, Wallace G G, et al. 2011. Torsional carbon nanotube artificial muscles. Science, 334: 494-497.

Ghosh S, Sood A K, Kumar N. 2003. Carbon nanotube flow sensors. Science, 299: 1042-1044.

Kral P, Shapiro M. 2001. Nanotube electron drag in flowing liquids. Physical Review Letters, 86: 131-134.

Kwak S S, Lin S S, Lee J H, et al. 2016. Triboelectrification-induced large electric power generation from a single moving droplet on graphene/polytetrafluoroethylene. ACS Nano, 10: 7297-7302.

Xu Y F, Chen P N, Zhang J, et al. 2017. A one-dimensional fluidic nanogenerator with a high power conversion efficiency. Angewandte Chemie International Edition, 56: 12940-12945.

Xue G B, Xu Y, Ding T P, et al. 2017. Water-evaporation-induced electricity with nanostructured carbon materials. Nature Nanotechnology, 12: 317-321.

Xue J L, Zhao F, Hu C G, et al. 2016. Vapor-activated power generation on conductive polymer. Advanced Functional Materials, 26: 8784-8792.

Yin J, Li X M, Yu J, et al. 2014a. Generating electricity by moving a droplet of ionic liquid along graphene. Nature Nanotechnology, 9: 378-383.

Yin J, Zhang Z H, Li X M, et al. 2014b. Waving potential in graphene. Nature Communications, 5: 3582.

Zhang Y, Tang Q W, He B L, et al. 2016. Graphene enabled all-weather solar cells for electricity harvest from sun and rain. Journal of Materials Chemistry A, 4: 13235-13241.

Zhang Z H, Li X M, Yin J, et al. 2018. Emerging hydrovoltaic technology. Nature Nanotechnology, 13: 1109-1119.

Zhao F, Cheng H H, Zhang Z P, et al. 2015. Direct power generation from a graphene oxide film under moisture. Advanced Materials, 27: 4351-4357.

Zhong H K, Xia J, Wang F C, et al. 2017. Graphene-piezoelectric material heterostructure for harvesting energy from water flow. Advanced Functional Materials, 27: 1104226.

第五节　极限条件下水 / 二氧化碳转化与地外人工光合成材料

一、研究概述

太空探索已成为人类共同目标，重返月球、载人火星探索等人类历史上的重大里程碑任务已逐步实施（International Space Exploration Coordination Group，2018）。地外生存是人类实现长期太空飞行（地球和月球轨道任务、地球和火星长期飞行任务）、地外长期居住和地外移民（月球和火星基地）的基本能力。人类的氧气消耗约 1 kg/（d·人），人类的生存一刻也离不开氧气的供应。人类脱离地球，开展太空探索，必须具备氧气的长期持续供应能力。将人类呼吸产生的二氧化碳转换为氧气，实现密闭空间的废弃物原位资源再生循环，可大大降低载人空间站、载人深空飞船的物资供应需求。更为重要的是，利用火星等地外大气环境中丰富的二氧化碳和水原位资源生产氧气与燃料，是实现人类在其他天体上长期生存和深空往返推进运输，支撑可承受、可持续的载人深空探索任务的重要基础。因此，水 / 二氧化碳转化有望在解决上述问题中发挥重要作用。由 20 世纪 60 年代发展至今，萨巴捷反应与博施（Bosch）还原法是二氧化碳还原技术中两种比较主流的方式，水电解法也得到广泛应用。近年来，随着光催化技术的快速发展，人工光合成也有望在此领域得到应用。太空探索活动中面临低微重力、强辐射、极端温压等特殊环境，给地面水 / 二氧化碳转化技术的空间应用带来一系列挑战，极限条件下水 / 二氧化碳转化与地外人工光合成将成为未来的前沿研究方向。

二、历史及现状

美国等航天强国为解决载人空间站和深空探索关键问题，在继承传统地面技术的基础上，持续开展了水 / 二氧化碳转化技术的研究。例如，为解决国际空间站的氧气供应难题，采用电解水的方式为宇航员补充氧气；为实现宇航员排出的二氧化碳再利用，美国 NASA 和日本宇宙航空研究开发机构（JAXA）计划在空间站上开发一套二氧化碳还原和氧气获取装置，其中二氧化碳还原是利用萨巴捷反应通过氢气将二氧化碳转化成甲烷和水，氧气通过电解水装置获得。萨巴捷反应过程为气固两相过程，核心装置温度为 250～450 ℃，气体最小压力为 55kPa，其中地面实验装置质量约 41 kg，总功率超过 100 W。这一系统已经于 2010 年 10 月完成了空间站测试，如图 4-18 所示（Junaedi et al., 2011）。JAXA 也正在开发新一代空间电解水装置，如图 4-19 所示（Sakurai et al., 2015）。采用阴极给水设计并对质子交换膜电解池进行优化，有望提高能量转换效率。在国际空间站上，通过上述化学反应过程可实现水、氧气等物质的循环利用，更好地支持环境控制和生命保障（ECLS）（图 4-20）。对于微重力环境的影响，JAXA 的研究人员利用抛物线飞行和落塔试验对水电解装置进行了大量的研究与改进工作，包括电解池工作温度、气液分离膜压强、电解质组分、工作电压和电流等。然而，即使经过了各项优化工作，该装置在微重力条件下的工作效率仍不足一般重力环境下的 1/3。研究表明，过溶气体在电极表面附近聚集形成的溶解气体分子的过饱和层，对物质输运速率和反应效率具有非常重要的影响。松岛永佳（Hisayoshi Matsushima）等发现，微重力环境下电极的润湿性对气泡的生长和电化学反应都有显著的影响（图 4-21）。但由于微重力实验和相关理论研

图 4-18　国际空间站上搭载的萨巴捷反应系统核心单元

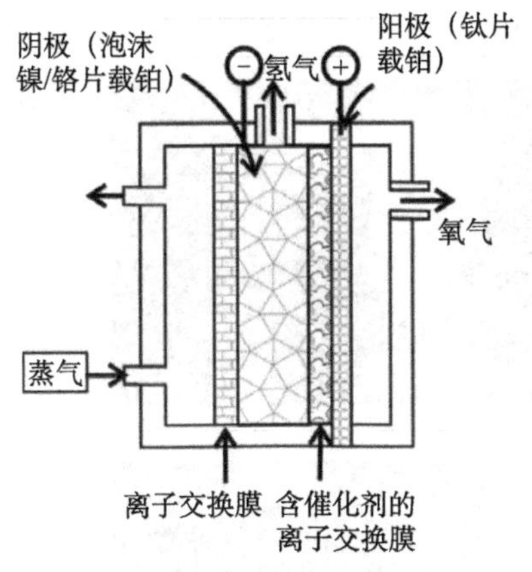

(a) 电解水装置原理图

(b) 电解水装置外观图

图 4-19　JAXA 电解水装置原理图与外观图

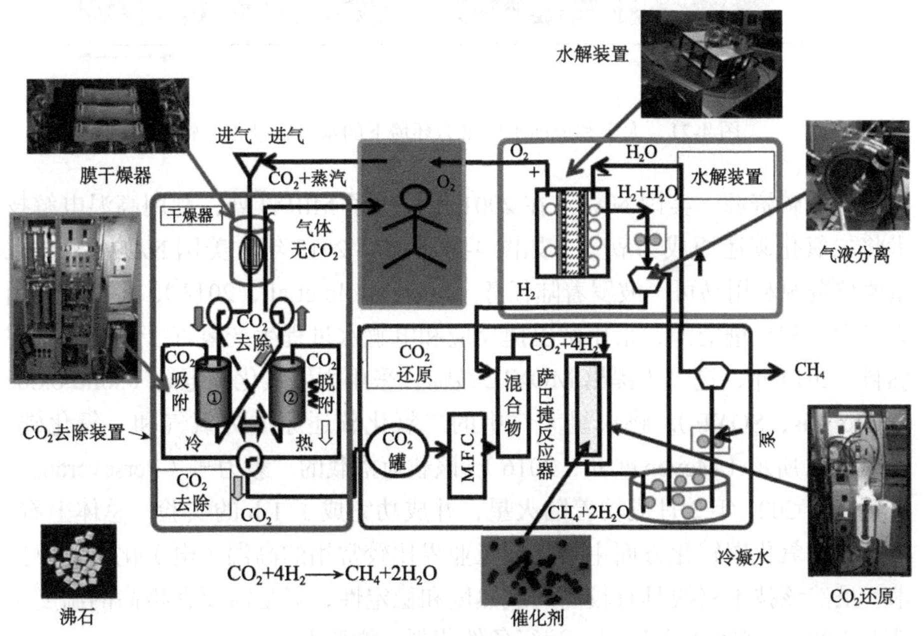

图 4-20　国际空间站新概念 ECLS 系统（Sakurai & Terao，2016）

究的缺乏，微重力条件下的多相催化反应涉及的气泡生成、演化与脱离、反应界面多场耦合机理等关键问题亟待解决。

对于更具挑战性的月球和火星等载人深空探索任务，美国最早提出利用原位获取的水和二氧化碳等资源来原位制造氧气和燃料。针对火星上大气中

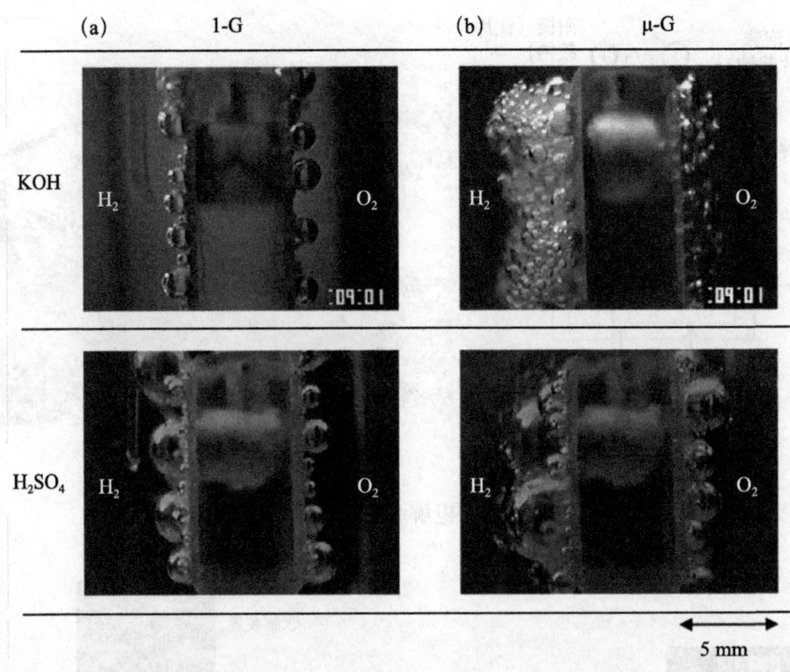

图 4-21 太空和地面不同重力环境下的水电解过程比较

的二氧化碳资源，美国 NASA 在 2001 年提出了 MIP 计划，利用高温电解技术将二氧化碳还原成为氧气，如图 4-22 所示。2013 年，美国 NASA 提出火星原位资源利用马可·波罗着陆任务（Interbartolo et al., 2013），将综合利用火星大气和星壤资源，采用萨巴捷反应和电解水过程生产氢气、氧气和甲烷燃料。2014 年，进一步提出 MOXIE 载荷，采用固体氧化物电解（solid oxide electrolysis，SOXE），将火星大气中的二氧化碳还原成为氧气和一氧化碳，如图 4-23 所示（Meyen et al., 2016）。该载荷搭载的"毅力号"(Perseverance）火星车于 2021 年 2 月成功着陆火星，并成功完成了 1 h 的实验。总体上看，美国在二氧化碳转化方面主要采用工业界比较常用的高温（电）化学转换技术。虽然该技术路线具有较高的成熟度和稳定性，但是需要在极高的温度条件下（900～1600 ℃）进行，运行条件苛刻，能耗大。

人工光合成是模拟地球上绿色植物的自然光合作用，通过光电催化加速，可控地将二氧化碳转化成为氧气和含碳燃料的化学过程。1972 年，日本东京大学的藤岛昭和本多健一报道了二氧化钛电极在紫外光照射下使水发生光解产氢，拉开了人工光合作用研究的序幕（Fujishima & Honda，1972），他们又在 1979 年报道了光解二氧化碳的研究工作。在随后的十几年中，包括

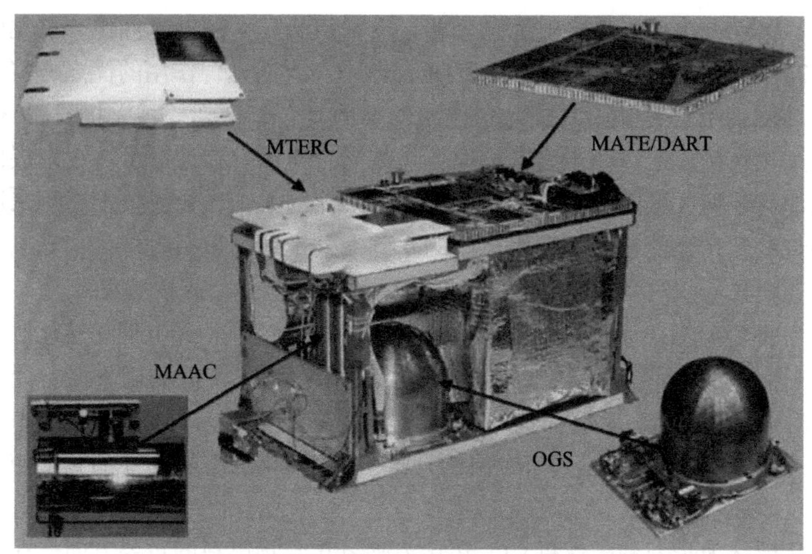

图 4-22　MIP 计划中的 CO_2 转换系统

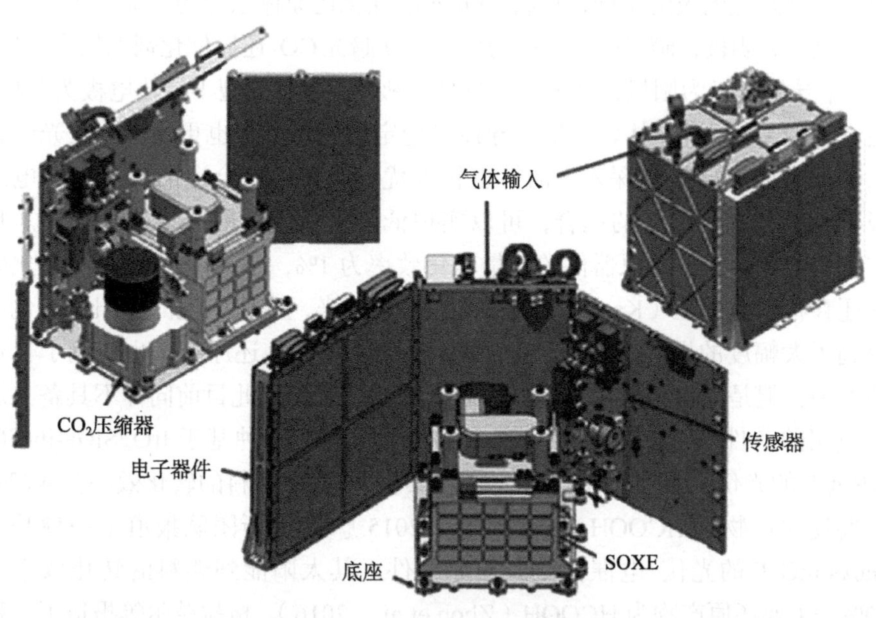

图 4-23　MOXIE CO_2 转换系统

巴德、堂免一成、安保正一（Masakazu Anpo）等国外科学家都进行了大量的研究工作。2001 年，邹志刚提出了调控光催化材料能带结构的新理论和新方法，发展了新一代可见光响应型光催化材料，拓宽了光催化材料的响应范围，实现了可见光下水的完全分解，进一步研究了可见光催化二氧化碳还原。2016 年，杨培东团队结合人工光合作用和天然光合作用的优点，将光吸收半导体纳米材料和不具备光合作用的细菌复合，得到一种无机材料-微生物的杂化系统，通过两步法来模拟自然的光合作用。在地球可持续发展面临的能源与环境关键问题的推动下，利用太阳能的人工光合成技术近年来取得重要进展。在光催化材料取得突破的基础上，发展了包括光催化、光电化学电池、串联-光电化学电池和光电池-电解池等光化学合成技术路线。

当前 CO_2 光还原技术主要包括粉末体系 CO_2 还原、光电催化转化法、光伏-电催化还原等。粉末光催化 CO_2 还原体系存在太阳能利用率低、光生电子-空穴对易复合、能带位置不匹配、产物选择性较差等问题，多集中于基础研究领域。与光催化转化法相比，光电催化转化法施加了外部偏压使光生电子发生定向移动，降低了光生电子-空穴对的复合概率，因此，光电催化转化法被认为是一种高效将 CO_2 转化为碳氢化合物燃料的重要方向。在光电催化还原 CO_2 的体系中，阴极发生 CO_2 的还原反应，阳极发生 H_2O 的氧化反应。根据感光电极的不同，可将 CO_2 光电催化还原体系分为三类：① p 型半导体电极为阴极，暗光析氧电极为阳极；②暗光 CO_2 还原催化剂电极为阴极，n 型半导体电极为阳极；③ p 型半导体电极为阴极，n 型半导体电极为阳极。通常来说，n 型半导体较 p 型半导体光稳定性好，价格也更便宜，在光电解水产氧中研究已较为深入，活性也较为优异，若配合合适的 CO_2 还原电阴极催化剂，实现两者的耦合，可以实现低能量输入下的 CO_2 高效还原。目前，CO_2 光电催化还原器件的能量转化效率为 1%，距离 10% 的目标转化效率还存在较大差距（Kang et al.，2015）。近年来，光伏电池的太阳能转化率得到了大幅度的提高，受益于此，光伏-电催化 CO_2 还原技术也得到了长足的发展，但是光伏材料成本高昂、制作工艺复杂，因此目前同样不具备大规模应用的条件。荒井健男（Takeo Arai）等报道了一种基于 IrO_x/SiGe-jn/CC/p-RuCP 的光伏-电催化 CO_2 还原器件，其太阳能到燃料的转化效率为 4.6%，主要还原产物为 HCOOH（Arai et al.，2015）。路易斯团队报道了一种基于 GaAs/InGaP 的光伏-电催化 CO_2 还原器件，其太阳能到燃料的转化效率为 10%，主要还原产物为 HCOOH（Zhou et al.，2016）。格拉兹尔等报道了一种基于 $CH_3NH_3PbI_3$ 的光伏-电催化 CO_2 还原器件，其太阳能到燃料的转化效率为 6.5%，主要还原产物为 CO（Schreier et al.，2015），2017 年，进一步报道了一种基于 GaInP/GaInAs/Ge 的光伏-电催化 CO_2 还原器件，其太阳能到燃料的转化效率为 14.4%，这也是目前文献报道的最高值，其主要还原产物为

CO（Schreier et al.，2017）。

当前 H_2O 光还原技术主要包括粉末光催化、光电化学、串联-光电化学池和光电池-电解池等光化学合成技术路线。在粉末光催化技术中，利用半导体材料吸收太阳光，在材料的导带和价带上分别产生电子与空穴，利用电子的还原性将 H_2O 还原为 H_2，并在此过程中将太阳能转化为化学能储存。另外，利用价带上的空穴将 H_2O 氧化生成 O_2。但是由于电子和空穴的非定向迁移，其太阳能利用率较低。堂免一成等报道了一种基于 La 和 Rh 掺杂的 $SiTiO_3$ 与 Mo 掺杂的 $BiVO_4$ 组成的 Z 型异质结，其太阳能到氢能的转化效率达到 1.1%，这也是粉末体系中公认的比较高的太阳能转化率（Wang et al.，2016）。目前的光电池-电解池系统中，叠层太阳能电池可实现较高的太阳能转化率，但结构复杂，电极制备成本相对较高。早在 1998 年，特纳（Turner）等通过构建 $GaInP_2$/GaAs 叠层电池，实现了高达 12.4% 的太阳能到氢能的转化效率（Khaselev & Turner，1998）。此后，马赛亚斯·梅（Matthias M. May）等设计了 GaInP/GaInAs 叠层电池，实现了 14% 的太阳能转化效率（May et al.，2015）。托德·多伊奇（Todd G. Deutsch）等同样采用 GaInP/GaInAs 叠层电池实现了 16% 的太阳能转化效率（Young et al.，2017）。哈拉米洛等构建了 InGaP/GaAs/GaInNAs(Sb) 叠层太阳能电池（图 4-24），实现了 30% 的太阳能转化效率，这也是当时叠层太阳能电池最高的太阳能转化效率（Jia et al.，2016）。通过 p 型光阴极和 n 型光阳极组成的串联结构，不仅拥有较高的理论转化效率（约 28%），而且成本相对较低，是理想的器件组成，但是目前能够实现的转化率依然较低。目前 p 型光阴极和 n 型光阳极组成的串联结构

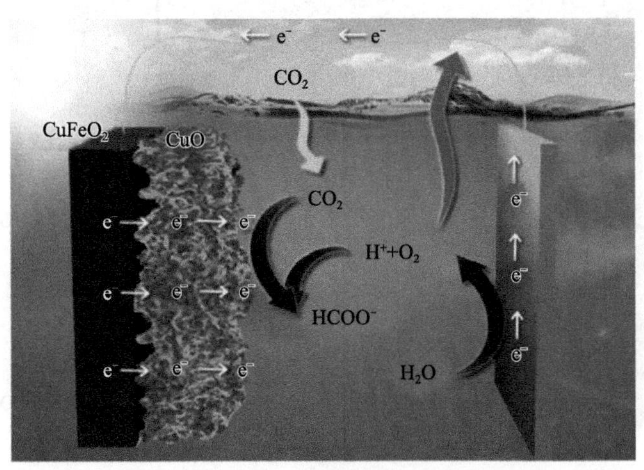

图 4-24　CO_2 光电催化还原示意图

中，波士顿学院的王敦伟团队通过 Fe_2O_3 基光阳极和非晶 Si 基光阴极组成的器件，获得了当时最高的（约 0.9%）太阳能到氢能的转化效率（Jang et al., 2015）；天津大学的巩金龙团队通过在 CIGS 上构筑 p-n 结，将其与 $BiVO_4$ 光阳极组成器件，获得了约 1.01% 的太阳能转化效率（Chen et al., 2018）；格拉兹尔、罗景山团队利用 Cu_2O 作为光阴极，通过 Ga_2O_3、TiO_2、NiMo 等多层结构调控后，与 $BiVO_4$ 光阳极组成器件，最终获得了约 3% 的太阳能转化效率（图 4-25）（Pan et al., 2018）；堂免一成团队调控 CIGS 中元素的比例，将获得的高性能 $CuIn_{0.5}Ga_{0.5}Se_2$ 光阴极与 $BiVO_4$ 光阳极串联，获得了约 3.7% 的太阳能转化效率，这也是当时获得的最高效率（Kobayashi et al., 2018）。p 型光阴极和 n 型光阳极组成的串联结构中，太阳光从 n 型光阳极侧照射，光阳极吸收短波长的光，长波长光穿透光阳极被后侧的光阴极吸收。在光阳极上发生 H_2O 的氧化反应，在光阴极上发生 H_2O 的还原反应（图 4-26）。为了使得此器件能够无偏压工作，就需要光阳极和光阴极的光电流相互匹配，有交点电流（J_{op}）器件的太阳能转化效率 $\eta = J_{op}$（1.23 V）$/P_{in}$，P_{in} 为入射光

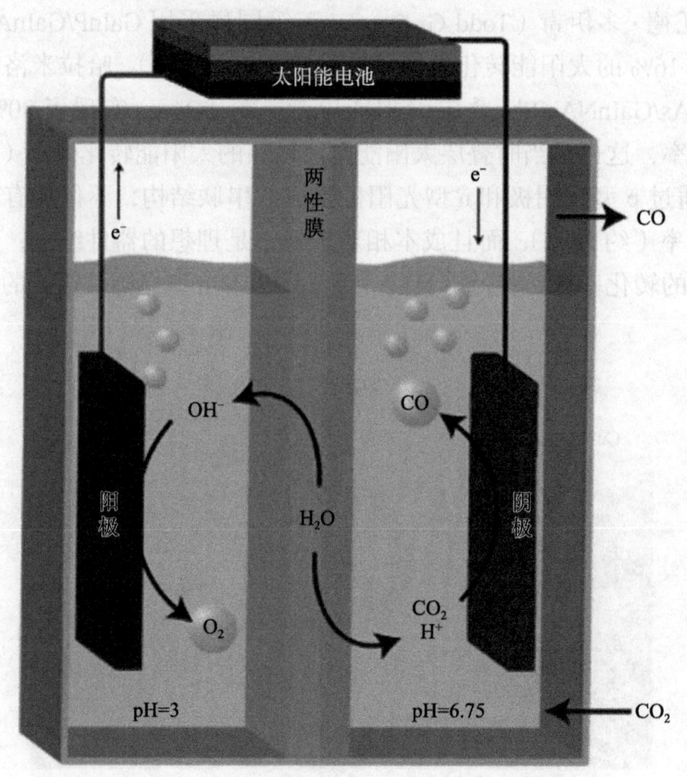

图 4-25 光伏－电催化 CO_2 还原技术示意图

强度 (mW/cm^2)，因此两个电极的交点电流决定了器件的最终效率（图 4-27）。为了解决 p 型光阴极和 n 型光阳极组成的串联结构中交点电流较低的问题，部分研究者通过将光伏电池与光电极结合来提高整个器件的工作效率（图 4-28）。韩国浦项科技大学、德国亥姆霍兹柏林材料与能源中心和韩国蔚山科学技术院的研究者将 Fe_2O_3 和 $BiVO_4$ 光阳极并联，并与 c-Si 太阳能电池串联，构建了无偏压的分解水器件，获得了约 7.7% 的太阳能转化效率（图 4-29）（Kim et al., 2016）。王连洲、刘岗团队等将双 $BiVO_4$ 光阳极并联，并与钙钛矿太阳能电池串联构建了无偏压的分解水器件，获得了约 6.5% 的太阳能转化效率（Wang et al., 2018）。

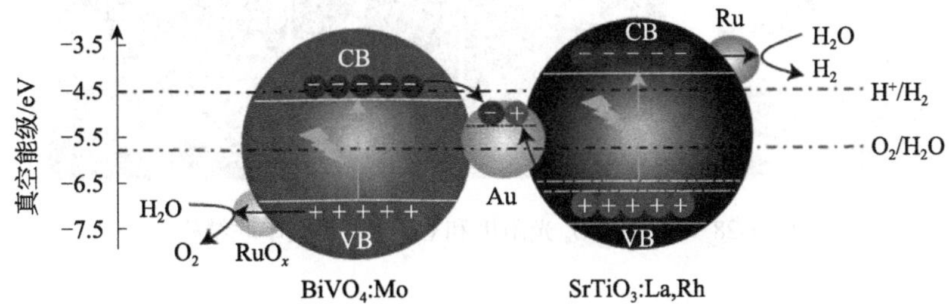

图 4-26　粉末体系光催化分解水示意图

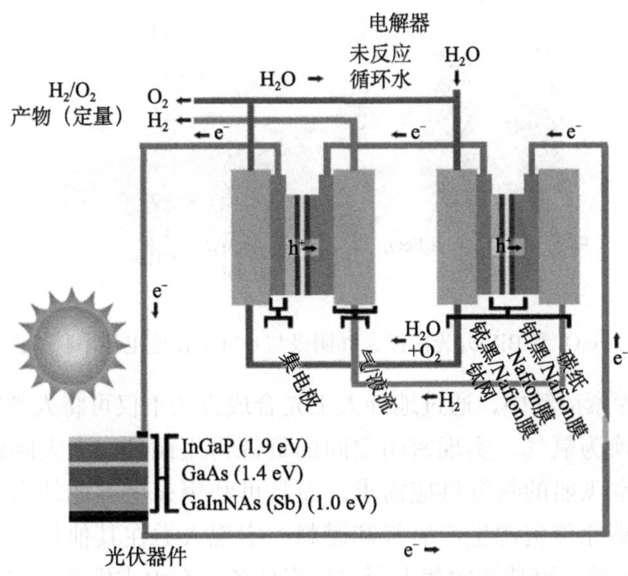

图 4-27　InGaP/GaAs/GaInNAs(Sb) 叠层太阳能转化

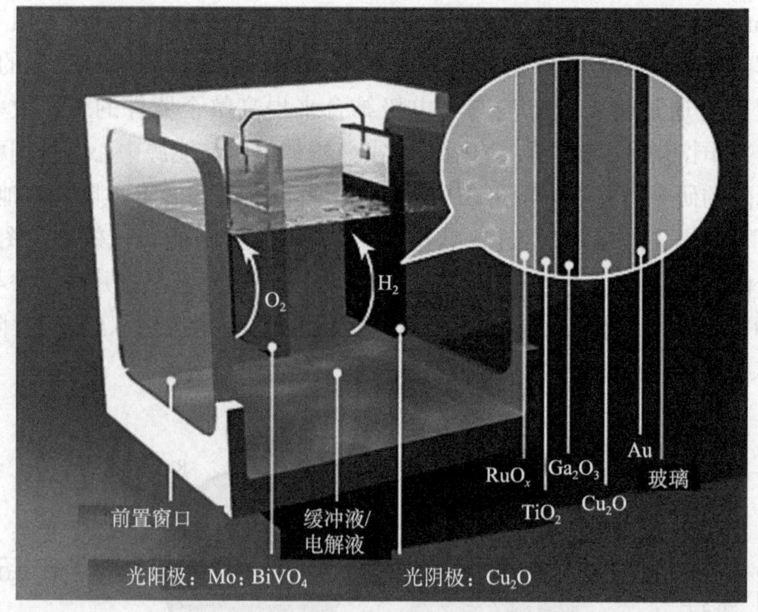

图 4-28　Mo:$BiVO_4$ 光阳极和 Cu_2O 光阴极串联结构

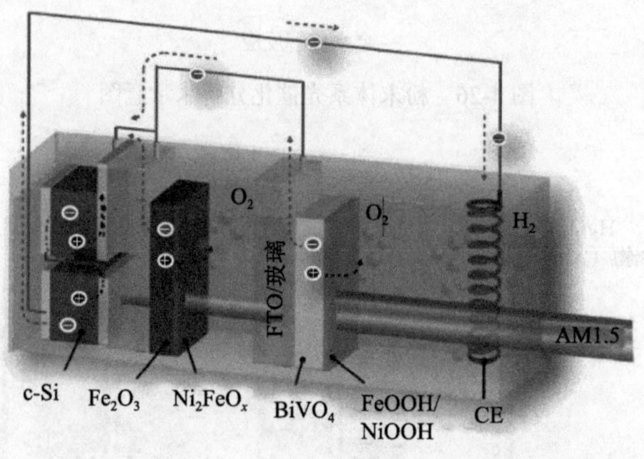

图 4-29　Fe_2O_3 和 $BiVO_4$ 光阳极、Pt 阴极与 c-Si 太阳能电池串联结构示意图

在太空探索活动中，通过地外人工光合成方法不仅可将人类呼吸产生的二氧化碳转换为氧气，实现密闭空间的资源再生循环，大大降低载人空间站、载人深空飞船的物资供应需求，而且可利用火星等地外大气环境中丰富的二氧化碳原位资源生产氧气和燃料，实现人类在其他行星上的地外生存，支撑可承受、可持续的载人深空探索任务，有力支撑我国载人航天和后

续深空探测的发展。与传统高温（电）化学转换技术相比，地外人工光合成可在常温下实现水 / 二氧化碳转换，可解决运行条件苛刻（高温、高压条件）、能耗高等问题，是太空探索的核心能力。2014 年，钱学森实验室结合我国太空探索任务，提出了地外人工光合成概念并逐步开展材料与器件研究。2015 年，美国《NASA 技术路线图》提出了微生物辅助人工光合成概念。2018 年，钱学森实验室和南京大学联合开展地外人工光合成装置研究与空间搭载实验，通过与微流芯片的结合（图 4-30 和图 4-31），利用液体剪切力迫使生成的气体脱离电极表面，可防止微重力条件下过溶气体在电极表面附近聚集形成溶解气体分子的过饱和层，提高物质输运速率和反应效率；通过太阳能聚光及光热电协同转化，可进一步提高系统对整个太阳光波谱的能量利用效率。该系统研制将为深空探测和载人航天的后续任务提供重要手段。

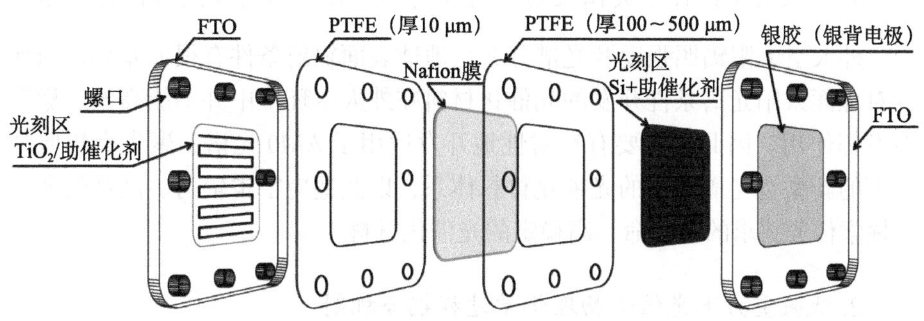

图 4-30　微流芯片结构示意图

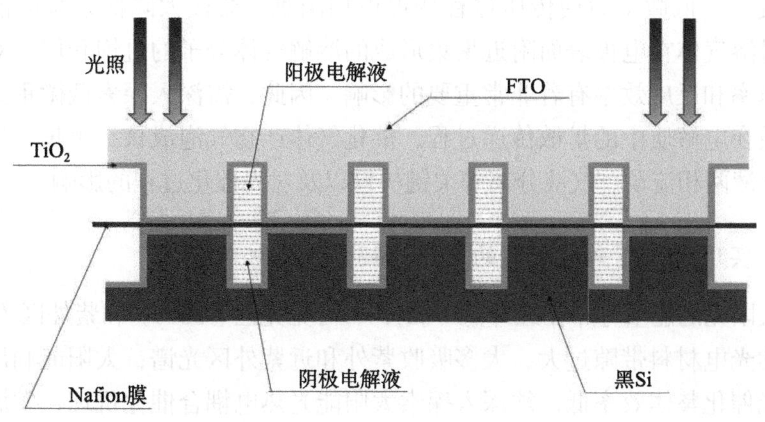

图 4-31　微流芯片光合成反应器

三、目前面临的重大科学问题及前沿方向

（一）重大科学问题

面对全球探索深空的共同目标和我国航天强国建设的重大任务，如何使人类具备脱离地球的生存能力，真正实现可承受、可持续的太空探索，面临极限条件下水/二氧化碳转化重大挑战，将迎来地外人工光合成前沿方向的发展。由于地球大气层的作用，地外空间太阳辐射光谱与地球表面有显著差异；由于围绕太阳的公转半径有很大区别，不同天体受到的太阳辐射强度差别很大。太空低微重力影响传递和输运过程，极限真空、低温和强宇宙辐射环境对转化过程有重要影响。极限条件下水/二氧化碳转化和地外人工光合成面临一系列重大科学问题。

1. 适应太空条件下太阳辐射光谱和高辐照强度的新型光催化材料体系

外太空太阳辐照强度及光谱分布与地球表面辐射条件有很大差异，当前针对地面太阳光谱条件开发的光催化材料在外太空环境中并不能高效、稳定地发挥作用。因此，需要有针对性地开发适用于 AM0 光谱、聚光高辐照强度下的新型宽光谱响应的光催化材料体系，以及适应宇宙辐射、满足长期太空探索任务要求的长寿命、高稳定的光催化材料。

2. 低微重力下光催化物理化学过程耦合机制

光催化过程产生的气泡引起的界面电阻（欧姆降）增加，将影响电极的表面覆盖，低微重力使传质过程变得更加困难，会极大降低系统的能量效率；过溶气体在电极表面附近聚集形成的溶解气体分子的过饱和层，对物质输运速率和反应效率有着非常重要的影响。因此，需深入探索低微重力下反应工质在电解质中的扩散传递过程，催化气体产物气泡成核、生长、界面脱离、气液两相流动、气液分离等关键机理以及对光催化过程的影响。

3. 实现太阳能全谱利用的光热电耦合催化机理

太阳光的能量分布在红外区（43%）、可见光区（50%）和紫外区（7%）。现有的光电材料带隙过大，大多吸收紫外和近紫外区光谱，太阳能利用率不高，光催化整体效率低，须深入探索太阳能光热电耦合催化机理，发展多谱段吸收、全谱段利用的复合光催化能源材料体系，提高光合成反应速率和光

化学转化效率。

（二）前沿研究方向

（1）适应太空环境的新型光催化材料。开发包括 AM0 下宽谱响应、高转换效率、高选择性的光催化材料（STF>10%），太空辐射和高光强条件下长期稳定的光催化材料（寿命 >3 年）。

（2）太空环境下太阳能全谱利用的光热电耦合催化复合材料。开发包括半导体能带调控实现紫外-可见光谱的高光电响应材料、实现可见-红外光谱热电子激发的局域等离激元结构材料、红外高吸收光热转换材料等，实现 STF>20%。

（3）低微重力下亲疏液微纳界面功能材料。开发促进反应工质扩散传递，反应产物气泡脱离、分离的亲疏液界面材料；开发促进质子输运的新型膜材料。

四、研究总结与展望

如何将人类呼吸产生的二氧化碳转换为氧气？如何利用火星等地外大气环境中丰富的二氧化碳和水原位资源生产氧气与燃料？这是人类实现地外生存、开展可持续探索的核心基础问题。美国等航天强国在继承传统萨巴捷反应、博施还原法等二氧化碳还原和电解水等地面技术的基础上，持续开展了太空极限条件下水 / 二氧化碳转化技术的研究，但面临运行条件苛刻（高温、高压条件）、能耗高、微重力环境下性能降低等突出问题。

近年来，在地球可持续发展的推动下，地面光催化技术得到快速发展。在太空探索活动中，通过地外人工光合成可在常温下实现将原位获取的水 / 二氧化碳转化为人类地外生存所需的基本物资，将成为太空探索的核心能力。我国科学家提出了地外人工光合成的概念，并率先开展了地外人工光合成装置的研制和空间实验，这将大大推动该领域的发展。地外人工光合成将成为新能源领域的新兴学科方向和前沿研究方向。

五、学科发展政策建议与措施

结合我国深空探测和载人航天后续发展需求，我国科学家首次提出地外人工光合成的概念并率先开展地外人工光合成装置研制。建议紧密结合全球太空探索共同目标和航天强国建设重大需求，加快推动地外人工光合成材料开发、系统研制和空间实验的全面发展，并以此为契机，带动地外人工光合成新学科发展，尽快取得具有重要国际影响力的研究成果，形成中国原创、国际引领的新能源领域的新学科方向。

张　策（中国空间技术研究院）、姚　伟（中国空间技术研究院）

本节参考文献

Arai T, Sato S, Morikawa T, 2015. A monolithic device for CO_2 photoreduction to generate liquid organic substances in a single-compartment reactor. Energy & Environmental Science, 8: 1998-2002.

Chen M X, Liu Y, Li C C, et al. 2018. Spatial control of cocatalysts and elimination of interfacial defects towards efficient and robust CIGS photocathodes for solar water splitting. Energy & Environmental Science, 11: 2025-2034.

Fujishima A, Honda K. 1972. Electrochemical photolysis of water at a semiconductor electrode. Nature, 238: 37-38.

Interbartolo M A, Sanders G B, Oryshchyn L, et al. 2013. Prototype development of an integrated Mars atmosphere and soil-processing system. Journal of Aerospace Engineering, 26: 57-66.

International Space Exploration Coordination Group. 2018. The Global Exploration Roadmap.

Jang J W, Du C, Ye Y F, et al. 2015. Enabling unassisted solar water splitting by iron oxide and silicon. Nature Communications, 6: 7447.

Jia J Y, Seitz L C, Benck J D, et al. 2016. Solar water splitting by photovoltaic-electrolysis with a solar-to-hydrogen efficiency over 30%. Nature Communications, 7: 13237.

Junaedi C, Hawley K, Walsh D, et al. 2011. Compact and lightweight Sabatier reactor for carbon dioxide reduction. 41st Int Conf Environ Sys, 1: 5033.

Kang U, Choi S K, Ham D J, et al. 2015. Photosynthesis of formate from CO_2 and water at 1% energy efficiency via copper iron oxide catalysis. Energy & Environmental Science, 8: 2638-2643.

Khaselev O, Turner J A. 1998. A monolithic photovoltaic-photoelectrochemical device for hydrogen production via water splitting. Science, 280: 425-427.

Kim J H, Jang J W, Jo Y H, et al. 2016. Hetero-type dual photoanodes for unbiased solar water splitting with extended light harvesting. Nature Communications, 7: 13380.

Kobayashi H, Sato N, Orita M, et al. 2018. Development of highly efficient $CuIn_{0.5}Ga_{0.5}Se_2$-based photocathode and application to overall solar driven water splitting. Energy & Environmental Science, 11: 3003-3009.

May M M, Lewerenz H J, Lackner D, et al. 2015. Efficient direct solar-to-hydrogen conversion by *in situ* interface transformation of a tandem structure. Nature Communications, 6: 8286.

Meyen F E, Hecht M H, Hoffman J A, et al. 2016. Thermodynamic model of Mars oxygen ISRU experiment (MOXIE). Acta Astronautica, 129: 82-87.

Pan L F, Kim J H, Mayer M T, et al. 2018. Boosting the performance of Cu_2O photocathodes for unassisted solar water splitting devices. Nature Catalysis, 1: 412-420.

Sakurai M, Oguchi M, Yoshihara S. 2008. Water electrolysis cell that are free liquid-gas separation system for microgravity conditions in order to establish circulated life support system. J.Japan Soc. Microg. Appl., 25: 653-656.

Sakurai M, Terao T, 2016. Study of water electrolysis under microgravity conditions for oxygen generation: applied to a ground demonstration system and development of new systems. Proc 46th Int Conf Environ Syst, 1: 204.

Sakurai M, Terao T, Sone Y, 2015. Development of water electrolysis system for oxygen production aimed at energy saving and high safety. Proc 45th Int Conf Environ Syst, 1: 273.

Schreier M, Curvat L, Giordano F, et al. 2015. Efficient photosynthesis of carbon monoxide from CO_2 using perovskite photovoltaics. Nature Communications, 6: 7326.

Schreier M, Héroguel F, Steier L, et al. 2017. Solar conversion of CO_2 to CO using Earth-abundant electrocatalysts prepared by atomic layer modification of CuO. Nature Energy, 2: 17087.

Wang Q, Hisatomi T, Jia Q X, et al. 2016. Scalable water splitting on particulate photocatalyst sheets with a solar-to-hydrogen energy conversion efficiency exceeding 1%. Nature Materials, 15: 611-615.

Wang S C, Chen P, Bai Y, et al. 2018. New $BiVO_4$ dual photoanodes with enriched oxygen vacancies for efficient solar-driven water splitting. Advanced Materials, 30: 1800486.

Young J L, Steiner M A, Döscher H, et al. 2017. Direct solar-to-hydrogen conversion via inverted metamorphic multi-junction semiconductor architectures. Nature Energy, 2: 17028.

Zhou X H, Liu R, Sun K, et al. 2016. Solar-driven reduction of 1 atm of CO_2 to formate at 10% energy-conversion efficiency by use of a TiO_2-protected Ⅲ-V tandem photoanode in conjunction with a bipolar membrane and a Pd/C cathode. ACS Energy Letters, 1: 764-770.

关键词索引

H

J

K

L

Q

R

S

T

W

X

Y

Z

其 他